Mathematics in Philosophy

Mathematics in Philosophy

SELECTED ESSAYS

CHARLES PARSONS

Cornell University Press

ITHACA, NEW YORK

Cornell University Press gratefully acknowledges a grant from the Andrew W. Mellon Foundation that aided in bringing this book to publication.

First published 1983 by Cornell University Press.
Published in the United Kingdom by Cornell University Press, Ltd., London.

International Standard Book Number 0-8014-1471-7
Library of Congress Catalog Card Number 83-45153
Printed in the United States of America
*Librarians: Library of Congress cataloging information
appears on the last page of the book.*

*The paper in this book is acid-free and meets the guidelines
for permanence and durability of the Committee on Production
Guidelines for Book Longevity of the Council on Library Resources.*

To the memory of my father

Contents

Contents

Preface

This book contains the most substantial philosophical papers I wrote for publication up to 1977, with one new essay added.[1] I distinguish philosophical work from work in mathematical logic; I have not included any of my technical papers.[2] The collection is unified by a common point of view underlying the essays and by certain problems that are approached from different angles in different essays. Most are directly concerned with the philosophy of mathematics, and even in those that are not, such as Essay 9 on the liar paradox, the connection between issues discussed and mathematics is never far from the surface, and the issues are approached "from a mathematical point of view." In the Introduction I articulate some elements of my general point of view and point out some connections between the essays. As I try to make clearer there, I hope that the essays reveal not just a specialist's concern with problems about the foundations of mathematics but also get across my sense of the central role of mathematical thought in our thought in general.

Along the way in the Introduction, something will be said about the importance I attach to certain historical figures. I will also offer the reader some guidance and assistance with the technical background of some of the papers.

Of the essays in Part One of the book, the first two concentrate on problems of ontology, the first in relation to elementary and

[1] The collection does not include reviews, short commenting papers, encyclopedia articles, and two essays: "Was ist eine mögliche Welt?" and "Some Remarks on Frege's Conception of Extension." More recent work, notably "Objects and Logic" and "Mathematical Intuition," is intended to form another book when combined with other work in progress or projected.

[2] This principle is compromised to some extent in Essay 11, in which I report some technical work that grew directly from the philosophical work of Essay 10.

constructive mathematics, the second in relation to substitutional quantification. The third discusses a number of issues about informal and formalized axiomatic theories, Gödel's theorem, and Tarski's theorem on the indefinability of truth. It could serve to introduce the themes of Part Three.

Part Two consists of essays on two important historical figures, Kant and Frege, and one contemporary, W. V. Quine.

Part Three consists of essays on the notions of set, class, and truth, notions among which I find crucial connections. The distinction between the notion of set and the notion of class is problematic; that is the main subject of Essay 8. The connection of the notion of class with satisfaction and truth, remarked on in Essay 8, looms large in Essay 9 on the liar paradox, which is also concerned with showing connections between the semantical and set-theoretic paradoxes. Essay 10, mainly concerned with a commonly used intuitive explanation of the universe of sets, returns to the notion of class in set theory and to the problem of speaking about *all* sets. The ideas of Essay 10 raise questions about reference to sets in modal contexts; some of the loose ends of the discussion are pursued in Essay 11, which also examines some axiomatic set theories suggested by the ideas of Essay 10 and some differently motivated modal set theories.

With some minor exceptions noted below in opening footnotes, the essays have not been revised, even where I now think their arguments unclear or mistaken. I have added to the footnotes in some places, either to clear up purely technical confusions or because I was not able to resist the temptation to add new comments or references. Such additions are enclosed in brackets. Postscripts have been added to Essays 1, 5, 6, and 9. Essay 5 generated some controversy about Kant's conception of intuition. Especially since I have never replied to Hintikka's criticism,[3] I now comment on that controversy and on some other issues about Kant's philosophy of arithmetic. The publication of Essay 9 was followed by important technical work on semantical paradoxes by others, notably Saul Kripke. Here the main purpose of the Postscript is to say how that work affects the viewpoint of my own paper on the subject.

In editing the published texts, I have put the references into a uniform style and updated some of them. Citations are given only

---

[3]"Kantian Intuitions."

by author and title, the latter often abbreviated. Full bibliographical information is given in the Bibliography at the end of the book.

Collecting the work of nearly twenty years offers an occasion to thank some of the many people who have offered me instruction, stimulation, and helpful criticism over the years. I must begin with my teachers at Harvard University. The intense activity in philosophy and logic at Harvard led me into both subjects. Burton Dreben, my principal teacher in mathematical logic, influenced my philosophical development in many ways, perhaps first by wisely urging me to postpone my ambitions in the philosophy of mathematics and do research in mathematical logic. His historical understanding of logic and of the analytic tradition in philosophy and his skeptical questioning of philosophical ideas have offered me both education and challenge.

It should be obvious to my readers that I owe much to the writings and teaching of W. V. Quine. Most apparent is my debt to Quine's *philosophy*, even where I disagree. From him I also first learned logic. In that subject he provided a model of clarity and philosophical conscience, although my own logical research has been in a different tradition. For my generation, especially at Harvard, he has been a model of a philosopher-logician.

Although Hao Wang was my teacher at Harvard for only one of my graduate student years, he did provide indispensable guidance into the work of the Hilbert school in proof theory, at that time little known in America. Over the years I have had valuable discussions with him on many subjects.

Though he was not formally my teacher, and our interchanges mainly concerned technical matters, I also learned much from Georg Kreisel. Proof theory, an important background for my philosophical work, owes its transplantation from Germany to the English-speaking world in large measure to him. Others from whom I have learned much about this subject are W. W. Tait, W. A. Howard, Nicolas D. Goodman, Warren D. Goldfarb, and Solomon Feferman.

The philosophy of Kant, phenomenology (especially Husserl), and Brouwer's intuitionism were formative influences on my philosophical work which did not come from teachers at Harvard. Though my study of Kant was begun with A. C. Ewing, my real education on the subject came during my graduate student years at Harvard from fellow students, who were former students of C. I. Lewis (by then retired). The "Kant group" of S. F. Barker, Hubert

Dreyfus, Samuel Todes, Robert Paul Wolff, and later Ingrid Stadler has already been chronicled in print.[4] To Dreyfus, Todes, and Dagfinn Føllesdal I owe what little understanding I have of Husserl and phenomenology. The stimulus to study Brouwer came from Margaret Masterman in Cambridge in 1955; Richard Braithwaite was a helpful guide in my first study of his difficult writings.

Since 1965 Columbia University has been my academic home, and my work owes much to the environment provided by colleagues and students there. Deserving of special mention are Sidney Morgenbesser and Isaac Levi for their friendship, their knowledge of many things I do not know, their philosophical acumen, and their readiness for discussion. Among others at Columbia (now or previously) with whom I have discussed matters connected with these essays, I should thank George Boolos, Raymond Geuss, Dieter Henrich, James Higginbotham, Ernest Nagel, Wilfried Sieg, Howard Stein, and Mark Steiner.

The secretarial staff of the department, presided over by Sheila Farrelly Sheridan, has been of great assistance. In particular, Michael Laser did much of the typing and photocopying for this book.

In the academic year 1979–1980 I spent a sabbatical year at Oxford as Visiting Fellow of All Souls College and as a Fellow of the National Endowment for the Humanities. Though no word of this book was written during that time, the conception of it resulted from the project on which I was then working, and the new material (particularly the Introduction) owes something to that work. Accordingly, acknowledgment is due to these institutions.

At an early stage, my work benefited crucially from the freedom and stimulation provided by a Junior Fellowship of the Society of Fellows at Harvard, which I held from 1958 to 1961.

Without the interest and assistance of Bernhard Kendler of Cornell University Press, this volume would not have come into being. I also thank Allison Dodge, who has seen the book through the Press, and Richard Tieszen, who did most of the work on the index.

To my wife, Marjorie, and my children, Jotham and Sylvia, I owe more than can be said, not least for their efforts to instill in me that "robust sense of reality" which, according to Russell, is needed even for "the most abstract speculations."

This book is dedicated to the memory of my father, Professor Talcott Parsons. He was for me the first and most significant model of "science as a vocation," and he and my mother provided much

[4]Wolff, *Kant's Theory of Mental Activity*, p. x.

encouragement, support, and advice. More particularly, it was through him, beginning even with the books on his shelves in my boyhood, that I first sensed the importance of the German intellectual tradition and especially of Kant. My own concern with classic figures such as Kant, Frege, and Russell no doubt owes inspiration to his dialogue with Weber, Durkheim, Freud, Pareto, and the classical economists. I greatly regret that I did not have this book, or another, ready to present to him in his lifetime.

CHARLES PARSONS

New York, New York

Introduction

Most of the essays in this volume concern the philosophy of mathematics, and issues about mathematics figure prominently in the rest. Obviously a direct attraction to mathematics and its foundations has motivated much of my work. I have also held the conviction that mathematical thought is central to our thought in general and that reflection on mathematics is accordingly central to philosophy. Few would dispute its importance in the history of philosophy. In some of the well-known arguments of Plato and Aristotle, for example, mathematics plays a strategic role. The scientific revolution of the seventeenth century consisted in the first instance in the creation of a mathematical science of the physical world, incomparably deeper than the pre-Copernican mathematical astronomy, in that it went beyond description of the phenomena to a theory of underlying structure and dynamics. Of course Euclidean geometry, one of the pillars of this structure, was inherited from the Greeks. In the history of modern philosophy, both rationalism and empiricism have supported their points of view by interpretations of this science. Seventeenth-century science and its historical successors are at once mathematical and empirical; in general the former characteristic was the basis for the rationalist's case and the latter the basis for the empiricist's. This meant neither that the rationalist could ignore experience nor that the empiricist could ignore mathematics. But the example of mathematics was used through much of the history of philosophy to support rationalistic views, and much argumentation by empiricists has been negative, either against the rationalist's conception of mathematics itself or against the inferences about other domains that he made with the help of a conception of mathematics.

I

Kant was expressing conventional wisdom when he maintained that mathematics is *necessary* and that it is *a priori*. He was expressing "intuitions" about mathematics that any philosophy has to take into account. It is certainly part of my own view that these intuitions cannot be ignored. The increasing rigor and abstraction of mathematics in the late nineteenth century brought to light another feature of mathematics that the rationalist could appeal to: its generality of content and application. So long as Euclidean geometry was the philosopher's paradigm of a mathematical theory and was interpreted as a theory of physical space, this generality could not have its due weight in philosophy. It came to be strongly emphasized by Frege as it was used by him to support his logicism. I will call this thesis, so far rather vaguely defined, the *formality* of mathematics; the idea this term conveys, that mathematics is if not part of logic, at least importantly like logic, is intended.

The traditional conception of mathematics as necessary and a priori, and the Fregean thesis of formality, had the effect of distinguishing mathematics sharply from empirical science. This traditional view has survived in twentieth-century philosophy, for example in the Vienna Circle's characterization of mathematics as analytic and science as synthetic. But this distinction of mathematics from science has been sharply challenged by Quine's critique of the analytic and the a priori. Quine presents a picture of mathematics as deriving its evidence from belonging to a body of theory which is tested by experience as a whole, so that in this testing either the mathematics or the specifically scientific part of the theory might be modified if experience proves recalcitrant. The line between logic and mathematics on the one side and natural science on the other is not sharp, either in itself or with respect to their basic epistemological or metaphysical character.[1] This assimilation of mathematics to natural science is in some respects pressed even further by Hilary Putnam.[2]

The traditional view that there is an important distinction between mathematics and empirical science is one of the guiding ideas of much of my writing. The line between pure mathematics and the special sciences seems to me to be clear enough.[3] But of course the real question is whether it is significant. A number of consid-

---

[1]For example, Quine, *Philosophy of Logic*, pp. 98-100.
[2]See especially "What Is Mathematical Truth?"
[3]See Essay 7, section V.

erations developed in these essays indicate to me that it is. First, in direct relation to Quine, I defend the necessity of mathematics and specifically the greater generality of mathematical over "physical or natural" modality (Essay 7). A second point is the intimate connection between the fundamental concepts of pure mathematics, such as number, set, and function, and the concepts of formal logic (Essay 6; Essay 7, section VI; Essays 8, 10). A third is the character of modern mathematics as a theory of *structures* of different types, where the theory is quite general with respect to the "internal constitution" of the objects and even of the relations making up the structure (Essay 1, section II; Essay 7, section IV; Essay 6, sections IV–V).[4] Scientific theories can be mathematical in form and can thus discern a mathematically describable structure in the real world, for example, a certain geometry in space-time. But the question whether such an attribution is right is different from the "internal" questions about the structure or the question whether it "exists" in the mathematically relevant sense. A fourth consideration is the gross difference in procedures of justification between mathematics and empirical science, in particular the role of proof in mathematics and of experiment and observation in science. This last point, most directly relevant to the idea that mathematics is a priori, is not much discussed in these essays.

It is one thing to claim a philosophically important distinction between mathematics and empirical science, another to defend the characterization of it given by one or another earlier philospher. Kant virtually identified the necessity and apriority of mathematics, but different considerations can be brought to bear supporting each. Moreover, two considerations prominent in contemporary philosophy argue for separating them. One is Quine's critique of the a priori. The other is the examples offered by Saul Kripke of statements that would have one of these characteristics but not the other.[5] The necessity of mathematics is defended against Quine in Essay 7, but I have not worked out a position on the question of the a priori. A priori knowledge is discussed only tangentially and cautiously in the essays in this book.[6] Quine's critique and its de-

---

[4]This point and its implications for the notion of mathematical object are discussed more fully in my work in progress.

[5]*Naming and Necessity*, lecture I. In using the word "statement," I do not want to commit myself on the question *what is said* by Kripke to be necessary and not a priori, or vice versa. A more careful formulation of what he can be taken to be denying is given in Essay 7, section I.

[6]See in particular Essay 7, section I and end of section V. See also "Mathematics, Foundations of," pp. 192-200, where the discussion is more extensive but still inconclusive.

velopment by other philosophers are doubtless responsible for this lack. But because he defined his position in relation to that of the Vienna Circle, particularly Carnap, Quine's attack on the a priori is primarily an attack on the analytic-synthetic distinction and the use of the concept of analyticity to explain instances of a priori knowledge. I am persuaded by arguments against the analyticity of mathematics coming from Quine and from other sources, some going back to Kant. But just as the a priori needs to be distinguished from the necessary, so it needs to be distinguished from the analytic. The possibility is thus open of a defense of the a priori character of mathematics independent of the thesis that it is analytic and even of the analytic-synthetic distinction. I am inclined to think that such a defense is possible, but it is not offered in these essays or elsewhere in my writing.

Such a defense might begin with the observation that the generality of mathematics and logic is of a different kind from that of laws in other domains of knowledge. Conceived as a theory of cardinal and ordinal number, arithmetic numbers objects *in general.* Set theory is a theory of sets of similarly arbitrary objects. This generality is obviously closely related to the generality of formal logic, in which a valid formula of, say, first-order quantificational logic is true when its quantifiers are interpreted as ranging over *any* domain of objects and its predicates are interpreted in any possible way as predicates of these objects. Husserl characterized generality of this kind as "formal" and contrasted it with the generality of the laws of particular "regions" of being, such as the physical world.[7] Formal generality cuts across such distinctions as that between the physical and the mental, living and nonliving, even abstract and concrete. Husserl's view that formal generalization (*Formalisierung*) is fundamentally different from ordinary generalization (*Generalisierung*) has much to be said for it. It finds an echo in the distinction made at one time by Carnap between *Formalwissenschaft* and *Realwissenschaft,*[8] and even in Quine as concerns the laws of *logic,* when he says that they can be expressed only by semantic ascent.[9] Husserl's distinction is not directly explored in these essays, although the "systematic ambiguity" found in certain generalizations involving such notions as set, class, proposition, and truth is a related phenomenon (see Essays 8, 9, 10).

Of the historic views in the foundations of mathematics, the one

---

[7]*Ideen,* § 13.
[8]"Formalwissenschaft und Realwissenschaft."
[9]*Philosophy of Logic,* pp. 11-12.

that has made most of the formality of mathematics and that has perhaps done it the most justice is logicism. The discussion of logicism in Essay 6 is critical, and it has not found many defenders recently. I never meant to deny the intimate connection of mathematics and logic (see Essay 7, section I, and Essay 9). One aspect of logicism that I do reject is Frege's conception of logical objects. It is in its commitment to mathematical objects that mathematics first goes beyond logic. The notion of object in mathematics is one that has seemed to me to need a lot of analysis; thus ontology has been one of the main themes of my writing. Since I accept set theory and believe, with Quine, that one who assents to statements such as "there are nondenumerable sets of real numbers" is committed to the existence of sets, I must count as a platonist in some sense. Much philosophical writing about platonism in mathematics, however, attributes to the platonist a very naive picture of a "realm" of mathematical objects, which results from pressing analogies between mathematical objects and everyday things or the objects of physics. A juster appreciation of what it means to talk of a domain of mathematical objects was pointed to by Frege (see Essay 6, se-citons II–IV) and developed by Quine (cf. Essay 7, section IV). The "concept of an object in general" comes from formal logic and its apparatus of singular terms, identity, and quantifiers. It is not required of an object as such that it should enter into causal relations, be an object of perception or something analogous to it, or be in space and time.[10]

Unlike Quine, I have undertaken to give a sort of phenomenology of different kinds of mathematical objects: numbers and formal expressions in Essay 1, sets in Essays 8 and 10, classes as extensions of predicates (potentially distinguishable from sets) in Essays 2, 8, and 10. The minimal general characterization of the notion of object which reflection on logic offers needs to be fleshed out in individual cases. There is also an epistemological motive to say something about how objects of these domains are known to us. There is something puzzling about abstract objects, especially the "pure" abstract objects of arithmetic, analysis, and set theory, which seem to lack any intrinsic connection with the concrete. The kind of explication that has interested me could be pressed in the direction of *reduction*. Thus the "phenomenology" of Essay 1 could be offered as a reduction of expression-types and numbers to

[10]A new, fuller presentation of this general view of the concept of object is in Parsons, "Objects and Logic."

expression-tokens, in the context of a modal language.[11] Substitutional interpretations of number theory and of classes have been seen by other writers as *eliminations* of these objects.[12] The understanding of mathematics as about structures also suggests a reductive program, since it suggests that we take a mathematical theory that talks of certain objects, say numbers of one of the number systems, as about *any* system of objects and relations which satisfies the laws characterizing the number system in question. As it stands, this procedure leaves unreduced the objects of set theory or some other theory that can talk of "systems of objects" and their "relations." For example, this is true for the structuralist treatment of the natural numbers in Dedekind's classic *Was sind und was sollen die Zahlen?*[13] The same "ontological relativity" about numbers that has been much discussed in philosophical literature also affects sets, classes, functions, and relations.[14] But if set theory is also a theory that does not in the end characterize its objects in more than structural terms, then an eliminative program will have to be more radical. One possibililty is what has been called "if-thenism" or "deductivism": the *mathematical* truth of a statement in a mathematical theory would be identified with the *logical* truth of its implication by the axioms characterizing the relevant structure. This

[11]Thus Chihara, *Ontology and the Vicious Circle Principle*, sees the possibility of interpreting parts of mathematics in a modal language talking about expression-tokens as part of a case for nominalism. In view of the admission of modality, it has to be a qualified nominalism; it might be called "modal nominalism." (Cf. Essay 1, section III.)

[12]Gottlieb, *Ontological Economy*. I comment on this subject in my review article "Substitutional Quantification and Mathematics."

[13]See esp. par. 71. It is scandalous that Dedekind's treatment has been ignored in nearly all the recent discussion of the implications of ontological relativity for the concept of number and of the structuralist view of mathematical theories, including my own previously published remarks on this subject. An exception is Kitcher, "The Plight of the Platonist," p. 123. In "What Numbers Are," Nicholas White develops a position that is (as Kitcher notes) close to Dedekind's, apparently unaware of the connection.

[14]Instances in recent literature of neglect of this obvious fact are White, "What Numbers Are," and Jubien, "Ontology and Mathematical Truth." Much of the discussion goes back to Benacerraf, "What Numbers Could Not Be." Benacerraf seems to acknowledge the ontological relativity of set theory (p. 68), but still talks of "systems of objects" in expounding his thesis that numbers are not objects (p. 70), and also of "relations" (ibid.); whether he holds that such entities are after all exempt from ontological relativity, or envisages some way of treating them as not really objects, he does not say. My own view is that the case of sets shows that for mathematical objects, ontological relativity and "incompleteness" have to be lived with, and therefore the considerations Benacerraf adduces about *numbers* do not show them not to be objects. This conclusion is just what would be expected on Quine's view of these matters. (See Essay 7, section IV.)

view faces serious difficulties.[15] A nominalist view might regard a mathematical theory as admissible if a model can be specified in which the objects of the model are physical or in some general sense concrete. Where an infinite domain is required, as is already the case with elementary arithmetic, it is questionable that such a model can be given independently of physical science, which depends on mathematics. This objection may not be decisive, but the hypothesis that there is such a model is stronger and more specific than is needed for mathematics and is open to question on mathematically irrelevant grounds.

A more promising eliminative strategy is suggested by Hilary Putnam's idea of "mathematics as modal logic."[16] A difficulty of ifthenism is that if a system of axioms has *no* model, then the conditional of the axioms and any statement is vacuously valid. To this difficulty one is inclined to respond with Putnam[17] that it is surely sufficient that the axioms *might* have a model. Taking this seriously as a statement of possibility, we are led to interpret mathematical statements as statements of what necessarily holds if the conditions defining the relevant structure obtain. The necessity and possibility involved here are more general than "physical" or "natural" modalities; otherwise the same objection could be made as is made to the nominalist.[18] For reasons like those developed in connection with Kant in Essay 4, I do not think it helpful to construe these modalities in terms of capacities of the mind. The modalities involved could be the "metaphysical" modalities of such writers as Kripke. In contrast to the applications of these notions in Kripke's writings and those of modal realists such as Plantinga, what is required is purely formal; what we have to do with is the possibility

---

[15]Putnam, "The Thesis That Mathematics Is Logic"; Resnik, *Frege and the Philosophy of Mathematics*, chap. 3. For Putnam, note the footnotes added in *Mathematics, Matter, and Method*, and compare the subsequent paper, "Mathematics without Foundations."

[16]"Mathematics without Foundations." My use of Putnam's idea as an eliminative strategy seems not in accord with his disclaimer that his idea offers a "foundation" for mathematics (ibid., p. 57). In view of my conclusion, this use is only for the sake of argument. In fact, I am much attracted by his own conception of the "modal logic picture" and the "mathematical-objects picture" as *complementary* descriptions,, with the difference noted below concerning the generality of the notion of object required by set theory. (See Essay 7, note 32.) There is also a problem as to how a modal formulation treats second-order notions.

The appeal to modality has been taken as eliminative by others, such as Chihara (note 11 above), Jubien, "Ontology and Mathematical Truth," and Harold Hodes (in unpublished work).

[17]Already in "The Thesis That Mathematics Is Logic," pp. 32-33.

[18]See Essay 7, section IV–V.

of *structures*. I prefer to speak, with Putnam, of *mathematical* modalities. I do not think that accepting them commits one to the meaningfulness of the "material" modal questions discussed in the modal realist literature.

But an important difference still emerges, such that arithmetic falls on one side and higher set theory on the other. The modal-logical picture does not eliminate the notion of the *possible* existence of objects. For arithmetic, we do not need the possible existence of objects essentially different from those we encounter in everyday experience; geometry and analysis would stretch the notion of object further, but it still would not go beyond what is postulated in current or even classical physical theories. In set theory, however, we have structures that for cardinality reasons are richer than any we have any reason to suppose that the physical world, or any other aspect of the actual world, instantiates. On any modal interpretation of higher set theory, we are concerned with the possible existence of *objects*, where, however, we do not have any reason to suppose that such objects are possible which have any of the characteristic marks of concreteness. It follows that the very general notion of object which the "ordinary" notion of mathematical object requires will not be eliminated by a modal interpretation. (See Essay 7, section V.) For this reason it seems to me that Putnam's idea fails at this point as an eliminative strategy. The reduction of statements of the *existence* of mathematical objects to statements involving commitment to the *possible existence* of "objects" which do not lack the features that made mathematical objects puzzling may then seem trivial; however, it does at least have the moral that mathematical existence is not a kind of *actuality* (as some writing about platonism seems to claim it should be); we might parody Meinong by saying that mathematical existence is "beyond possibility and actuality."

II

Even if no more is required for the *existence* of mathematical objects belonging to an abstract structure than for the *possibility* of the objects which would be in an instance of the structure, the question still remains how this possibility is made out. In my attempts at a "phenomenology" of strings of symbols and of natural numbers, I claimed that these objects are objects of *intuition* in a sense like Kant's. (See especially Essay 1, section III; also Essay 5, esp. section VII; Essay 6, section IX.) Few philosophical terms have

been surrounded by more confusion than "intuition." To begin
with one must distinguish intuition of *objects*, which is what is here
in question, from intuition of *truths*. Kant's term *Anschauung* is used
for intuition of objects, and the intuition of truths that he admits,
for example intuitive knowledge in geometry, is evidently derivative
from intuition of objects. Most of the everyday use of "intuition,"
on the other hand, is for intuition of truths, or at least intuition as
a propositional attitude. The two would on any account be intimately connected, however, and therefore difficult to disengage.
In mathematical writing, for example, intuition is very often just
a nose for conjectures that are likely to turn out to be true; this
certainly falls on the propositional attitude side. Such a nose for
the truth, however, will go with the ability to visualize the mathematical structures in question (most literally in geometry or topol-
ogy). This falls on the side of intuition of objects.[19]

My own attempts have been to explicate what would be intuition
in what I take to be Kant's sense.[20] I should remark that a roughly
Kantian conception of intuition is prominent in early twentieth-
century work on the foundations of mathematics. This is perhaps
clearest in Hilbert's conception of *metamathematics*. Brouwer explicitly connects his own ideas with Kant's in his early address "Intuitionism and Formalism." His later, more systematic philosophical
writings show that his general philosophy was not very Kantian in
spirit, but a Kantian conception of mathematics as having to do
with the *form* of a certain kind of experience survives.[21] More recent
philosophical writing on the foundations of intuitionistic mathematics has undertaken to do without such a conception of intuition,
substituting what might be called a "meaning-theoretic" approach.[22]

My present view of this matter is that what is accomplished by
the concept of intuition developed in these essays and elaborated
in "Mathematical Intuition" is quite limited, but it would have to
have a place in the meaning-theoretic approach as well. In several
places, I make a distinction between intuition of objects such as

[19]On this distinction see Essay 10, p. 278 below and note 12. It is elaborated in
"Mathematical Intuition."

[20]See Essay 5, esp. section I and Postscript.

[21]See "Mathematik, Wissenschaft, und Sprache" and its expanded Dutch version,
"Willen, weten, spreken," and "Consciousness, Philosophy, and Mathematics."

[22]The philosophical aspects of this approach are most fully developed in writings
of Michael Dummett; see especially "The Philosophical Basis of Intuitionistic Logic."
Technical work developing intuitionistic mathematics from a point of view of this
kind has been done by W. A. Howard, Per Martin-Löf, Dag Prawitz, W. W. Tait,
and others.

geometic configurations or strings of symbols and whatever confers evidence on mathematical induction and other principles of the same kind.[23] The latter could not be a kind of intuition giving new *objects*, since it concerns general propositions about objects that could be objects of intuition in the first sense. From reflections of Bernays and Gödel on the difference between the kind of evidence Hilbert claimed for finitist mathematics and constructive evidence generally, it follows that induction as a general principle, even applied to objects of intuition such as strings, is *not* intuitively evident. The reason is that it involves the "abstract" general notion of predicate, property, or class.[24] Hilbert, however, claimed intuitive evidence for individual instances of induction where the predicates involved are of the right kind (do not themselves involve "abstract concepts"), in practice primitive recursive. There is still something involved which is not involved in singular propositions, namely, understanding of the predicate such as "is a string (from the given alphabet)" which gives the domain for the general proposition. This fact has made me doubtful about Hilbert's claim, but even if it is accepted, the *explanation* of such intuitive evidence has to be different from that in the singular case. I think the most promising line for the Hilbertian to take is that induction is like a logical principle and in particular that it is *conservative* of intuitive evidence, that is, that it leads from intuitively evident premises to intuitively evident conclusions (at least in the kind of case Hilbert admits).

At all events, one comes very quickly to cases involving "abstract concepts." Although perhaps (as I claimed in section V of Essay 1) one can still go a long way without appealing to *objects* that are not given by Hilbertian intuition, the conceptual demands are increased further by the introduction of successive stages of semantic reflection. At this point, the Kantian conception of pure intuition is not very much help. But the reader may be reminded of Gödel's thesis that we have "something like a perception of the objects of set theory," which he is prepared to call mathematical intuition.[25] A number of indications suggest that Gödel has in mind here something of a more intellectual nature, not concerning a "form of sensibility" in Kant's sense. My conjecture is that Husserl's and not Kant's conception of intuition is his model. But in spite of the

---

[23]See the end of Essay 5; also "Mathematical Intuition," pp. 164-165.

[24]Bernays, "On Platonism," pp. 280-282; Gödel, "Über eine bisher noch nicht benützte Erweiterung," p. 281.

[25]"What Is Cantor's Continuum Problem?" p. 271.

undoubted interest of Husserl's ideas, I have not arrived at a satisfying understanding of such a stronger conception of intuition,
and both in these essays and elsewhere I am skeptical about Gödel's
view of our knowledge of set theory.

The notion of intuition that I do defend raises problems when
it is confronted with the formality of mathematics, discussed above.
The objects of such intuition, such as geometric figures and Hilbert's strings, are abstract objects of a kind I call "quasi-concrete."
That is, they are determined by intrinsic relations to concrete objects. Strings and expressions are the clearest case: these are types
that are instantiated by concrete objects (tokens), and what object
a type is is determined by what is or would be its tokens. Obviously
this concrete representation is something about these objects that
goes beyond the formal structure, say that of an ω-sequence, that
they instantiate. Moreover, the manner in which I conceive intuition
of such objects as "founded" on ordinary perception and imagination means that the representation is essential to the intuitability
of the objects. It follows that we have here a domain of mathematical
objects that is not characterized in purely structural terms. The
case is like that of the classical geometry of space.

To be sure, the construction of Essay 1, section III, was intended
to show that the natural numbers, which are certainly not quasi-
concrete, are intuitable. But I do not now regard that attempt as
successful.[26] In any case, the other examples would retain their
force even if natural numbers were shown to be objects of intuition.

The familiar idea that a set is constituted by its elements should
remind us that something like this relation of "representation"
arises for objects that undoubtedly belong to pure mathematics.
Even if in the end the "universe of sets" is regarded as an abstract
structure, such an idea is essential to the formation of our conception of this structure and to our convincing ourselves of its possibility. As a residue of the fact that we begin with sets of objects
that are not sets, the universe of sets (or a model of set theory)
contains an underlying domain of individuals that has to be given
from outside. The abstract set theorist can suppose this domain
empty, but the generality of pure set theory as a theory of sets of
any kind of object calls for no restriction on the individuals, except
perhaps that they constitute a set.

---

[26]Cf. "Mathematical Intuition," pp. 163-164. The issue is discussed further in my
work in progress. I should add that the quasi-concrete character of some mathematical objects does not nullify their incompleteness; on this point see ibid., pp.
161-162.

One might indeed see intuition of a set as founded on perception of its elements in a way analogous to that in which intuition of a type is founded on perception of its tokens. In the case of finite sets such a picture has considerable plausibility, but it still has difficulties.[27] At all events a set whose elements are concrete seems reasonably called quasi-concrete. There is still an important difference between this case and that of expression-types or geometric figures, however, in that the concept of set, and the principles of set theory, provide for sets of objects *in general,* as is shown by the arbitrariness of the domain of individuals. Thus the conception of mathematical intuition that I defend involves a qualification of the formality of mathematics that set theory would not lead to. This would also affect the kind of case for the apriority of mathematics based on formality that I think might be made, since even for number theory and set theory intuition plays a role in yielding the possibility of the structures involved.

III

Both philosophical discussion of mathematics and a considerable amount of technical work in foundations have been guided by what can be roughly called issues about platonism. By this I mean in the first instance epistemological realism about mathematics, though questions of the existence of sets and classes, more immediately suggested by the name, are bound up with it. A remarkable outcome of the classic work in foundations of the early twentieth century, particularly that of Poincaré, Russell, and Brouwer, is that realistic and antirealistic assumptions offer strong motivations for *mathematically* substantive principles. This is a many-times-told tale, but I will say something about it below for the reader's orientation. The reader of these essays may be disappointed at the absence of a definite taking of position between platonism and one or another version of constructivism. An eclectic attitude in these matters is in fact widely shared. I shall try to say briefly why I hold it and simultaneously how these essays relate to issues of this kind.

The critique by Poincaré, Russell, and others of "impredicative definitions" is a classic example of a philosophical argument that

[27]Cf. ibid., p. 164. The thesis that finite sets of objects of intuition are intuitable is discussed at length in my work in progress, where, however, it is defended only in a much qualified form. A similar thesis was developed and defended by Husserl in *Philosphie der Arithmetik.*

would directly affect what principles are acceptable in analysis and set theory. Poincaré argued that it would be viciously circular to define a set of natural numbers by a formula containing a quantifier that would have in its range *all* sets of natural numbers. Taking this idea strictly, as for example Hermann Weyl did for a time,[28] was thought to involve not only abandoning set theory as developed by Cantor, but modifying elementary theorems of analysis. Efforts to develop mathematics on a "predicative" basis have continued down to our own day, and methods have been found for preserving much more of ordinary analysis.

Arguments that such impredicative definitions of sets of natural numbers involved a vicious circle presuppose that sets must be given by properties or predicates. It then seems evident that the totality of such properties does not exist independently of our means of expressing them by language, or at least of our thought. But conversely, there should be no vicious circle if one admits that the totality of *sets* of natural numbers exists independently of our means of defining them; on this view, there should be no more objection to defining a set of natural numbers by a predicate containing a quantifier over all such sets than there is to defining a number by a definite description containing a quantifier over all numbers.[29] For then a statement involving quantification over sets of natural numbers will have a definite truth-value, and an open sentence involving such quantification will have a definite extension.

There are not many confirmed predicativists nowadays, if one means by that someone who simply rejects all mathematical arguments that involve impredicative definitions or other forms of impredicativity. There is no doubt that impredicativity marks one of the watersheds among mathematical principles with respect to intuitive evidence and clarity, even though a form of impredicativity enters into our specification of what a natural number is. A symptom of this, discussed in Essay 6, is the fact that impredicative second-order logic is required for Frege's derivation of mathematical induction from his definition of the predicate 'natural number'. Once one accepts the natural numbers (as Poincaré and Weyl did[30]), the assumption that *sets* or classes of natural numbers offer

---

[28]*Das Kontinuum*, "Der *circulus vitiosus*," "Über die neue Grundlagenkrise."

[29]Cf. the analysis of the vicious circle principle by Gödel in "Russell's Mathematical Logic."

[30]One reason for Russell's adoption of the compromise embodied in the ramified theory of types with the axiom of reducibility rather than following Poincaré in rejecting impredicative definitions outright was the needs of the logicist construction of the natural numbers. The derivation of mathematical induction in the system of

a definite domain for quantification, which is what is required for the stock examples of impredicative definitions in analysis, is clearly another conceptual leap. From my point of view, it is important to point out that it *is* a leap, even if it is not (as Poincaré seems to have thought) the source of the paradoxes. The trouble it leads to is not contradictions but questions that current or clearly foreseeable conceptual resources cannot resolve. For example, once one treats sets of sets of natural numbers in the same realist fashion, one can pose the continuum problem: whether there is a one-one mapping of the order types of well-orderings of the integers onto the set of sets of integers. This problem cannot be decided in standard set theory or in any known extension of it whose principles set theorists find intuitively convincing.

The limits of predicativity given the natural numbers were described convincingly in work of the 1960s of Feferman and Schütte.[31] Since that time the attention of proof theorists has tended to center on extensions of predicative mathematics as thus defined which still fall short of full second-order arithmetic, which is the theory that arises most directly from a realistic conception of sets of natural numbers. That such extensions might be motivated independently of such a conception of sets follows from the fact that intuitionistic analysis is already no longer predicative in this sense. The general logical scheme that the intuitionistic conception of ordinal embodies is that of "generalized inductive defintions," by which a predicate is introduced by rules that say (roughly) that it holds in certain initial cases and that the predicate holds in a given case if it holds in a (generally infinite) class of other cases. Since the idea is that the predicate is of minimal extension so as to satisfy these conditions, there is then an induction rule to the effect that any formula true of the initial cases and for which the inferences hold is true of everything for which the predicate holds. An example is the accessibility predicate for an ordering of the natural numbers, discussed in Essay 3, section III.

Proof theory has treated theories in which generalized inductive

---

Principia requires reducibility for the same reason that Frege's derivation requires impredicative second-order logic.

It should be pointed out that Russell was less ready to bend mathematics to his philosophy than especially the Weyl of 1918-1921. He repeatedly defended the axiom of reducibility on the grounds that it was needed to derive extant mathematics.

[31]See Feferman, "Sytems of Predicative Analysis," and the papers of Schütte cited in Essay 1, notes 20-21. Feferman has given newer formulations in later papers; see especially "A More Perspicuous Formal System for Predicativity" and "Gödel's Incompleteness Theorems and the Reflective Closure of Theories."

definitions, even infinitely iterated, are added to predicative theories based on intuitionistic logic as well understood and has used them as the basis for proof-theoretic analysis of other theories, particularly subsystems of classical second-order arithmetic.[32] A great deal more could usefully be said philosophically about generalized inductive definitions, and indeed about the intermediate realm between predicative and set-theoretic mathematics. The subject is only lightly touched on in these essays.

The absence of any fatal trouble with impredicative theories, set theory in particular, and the occurrence of impredicativity in the explanation of the notion of natural number, seem to me (as they have to many others) decisive reasons against the thesis that mathematics ought to be confined to what can be developed predicatively relative to the natural numbers. Moreover, one can add the further consideration that generalized inductive definitions and other apparatus of intuitionistic analysis constitute a smaller leap than that involved in full second-order arithmetic.[33] But the critique of impredicativity retains its interest because of its role in marking conceptual distinctions between different levels of mathematics.

The concept of set as it is developed in full-blown set theory is the subject of Essays 8, 10, and 11 and also enters into Essays 3, 7, and 9. Even here, however, predicativity plays a role. The rejection of impredicative specifications of sets results in my view from the conception of sets as classes (in the terminology of Essay 8), that is, objects associated with predicates with extensional indentity conditions. But then if one accepts set theory, one can turn such considerations around and question that conception of what sets are (see Essay 8). Moreover, predicativity relative to the universe of sets is the appropriate conception for nearly all uses in set theory of the notion of classes over and above sets.

The other major cleavage distinguishing set-theoretic from the most elementary mathematics that was central to the early twentieth-century controversies is that between constructive and nonconstructive. This issue, though related to that of predicativity, is

[32] A comprehensive recent presentation of such work is Buchholz, Feferman, Pohlers, and Sieg. The Introduction by Feferman and Sieg contains a helpful general exposition on generalized inductive definitions.

[33] A further complication of the picture arises from the development of theories based on intuitionistic impredicative higher-order logic and even of set theories based on intuitionistic logic. The problem of understanding these theories bears more on the question of what is constructive. I do not undertake to discuss the issues here. The existence of controversey and uncertainty here, however, may indicate unclarity in the idea of constructivity, or in the conception of an "antirealist" understanding of mathematical statements.

not the same. Generally speaking, a mathematical proof is constructive if it contains an explicit construction of any objects asserted in the proof to exist. "Constructivism" would then be a methodological view according to which proofs are required to be constructive. The set-theoretic foundations given to the theory of real numbers in the late nineteenth century violated this requirement. How fundamental a point it was was shown early in this century by Brouwer, who in effect showed that the use of the law of the excluded middle in mathematical arguments led to conclusions that *could not* be proved in a strictly constructive fashion.[34] Brouwer's "intuitionism" was an attempt to reconstruct mathematics in such as way as to rule out nonconstructive proofs.

The philosophical interest of constructivism is that it is the expression of a thoroughgoing idealist or antirealist conception of mathematical existence and truth. Brouwer interpreted his own work in terms of a highly idiosyncratic subjectivist philosophy, and the ideas that mathematical constructions are mental, that mathematical objects are constructions of the mind, and that the last arbiter of mathematical truth is such mental construction are generally taken as part of the philosophy of intuitionism in a more general sense. This view has to be distinguished from the view that mathematical assertions should be limited to what is intuitively evident. The latter view of course depends on a conception of the intuitively evident. A reasonably clear conception emerges from Hilbert's ideas about the methods of proof theory (on which the conception of mathematical intuition discussed above is partly based). Bernays pointed out long ago that intuitionistic methods went beyond the intuitively evident in this sense because of the use of "abstract" concepts of proof, function, and species (class).[35] What such a conception of intuitive evidence recommends is rather something closer to the finitary method of Hilbert. But it is very difficult to take seriously the claim that the limits of acceptable mathematics are that narrow, since very elementary and natural forms of reflection on finitary mathematics already go beyond it. If one takes the

---

[34] The most elementary type of counterexample, such as that of a real number that can be determined to be equal to or different from o only if some problem is solved that we have no idea how to attack, is well known and presented in almost any exposition of intuitionism. (See for example "Mathematics, Foundations of," p. 205.) The force of all such examples depends on the idea that a proof of a disjunction involves a proof of one or the other disjunct and the claim that it is senseless to talk of the *truth* of one of the disjuncts independently even of the possibility of knowing it (and therefore of knowing which one holds).

[35] "On Platonism," pp. 280-282.

notion of intuitive evidence in a less strict way to mean simply "obvious," it is by no means clear that the assumptions of intuitionistic mathematics are more evident than those of at least the weaker classical theories. Some other analysis is needed to make the case against classical mathematics, whether based on a different conception of intuitive evidence or on something else. What has been developed are philosophical arguments in favor of idealism or antirealism, at least with reference to mathematics. This line has been taken in recent years by Dummett.[36]

Heyting's development in 1930 of an intuitionistic logic transformed the foundations of intuitionism, and in more recent times the interpretation of intuitionistic logical connectives in terms of proof-conditions has been a focus of research and argument. Without going into the matter, I will say that the literature seems to me to show that the difficulties in the foundations of institutionism are no less great than those of classical mathematics. I am generally inclined to be skeptical of a philosophical construction that claims to be legislative for an established branch of science.[37]

The development of intuitionistic logic has another effect that over the years has worked toward the dissolution of intuitionism as a "school" and encouraged eclecticism. It made possible the construction of a variety of formal theories based on intuitionistic logic and, given the fertility of logicians' imaginations, led to interpretations of these theories from a variety of points of view, some of them depending essentially on classical concepts. Intuitionistic logic has come to have applications far removed from Brouwer's original motivations. In this way, intuitionism has had a success that worked to undermine the philosophy on which Brouwer's rejection of classical mathematics was based.

Nonetheless these essays do not enter into the issues raised by intuitionism, understood as an attempt at a systematic antirealist account of mathematical truth. Many readers will think that some of my ideas, such as those about intuition in Essay 1 and the analogy of "existence" in substitutional theories and idealistic conceptions in Essay 2, ought to be developed in that direction. I do not agree that they *require* such a development, but there is no doubt that an exploration of the issues is called for and lacking in these essays as they stand.

---

[36]"The Philosophical Basis of Intuitionistic Logic" and *Elements*, chap. 7.

[37]There is an evident connection between Dummett's attempt to develop an argument for the rejection of classical mathematics in favor of intuitionism and his belief in philosophy of language as "first philosophy."

IV

My essays vary in the demands they make on the reader's technical knowledge. Readers who are not well versed in mathematical logic might be helped by some guidance as to what parts of the book demand the least in this respect and also by suggestions of readings to fill in the gaps.

The historical essays 4, 5, and 6 require little beyond elementary logic and arithmetic, though Essay 6 requires a little second-order logic as well.[38] The same is true for Essay 1 (except for section V), much of Essays 2 and 9, and the earlier sections of Essay 7, except that Essays 1 and 7 involve elementary modal logic in a way now pretty standard in philosophical work on modality.

Tarskian semantics is the background of Essay 9, portions of Essays 3 and 8, and indirectly of Essay 2. Fortunately, through the work of Donald Davidson this has become better known to nonlogicians. Sections II and IV of Essay 3 could be regarded as concerned with the motivation and consequences of "truth theories" in Davidson's sense for elementary number theory and set theory. In this context, however, it is of first importance to keep in mind the distinction between a truth theory that merely serves to prove the "Tarski biconditionals" (of which ' "Snow is white" is true if and only if snow is white' is a paradigm instance) and a theory that postulates or proves the *laws* relating truth to the logical operators. (See Essay 3, pp. 78-79 below.)

Some of the mystery should have been taken out of Tarskian *definitions* of truth by the rigorous treatment of the notion of truth under an interpretation (or truth in a model) in the best recent logic textbooks. That of Mates, *Elementary Logic*, is especially to be recommended. Excellent, more informal treatments are to be found in Quine, *Philosophy of Logic*, chapter 3, and Field, "Tarski's Theory of Truth." These treatments define truth for sentences containing quantifiers by versions of Tarski's original method of a detour through the concept of satisfaction, rather than Mates's method, by which truth of a quantified sentence is characterized in terms of truth under interpretations in which the denotation of an individual constant substituted for the quantified variable is varied. The method of satisfaction fits better with one use I make of Tarskian semantics, to analyze the relation between the notion of truth

---

[38]Second-order logic is inadequately treated in textbooks, but there is now a little on the subject in Jeffrey, *Formal Logic*, 2d ed., chap. 7. The best introduction to the subject is probably still Church, *Introduction*, chap. 5.

and predicative notions of class (Essays 8 and 9). If truth has been defined by Mates's method, then satisfaction can be defined in terms of it.[39]

The essays concerned with set theory unavoidably presuppose some background in that subject. The reader of Essays 8, 10, and 11 and portions of Essays, 3, 7, and 9 might do well to look at the discussion of the axioms of ZF and of ordinals and cardinals in an introductory book on axiomatic set theory such as those by Enderton, Krivine, Lévy, or Suppes. George Boolos's "The Iterative Conception of Set" is an excellent introduction to the issues discussed in Essay 10.

Modal quantificational logic enters into Essays 10 and 11 in a more sophisticated way than in the earlier essays. The treatment of quantifiers in the literature on modal logic, especially textbooks, is not too satisfactory. Kripke's "Semantical Considerations on Modal Logic" (1963) and Thomason's "Modal Logic and Metaphysics" (1969) are still well worth reading, and Schütte's *Vollständige Systeme* is a superior treatment of the topics it covers. I have tried to incorporate a little of the necessary information into Essay 11. For higher-order modal logic, Gallin's *Intensional and Higher-Order Modal Logic* is an indispensable source.

In spite of its importance in my work in general, proof theory is directly appealed to only occasionally in these essays, notably in section V of Essay 1 and in Essay 3. Gödel's incompleteness theorem and related results based on the arithmetization of syntax and diagonal constructions are treated in a number of textbooks, perhaps most attractively for the philosopher in Boolos and Jeffrey's *Computability and Logic*.[40] Proof-theoretic consistency proofs and other results of "positive" proof theory growing out of the Hilbert program and the analysis of logical proofs by Herbrand and Gentzen are, by contrast, slighted in general books on mathematical logic. If the reader of Essay 3 wishes actually to see a proof of the consistency of first-order arithmetic by transfinite induction up to ϵ_0, he would probably do best to read Schwichtenberg's "Proof Theory: Applications of Cut-Elimination"; however, such a proof is also presented in the appendix to Mendelson's textbook (first edition

---

[39]Let f be a function that assigns to each variable x free in a formula A an object $f(x)$ in the domain of the interpretation in question. Then f satisfies A just in case, when we substitute for the free variables of A distinct new individual constants and interpret them so that the constant substituted for x denotes $f(x)$, the resulting sentence is true.

[40]Particularly the second edition.

only).[41] For a more serious study of the subject there is no substitute for the books of Takeuti and Schütte, both entitled *Proof Theory*. The background of section V of Essay 1 is the papers cited there. The proof theory of ramified analysis is also treated in Schütte's book.

[41]The consistency proof presented in Shoenfield, *Mathematical Logic*, section 8.3, uses instead Gödel's functional interpretation.

PART ONE

Mathematics, Logic, and Ontology

1

Ontology and Mathematics

The question I want to ask in this paper is what the more "elementary" or "constructive" parts of mathematics commit us to in the way of entities. The purpose of this is perhaps more to shed light on the general notion of mathematical existence than to give an inventory of the entities required by the theories I discuss.

I

For guidelines for this sort of ontological investigation, one naturally turns to the writings of Professor Quine. Indeed, what follows is perhaps largely a commentary on Quine. Quine tends to look at mathematics in terms of set theory, relying on the fact that classical mathematical theories can all be modeled in set theory. Thus for him, the question what there is in mathematics comes down essentially to the question what sets there are. And this is really the question what set theory to adopt. This is a question to which no one nowadays tries to give a unique answer. In his latest work on set theory, Quine tries to bring out minimal existence hypotheses for particular purposes.[1]

It is easy to see that the codification of mathematics in set theory

From *Philosophical Review*, 80 (1971), 151-176; reprinted by permission of the editors.

I am indebted to several persons who made helpful comments on earlier versions of this paper, among them Gilbert Harman, Sidney Morgenbesser, W. V. Quine, Hans Sluga, and John Watling. I also had the opportunity to see an early version of Charles Chihara's *Ontology and the Vicious Circle Principle*, which deals with matters related to those discussed here.

[1]*Set Theory and Its Logic.* It is generally held by researchers in set theory that the notion of set is sufficiently definite that no set theory which is incompatible with Zermelo-Fraenkel set theory could be accepted as a genuine set theory. Although I agree with this view, it falls far short of determining set theory uniquely.

forces it to have *some* ontic commitments, unless one rejects an application of Quine's "criterion of ontological commitment" which, it seems to me, hardly deserves to be controversial. If one does set theory in informal language, one finds among the principles of the subject assertions that there exist sets satisfying certain specified conditions. For example, for all objects x and y there is a set whose members are just x and y. Any formalized theory which we would normally call "set theory" has theorems of the form '$(\exists x)\, Fx$' where we read 'x' as ranging over sets, or '$(\exists x)\,(Mx \wedge Fx)$' where '$Mx$' is interpreted as '$x$ is a set'. Hence we conclude that set theory affirms the existence of sets and in fact of infinitely many sets.

Thus if one wanted to develop the idea of "mathematics without ontology" one could not take the translation into set theory as the final formulation of mathematical theories on the basis of which their ontic commitments will be determined. And it seems to me that, Quine's criterion apart, the best approach to the "question of the meaning of being" in the most elementary parts of mathematics is not by way of set theory. Just by its universality, set theory leads us to neglect essential distinctions.

There is another difficulty, more of principle, with a strictly Quinian approach to our problem. In effect, Quine recommends the following procedure for determining the ontology of a theory: translate it into an interpreted canonical notation based on *classical first-order logic* and observe what objects fall into the range of the variable of quantification. But the choice of classical first-order logic as the framework for canonical notation imposes some of the conceptual apparatus of classical set-theoretic mathematics on the theories which are the objects of our investigation.

The "official" justification of the laws of classical first-order logic is that they are valid, and validity here is a set-theoretic concept. No doubt, underlying this is a more primitive acceptance of inferences codified in formal logic, in particular in view of the fact that logic is needed in order to *show* that logical theorems are valid. Brouwer's critique of the law of excluded middle shows, however, that even on the level of unself-conscious acceptance, classical logic is bound up with principles which can be rejected. When the classical mathematician asserts that every real number is either rational or irrational and the intuitionist asserts that not every real number is either rational or irrational, it would be too naïve to take this as a straight disagreement about a single statement whose meaning is clearly the same. On the other hand, it would not do either to take the difference as "verbal" in the sense that each one can formulate what the other means in such a way that the disagreement will disappear. (At least if the intuitionist is Brouwer; now that he is dead there may be no one left who really rejects classical mathematics. Anyway, a version of intuitionism from which this rejection follows can be clearly formulated and is not especially difficult to understand.) Rather it is the kind of case which fits Quine's general views about meaning rather well: a theoretical disagreement which affects the greater part of the parties' use of mathematical language and in which differences in the meaning of words and differences about the truth cannot be clearly separated from one another.

This point is related to a problem about determining the ontology of a theory which Quine himself signalizes, that the translation into canonical notation is subject to the "indeterminacy of translation": it is possible to translate it in incompatible ways which, however, accord equally well with the possible ultimate evidence for such translation. So it is only relative to a set of "analytical hypotheses" that we can say what the ontology of a theory is. Quine emphasizes that the analytical hypotheses which one uses in translating an alien language into one's own are likely to assimilate the conceptual scheme of the "native" to one's own, and in Quine's case the latter importantly involves set-theoretic mathematics. This assimilation, however, will arise partly through the application of heuristic maxims which need not be applied where other strong complicating considerations are present. One of these maxims is "preserve logical truths," but this would be ill suited to the comparison of theories where other theoretical differences can be bound up with what seem to be differences in logic.

Quine takes up the matter of intuitionism and says that one might attribute to it a different "doctrine of being" from his[2] or a "deviant concept of existence."[3] In investigating the ontology of constructive mathematics, we shall differ from Quine's customary procedure in that we do not insist on an answer in terms of classical quantification theory, so that we shall make positive use of "deviant" concepts of existence. The most illuminating account seems to derive from setting the deviant concepts against Quine's.

II

Any mathematics worthy of the name will contain elementary number theory, and it seems evident that the ontology of elemen-

---

[2]"Existence and Quantification," p. 108.

[3]Ibid., p. 113. The former quotation suggests a difference of substantive doctrine, the latter a difference of meaning.

tary number theory involves commitment to numbers. In an earlier paper,[4] I tried to sketch an account of what is involved in this and more specifically in the idea of numbers as *objects*. This idea faces the obstacle that numbers and other abstract entities are not identifiable in quite the same way as ordinary physical objects and perhaps other entities which we are inclined to call "concrete objects" at least seem to be.[5] Not even small natural numbers, the simplest and most transparent mathematical objects that there are, are objects of sense perception in the ordinary sense, and they are therefore not identifiable by ostension in the ordinary sense.

Moreover, in the foundations of mathematics and to some extent in mathematics itself one finds oneself playing rather fast and loose with the truth-values of identity statements for mathematical objects of ostensibly different categories. Thus in a work on set theory one can assume certain axioms about sets and go on to identify numbers with certain sets by definition. But this can be done in incompatible ways in different treatments. For example, the von Neumann numbers are introduced by defining o as Λ and $n + 1$ as $n \cup \{n\}$, the Zermelo numbers by defining o as Λ and $n + 1$ as $\{n\}$. So it seems that it does not matter for the development of number theory whether one stipulates that $2 = \{\{o\}\}$ or $2 = \{o\} \cup \{\{o\}\}$.[6] This would suggest that antecedently to such stipulation the truth-value of these identity statements is undetermined. Moreover, one can go the other way and, taking number theory as given, identify finite sets built up from the null set with certain numbers (e.g., Λ with o and $\{x_1 \ldots x_n\}$ with $2^{x_1} + \ldots + 2^{x_n}$). (The cases are not quite symmetrical, since the usual set theories are much stronger than number theory and cannot be modeled in it, so that although one can start with set theory and build up number theory, if one starts with number theory one can build up only a limited part of set theory.)

In the case of number theory and set theory, one might be convinced that in the usage of mathematicians uncorrupted by mathematical logic, equations such as the above are false—that is, sets are essentially different from numbers. In other cases this is much more doubtful: rational numbers and (equivalence classes of) pairs of integers, real numbers and Dedekind sections, geometric spaces

---

[4] "Frege's Theory of Number" (Essay 6 of this volume).

[5] The relativity of reference which Quine associates with the indeterminacy of translation implies that for him this is only an appearance. He does not dispute my claims about abstract entities, but denies that the situation is essentially different for concrete entities. Cf. "Ontological Relativity."

[6] Λ = the null class. $\{x\}$ = the unit class of x. $\{x_1 \ldots x_n\}$ = the class whose members are $x_1 \ldots x_n$.

and sets of triples of real numbers, ordinal numbers and the sets
they are construed as in von Neumann's development, and so forth.
It is a commonplace of the working philosophy of modern mathematicians that they are concerned with *structures*—that is, systems
of objects and basic relations such that it does not matter what the
underlying objects are or, more precisely, it can matter only if the
structure is part of a larger structure. One might on this account
construe any mathematical discussion as taking place against the
background of an overarching structure (say, a model of set theory)
such that statements made will be invariant under isomorphisms
of the overarching structure with another.

If we go outside this structure it is not necessary even to regard
what we call identity as true identity. Two objects can be construed
as "identical" if they are "indiscernible" with respect to the basic relations. If set membership is the sole relation, x and y are indiscernible
if $(\forall z)\ (z \in x \equiv z \in y)$ and $(\forall z\ (x \in z \equiv y \in z)$. We could *define* '$x =
y$' as '$(\forall z)\ (z \in x \equiv z \in y) \land (\forall z)\ (x \in z \equiv y \in z)$'; the usual "axiom of
extensionality" just says that the first conjunct implies the second.

Another example is the construction of rational numbers within
the theory of integers. We can define in the latter theory an equiv-
alence relation $\langle x,y \rangle \equiv \langle u,v \rangle$ of pairs which holds if and only if $xv =
yu$ (i.e. if y, $v \neq 0$, $x/y = u/v$). Then the simplest way to proceed is
to construe the variable for rationals as ranging over pairs $\langle x,y \rangle$ with
$y \neq 0$ (themselves construed as integers) and equality as $\equiv$. More
puristically, we can pick out a representative of each equivalence
class of $\equiv$ —for example, the pairs $\langle x,y \rangle$ with $x \neq 0$, $y > 0$ and x
and y relatively prime, with the pair $\langle 0,1 \rangle$—and construe $=$ as $=$,
or use class variables and construe the rationals directly as equivalence classes. The same theory of rationals develops in each case
and the differences in simplicity are not of great importance.

We might digress for a moment to consider the relation of this
phenomenon to Quine's thesis of the indeterminacy of translation.
Gilbert Harman, with the approval of Quine, has used cases similar
to the above to illustrate that thesis.[7] And indeed it seems that we
have here a case of just that relativity of reference which is a striking
conclusion drawn by Quine in *Ontological Relativity* from the indeterminacy. (Thus he considers the apparent difference between
abstract and concrete entities to be *only* apparent.)

In the case of the above three interpretations of the theory of

---

[7]"Quine on Meaning and Existence"; "An Introduction to Translation and
Meaning."

rationals, the ontology comprises one class of integers on the first translation, another class on the second, and a certain class of *classes* of integers on the third. If all we know about someone's use of the theory of rationals is that he assents to theorems of this theory and dissents from negations of theorems, then the translations are equally defensible.

We do not, however, get in this case one of the features Quine claims for the indeterminacy of translation, that on different schemes of translation a sentence can get different truth-values. For we can show in *our* theory that, given any sentence of the theory of rationals, all three translations of it get the same truth-value.

In his discussion of the number theory-set theory case, Harman presupposes a first-order theory which talks about both numbers and sets (so that '3 ∈ 5' will be meaningful) but leaves their relation undetermined. This would be atypical of formalized mathematical theories: it would be more usual in such a case to use a two-sorted theory, so that sentences like '3 ∈ 5' would be meaningless. There would, however, be no difficulty in *formulating* the theory. For example, take a first-order set theory which permits *Urelemente*, introduce a predicate N (number), a singular term o, an operation symbol S (successor), and add a suitable version of Peano's axioms. It would then be easy to show that different explicit definitions of N, o, and S could be consistently added. Note, however, that if N is assumed to have an extension, a sentence A of nth-order number theory can be translated into a sentence A' of this theory in an obvious way, and if the set theory includes Zermelo's with the axiom of infinity, the two reductions of A' have the same truth-value. The same equivalence will hold for a much wider class of sentences. In fact, those sentences which fail to get the same truth-value are probably recognizable in advance as undetermined by our prior notions of number and set.

The following might seem to be a more substantial indeterminacy in the translation of mathematical theories. Suppose that all we know about our native's use of a certain *incomplete* theory—say, elementary number theory—is that he assents to theorems and dissents from their negations. We can certainly have translations which accommodate these data and translate the same sentence incompatibly: for any sentence which is neither provable nor refutable, there is a pair of translation sentences which will do this. The limitation of our information is quite artificial, however, since in a real case the native will have a conception of number-theoretic truth such that some non-theorems are true.

The most interesting type of case of indeterminacy arises when the sentences in question are undetermined by all possible evidence (in this case, proofs and whatever convinces us of the truth of statements which play the role of axioms) and yet are not "don't cares" whose truth-value can be seen to be left indeterminate by the conceptions involved. This situation may arise in set theory; in fact to some it seems to be the case for the continuum hypothesis and some large cardinal axioms.

III

I want now to consider two kinds of perception which serve to fill the gap left by the fact that there is not in the usual sense perception of abstract entities.

These are perceptions of linguistic expressions for them and perception of *concrete instances* of mathematical structures. Either one might play the role, for "deferred ostension"[8] of mathematical objects, which ordinary perception plays for direct ostension of physical objects. It appears that the latter, perception of instances of structures, must be primary, because perception of a symbol can only provide consciousness of its denotation if it has somehow *got* a denotation. Nonetheless, perception of expressions is worth dwelling on for a moment. For example, to see a configuration of black against a white background as a *word* involves seeing it as a token of a certain type—that is, as equivalent to other tokens of the same type and as fitting into a syntactic structure which would ordinarily be described in terms of such types. What matters in the perception of expressions is perception of form, where "form" in the most immediately relevant sense is determined by the type.

How much of this state of affairs can be described without commitment to expression types as abstract entities? The relation 'x is of the same type as y' is a relation of physical things (e.g., ink marks). It seems that it might be explicable independently of the notion of types, at least of types as a kind of entities. For example, in a single, standardized written notation, what it means for two primitive signs to be of the same type could be understood ostensively. (Perhaps this is not true of the phonemes of a natural language.) In general, two inscriptions in a linear notation are of the same type if they can be decomposed into sequences of primitive signs of the same

[8]Quine, *Ontological Relativity*, p. 40.

length such that corresponding signs are of the same type. We can think of a sequence of primitive-symbol inscriptions as itself a physical inscription, but it must have a distinguished decomposition. This could be taken as given by the "inductive" generation of sequences: the "expressions" are just those inscriptions which either consist of a primitive sign or are obtained by successive juxtaposition of primitive signs. Two such sequences are "of the same type" if they can be placed side by side so that their terms correspond one to one and so that the corresponding symbol inscriptions are of the same type.

This characteristic seems not to involve any nonphysical *entities*, but the *notions* of inscriptions which can be constructed by *successive* addition of signs, and of sequences *being capable of* being placed in a spatial one-to-one correspondence are perhaps peculiarly mathematical. It is noteworthy that the appeal to modality makes possible the avoidance of a second-order quantifier: one says of a specific relation that *it* can be brought about rather than saying that some one-to-one correspondence *exists*.

Let us consider now the matter of perception of instances of a structure. A set of n objects (perhaps with an ordering) is a structure associated with the number n. To see five objects of some kind (perhaps as in some order, either natural or arising from the movement of one's attention) is the sort of concrete perception which "represents" the number five. Kant, for example, thought of configurations such as the fingers of his hand (or, perhaps, the successive utterances 'one', 'two', 'three', 'four', 'five') as representing the number five in intuition. Having the notion of 5, however, involves perceiving such configurations or sequences as *fives*—that is, being able to recognize equivalence to other such configurations and the appropriate relations to fours, threes, twos, and ones. (Kant also emphasized the ability to construct such configurations in imagination, and indeed one of the features which marks the notion of 5 as mathematical is that an *imagined* sequence of five events serves just as well to represent it as a real one.)

The standard numerals of formal systems of number theory are representatives in this sense of the numbers they denote. For example, I, II, III, . . . have the property that the nth is a configuration of n objects, and moreover it is possible to reconstruct from it the sequence of the numerals which precede it. (o, So, SSo have the additional property that the initial term o is singled out by having an empty sequence of S's.) Now consider three inscriptions of standard numerals:

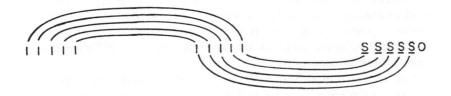

The correspondences mark them as representations of and therefore as numerals for the same number. But in the case of the first pair the correspondence also shows the inscriptions to be tokens of the same type.

Thus within a given system of numerals, what is involved in the notion "same expression" is just what is involved in the notion "denotes same number." But once one goes outside a single system, then the condition correspondences satisfy is more abstract in that symbols are no longer paired according to correspondence but rather according to role with respect to the rule for constructing numerals. We could easily imagine someone able to judge with respect to the first sort of correspondence and not with respect to the second. The principle of the first sort of correspondence implicitly contains the principle of the second, however, for it involves the notion of order-preserving correspondence of sequences. The first type of correspondence is simply that, plus the additional condition that corresponding terms should be tokens of the same symbol.

These remarks bear on the following suggestion of a "nominalist" treatment of the natural numbers. It seems that if we can construct those *expressions* (tokens) which we *call* "expressions for numbers" and recognize such expressions as standing for the *same* number, then we have all that we need for the theory of natural numbers. To say that *there is* a natural number having a property F is to say that *we can construct* a *perceptible* inscription such that it can be put into the empty place in an inscription of 'Fa' so that a truth results.[9]

[9]Modern nominalists will no doubt prefer "physical" to "perceptible." It will be necessary for number theory, however, that it should be possible to construct inscriptions where it seems clearly physically impossible that we should construct them and by no means clear that it is physically possible that they should exist. The notion of perceptibility is rather vaguer than the notion of physical possibility and is designed to signalize the fact that certain basic assumptions about what is constructible are justified by imaginative *Gedankenexperimente*. The fact (by no means obvious at the outset) that workable theories arise in this way is some argument for Kant's conception of a "form" of our "sensibility."

[This sentence (in the text) is ambiguous. Perhaps the most natural way of taking it would be to say that '$(\exists x)Fx$', where the variable "ranges over natural numbers," is *true if and only if* we can construct a perceptible inscription which can be put into

The kinds of arguments which are used in expositions of finitism and intuitionism to justify the elementary principles of number theory can be expressed in this vocabulary. Suppose our numerals are to be I, II, III, IIII, and so forth, and suppose (as fits the notation better) our theory is of *positive integers*.

We can say that the number 1 exists because we can construct an inscription equivalent to 'I'. We can say that every number has a successor, because given an inscription of the form I . . . I we can add another I to it. It is clear, moreover, that if two such inscriptions are of the same type, then so are their successors, so we have $x = y \supset Sx = Sy$.

If we have two inscriptions a and b such that aI and bI are of the same type, it is clear that a and b are also. Hence we have $Sx = Sy \supset x = y$.

This covers all the Dedekind-Peano axioms except induction. The intuitive justifications of induction offered in writings on constructive mathematics have for me an air of circularity, but at any rate in this formulation the argument is as good as it is anywhere. Suppose we have convinced ourselves of A (I) and A (a) $\supset A$ (aI). Consider an arbitrary numeral. Every such expression must have been constructed step by step starting with a I and successively adding I. But if b is the numeral constructed up to a given stage, at each stage we can convince ourselves of A (b) by using A (I) at the first step and A (a) $\supset A$ (aI) at each succeeding step.

We can also add to this explanation by using recursive rules of computation to give us addition, subtraction, multiplication, order, and other number-theoretic functions and relations. That is, we now interpret '$=$' as "reduces to" (which itself involves the idea of possible construction) and give instructions for reducing expres-

the empty place of an inscription of '*Fa*' so that a truth results. Properly, this should be recast as a modal statement about inscriptions. This truth-condition has a substitutional character, and then the resulting interpretation of the language of arithmetic is substitutional; inscriptions and constructions thereof are talked of only in the metalanguage.

Whether the metalanguage thus talks modally about inscriptions or other tokens, or in the usual way about expression types, this substitutional conception of the existence of numbers seems to me now of rather a different character from what I had primarily in mind in this section, according to which the numerals are thought of as *naming* objects whose identity conditions are as described in the last paragraph. These objects can be called "generalized types" (as in "Mathematical Intuition," pp. 163-164), and the account given here can be understood as a phenomenological description of how such objects are given in intuition.

On the substitutional conception, see Essay 2 below and Parsons, "Substitutional Quantification and Mathematics." Both conceptions and their limitations are discussed in detail in my work in progress.]

sions of such forms as | ... | + | ... | to numerals. (A numeral reduces only to itself.)

In order to say how much number theory we can get on this basis, we have to consider whether to restrict ourselves so as to be "finitist" or to use intuitionistic or classical logic. For any choice, the underlying assumptions would have to be clarified, and perhaps the use of full intuitionistic or classical logic would involve an expansion of our ontology. It is anyway not obvious that classical logic would square with the meaning suggested for the existential quantifier.

We can, however, discuss the claim of this conception to be "nominalist" independently of these decisions. From Quine's point of view, the formulation as it stands does not insure nominalism, and it is most doubtful that it can be altered so that it does. This might seem odd, because the only categorical existence statements are of inscriptions. The modal element in the interpretation of the existential quantifier (*can* be constructed) however, blocks the "naïve" translation into a standard first-order language in which the variables range over inscriptions. Given the Peano axioms, such a formulation would have to make assumptions about the actual existence of inscriptions which are out of place on this conception and which are anyway probably false. The language of our mathematics might be translated by a quantified modal logic, 'an *F* can be constructed' meaning 'possibly $(\exists x)\,Fx$'. But for Quine, the question of the ontology of a theory with that apparatus does not have a direct answer.[10]

---

[10]This formulation provides a simple model of the idea of "potential infinity." Consider a model system for a modal logic which contains $S4$, in which the possible worlds are all finite, but for each world there is one possible relative to it which contains at least one more member. Then for each n and for each world, there is a world possible, relative to it, which contains at least n more members.

Structures of this sort provide models for the following modal formulation of number theory, in which we have as primitive predicates 'Zx' (x is zero) and 'Syx' (x is the successor of y) and the following axioms:

$$\Diamond \ (\exists x)Zx$$
$$(\forall x) \ \Diamond \ (\exists y)Syx$$
$$(\forall x)(\forall y)(\forall z)(Szx \wedge Szy. \supset x = y)$$
$$(\forall x)(\forall y)(\forall z)(Syx \wedge Szx. \supset x = y)$$
$$(\forall x)(\forall y)(Sxy \supset \neg Zx)$$
$$(\forall x)(\forall y)(Zx \wedge Zy. \supset x = y)$$
$$(\forall x)(Zx \supset Fx) \wedge (\forall x)(\forall y)(Fx \wedge Syx. \supset Fy). \supset (\forall x)Fx.$$

In a [standard] model for this system, each possible world consists of the natural numbers up to some fixed number [or perhaps all of them]. (For first-order number theory, additional predicates need to be added to represent addition and multiplication.)

It would not help in the end to transfer the modal element in our conception from the quantifiers to the objects quantified over—that is, to take '$(\exists x)(Nx \wedge Fx)$' as saying not just 'possibly there is an inscription |. . . | such that "$F(|. . . |)$" is true' but "there is a possible inscription |. . . | such that "$F(|. . . |)$" is true'. This would give an ontology not of inscriptions but of possible inscriptions. It seems to me that a nominalist would probably not admit them into his ontology and that the notion of possible entities which *do* exist is inherently less clear than that of entities which *might* exist, but which of course would be actual if they did.

Although we expressed reservations about Quine's taking clascal first-order logic as the universal measure of ontology, we must

---

[The presentation of this system should have been more explicit about the underlying quantificational logic. If we add to S4 (or a normal extension) ordinary quantifier rules with free variable reasoning, then in the semantics we do not lose objects when we go to an alternative world, and therefore such a world will not have fewer numbers, in accord with the original conception. If we add the assumption that 'Z' and 'S' are persistent, i.e.,

$$(\forall x)(Zx \supset \Box Zx)$$
$$(\forall x)(\forall y)(Sxy \supset \Box Sxy),$$

the effect is that in an alternative world the same object represents each number. A model of the theory with this addition can be viewed as an intuitionistic Kripke model for the nonmodal language. However, it is not a model for intuitionistic arithmetic but only of its "negative fragment," that is, formulae without disjunction or existential quantification whose atomic formulae are prefixed by '$\neg\neg$'; this is in effect *classical* arithmetic. The condition on atomic formulae can be dropped if we also assume that '$\neg Zx$' and '$\neg Sxy$' are persistent (so that '$(\forall z)(Zx \vee \neg Zx)$' and '$(\forall x)(Sxy \vee \neg Sxy)$' hold in the model on the intuitionistic reading); then the model fails to be a model of intuitionistic arithmetic only because '$(\forall x)Zx$' and '$(\forall x)(\exists y)Syx$' may fail, being replaced by the weaker '$\neg\neg (\exists x)Zx$' and '$(\forall x) \neg\neg (\exists y)Syx$'.

(An excellent study of the quantified S4 we assume here and of its relation to intuitionistic logic is in Schütte, *Vollständige Systeme*. Kripke's original paper, "Semantical Analysis of Intuitionistic Logic," is still an excellent introduction to Kripke models; see also Dummett, *Elements*, chap. 5.)

Since this logic rules out the empty domain, its use has the unintended consequence that '$(\exists x)Zx$' is derivable. A free modal logic (such as the Q3 of Thomason, "Some Completeness Results") would allow us to keep the existence of all numbers, even zero, "merely potential," but if we still want to avoid losing numbers on going to alternative worlds, we need either to make the rather unnatural assumption that everything necessarily exists, or relativize our axioms to a predicate 'N' (natural number) with the axioms:

$$(\forall x)(Nx \supset \Box Nx)$$
$$(\forall x)(Zx \supset Nx)$$
$$(\forall x)(\forall y)(Ny \wedge Sxy. \supset Nx).$$

This construction was offered only as a simple model of the potential infinity of the number series. It was, however, a precedent for a development along similar lines of set theories; see my "Modal Set Theories" and Essay 11 below.]

agree with the conclusion that we attribute to him, that the conception we have sketched does not succeed in basing number theory on an ontology only of inscriptions. At any rate an important implication which one might expect is lacking. Any way I have of stating this seems objectionably vague. The essence of it is that the truth-value of statements is not a matter of the properties and relations of what there actually is. What actually exists in the way of inscriptions is not enough to make general number-theoretic propositions true. It is very likely still not enough even if we count as "what actually exists" also what will exist in the future and what has existed in the past; at any rate we are far from knowing that it is enough. (Perhaps what "actually exists" in the way of meanings and other intentions entertained by human beings *does* make statements of number theory true. But that is another matter.)

Thus it seems that although there is a possible sense of "ontic commitment" in which our conception is committed only to inscriptions and not to "abstract entities," this sense diverges from Quine's in a way which is significant. As one might expect from reflections on the foundations of mathematics, Quine's pegging of ontology to classical logic involves a realistic (as opposed to constructivistic or idealistic) conception of the existence of things.

IV

One might try to show that arithmetic (classical first-order arithmetic, not merely what admits the interpretation just sketched) has no ontological commitment, by observing that the quantifiers in a first-order formalism for number theory can be taken as substitutional quantifiers. For the idea that the variables "range over numbers" means that for every closed term t, there is a number n such that $t = \bar{n}$ is true ($\bar{n}$ = the nth numeral), and '$(\exists x)Fx$' is true just in case '$F\bar{n}$' is true for some numeral $\bar{n}$, and '$(\forall x)Fx$' is true just in case '$F\bar{n}$' is true for all numerals $\bar{n}$.

What Quine says about this is that for a theory based on substitutional quantification the question of ontic commitment does not arise (and therefore *not* that the theory has *null* commitment). The question of commitment has an answer only relative to a translation into ordinary objectual quantification. And then something suggests itself which in this context is useless, though it would not be elsewhere: the ontology of a substitutional theory can always be taken to comprise only natural numbers.[11]

Quine would no doubt respond in much the same way to a third suggestion for an ontology-free number theory: we can develop much number theory, including all "finitist" number theory[12] in a formalism which does not contain existential quantifiers at all. Generality is expressed by free variables which are really schematic letters for terms constructed from o, S, and other function symbols. We cannot say '$(\exists x)Fx$' but only things of the form Ft where t may contain parameters. In other words, we cannot merely say that there are numbers of such and such a sort; we must say in terms of given functions *what* they are.

Once again Quine would say that the question of the ontology of this theory is not directly answerable. We may suppose the background theory also contains numerals and expressions for recursive functions, so that a direct and obvious translation is possible. Still, we may suppose that the background theory does have quantifiers and a predicate 'x is a natural number' (Nx) so that if Ft is translated as $F't'$, $(\exists x)(Nx \wedge F'x)$ will follow in the background theory. This translation, then, does attribute to *PRA* an ontology of natural numbers. It translates the numerals of *PRA* as *names* of numbers, because the numerals of the background theory are such that the theory asserts $(\exists x)(Nx \wedge x = \bar{n})$ of them.

I might remark that I do not agree with the view Quine expresses in "Existence and Quantification" that substitutional quantification does not yield a genuine concept of existence. I take this matter up elsewhere.[13]

---

[11]"Existence and Quantification," pp. 106-107. There is an enumeration t_0, t_1, t_2, ... of the closed terms of the theory, and then the term t_n can be taken to denote the least number m such that $t_m = t_n$ is true. This requires a suitable interpretation of the function symbols.

[12]This is clear if we accept the idea that finitist reasoning does not take us essentially beyond primitive recursive number theory. This view is persuasively defended in Tait, "Constructive Reasoning." Tait's conception of finitism seems to agree with that of Gödel, "Über eine bisher noch nicht benützte Erweiterung," and, I should argue, with that of Hilbert. The alternative conception of Kreisel, "Ordinal Logics and the Characterization of Informal Concepts of Proof," yields a version of finitism which can only be reduced to free variable reasoning *ex post facto*, via some interpretation in a free variable formalism (e.g., with ordinal recursion). [See now also Tait, "Finitism." Gödel's remarks are considerably extended in "On an Extension of Finitary Mathematics Which Has Not Yet Been Used," notes (h), (i), (k).]

[13]"A Plea for Substitutional Quantification" (Essay 2 of this volume). [See now also "Substitutional Quantification and Mathematics."]

V

In some of the literature on finitist and constructive mathematics, a contrast is made between "abstract" concepts and those which involve only properties and relations of "concrete objects."[14] It seems that this difference is at least in part a difference of ontology: Gödel, for example, speaks of abstract concepts as applying to *Denkgebilde*, of which he gives proofs and meaningful statements as examples. It might seem therefore that the difference is between an ontology which involves only concrete entities in the normal philosophical sense (e.g., physical objects, or more generally objects of perception, with a real or apparent location in space and time) and one which involves abstract entities. This, however, could be true only if "ontology" is understood in a non-Quinean way such as that above, where the interpretation of quantification involves modal ideas. Gödel's example of a "concrete object" is a "sign combination," and it seems that for him concrete objects include signs as *types*. For him, however, the emphasis is not on ontology but on the contrast between "intuitive" (*anschaulich*) and abstract *evidence*, where the latter uses "insights . . .which arise not from the combinatorial (spatio-temporal) properties of the sign combinations representing them [proofs and statements], but only from their *sense*."

I think it is best to construe Gödel as meaning by "concrete objects" either spatiotemporal objects or *forms* of spatiotemporal objects, of which geometric figures would be a primary example.[15] We can attribute to him a version of Kant's theory of pure intuition relating to the spatiotemporal form of objects of perception.[16]

Thus insofar as the distinction Gödel is making is an ontological distinction, it would be that between forms of perceptible objects and the kind of entities he calls *Denkgebilde*. To view the ontology of our conception as comprising such forms has many advantages. It is compatible with Quine's conception of ontology but also with a corresponding one based on intuitionistic quantification theory. Moreover, it brings out accurately the relation to perception of our consciousness of such entities. The difference between perception of a physical inscription and "pure intuition" of a natural number

[14]See esp. Gödel, "Über eine noch nicht benützte Erweiterung des finiten Standpunktes."

[15]Some figures might require for their definition ideas which go beyond the notion of spatiotemporal form. That need not prevent them from *being* forms.

[16]Cf. "What is Cantor's Continuum Problem?" This need not extend beyond the combinatorial properties of objects (arithmetic); i.e., Gödel's remarks about finitary arithmetic do not commit him to a Kantian philosophy of geometry.

lies in the underlying equivalence relation which makes different perceptions or intuitions of the "same" object. In the latter case, of course, the relation does not fix for the object a spatiotemporal place, and propositions about numbers have a superior generality to everyday propositions about physical things. And the one-to-one correspondence of sequences which is an essential constituent of the equality relations both for numbers and expressions does not involve reference to space and time so that, to the extent that we have a conception of object not bound to space, time, or perception, the notion of one-to-one correspondence can be applied to it and thus gives the notions of number a wider application than in relation to physical objects, spatiotemporal forms, and other such entities.

To conclude this discussion I shall consider the question how far we can go in mathematics with the sort of ontology suggested by Gödel's remarks. Gödel seems to regard abstract *concepts* as concepts which must apply to *objects* of a different nature from spatiotemporal forms. The objects which would most readily come to mind from what he says are *intensional* objects—propositions in the first instance, then proofs, perhaps also attributes or "concepts." (He is talking about constructive proof theory and not about set theory.) It is far from clear, however, that the applications of abstract concepts he has in mind really require intensional *objects*. From examples, it seems that some at least of the uses of abstract concepts that he has in mind involve a certain semantic reflection (or, in Quine's phrase, "semantic ascent") in that one talks of the *truth* of statements or of predicates being true or false of objects. In this sense one must indeed have "meaningful statements," but one precisely does not need as an object the meaning which a statement has; that is, one never has to say when two statements (sentences or utterances) express the same *proposition*. So it seems that the introduction of abstract concepts can be carried out as an expansion of "ideology" beyond the intuitive-combinatorial without being an expansion of *ontology*.

Thus we can in constructive elementary number theory introduce a new function symbol $\phi(x_1 \ldots x_n)$ with defining equations, provided we can show that for any $m_1 \ldots m_n$, $\phi(\bar{m}_1 \ldots \bar{m}_n)$ can be reduced by means of the equations to a numeral, in the following sense. Let a *reduction* be a sequence of terms $t_1 \ldots t_m$ such that t_{i+1} results from t_i by replacing some part s of t_i by s', where $s = s'$ results from a defining equation by substituting numerals for variables. Then a term r can be reduced to a term t if a reduction can be constructed whose first term is r and whose last term is t.

Then we can justify on intuitive grounds the introduction of functions by primitive recursion. Consider, for example:

$$x + o = x \qquad x + Sy = S(x + y)$$

We can construct for any m, n a reduction which begins with $\overline{m} + \overline{n}$ and ends with a numeral.

If $\overline{n}$ is o, we have $\overline{m} + o, \overline{m}$.

If $\overline{n}$ is $S\overline{k}$, then from a reduction

$$\overline{m} + \overline{k}. . .\overline{l}$$

we obtain the reduction

$$\overline{m} + S\overline{k}, S(\overline{m} + \overline{k}) . . . S\overline{l}.$$

Thus we have

$$\exists \text{ reduction of } \overline{m} + \overline{k} \supset \exists \text{ reduction of } \overline{m} + S\overline{k}$$

that is, $(\forall r)(r \text{ reduction of } \overline{m} + \overline{k} \supset (\exists s)(s \text{ reduction of } \overline{m} + S\overline{k}))$

But now suppose we have the defining equations

$$[\phi \ (o,y) = 2y \qquad \phi \ (Sx,o) = 1 \qquad \phi \ (Sx, Sy) = \phi \ (x, \phi \ (Sx,y))]$$

(ϕ is not primitive recursive).[17]

Using the above we can construct for each $\overline{n}$ a reduction of $\phi \ (o,\overline{n})$ to a numeral, and trivially of $\phi \ (S\overline{m}, o)$ to a numeral. We should like to show by induction that for every n, $\phi \ (S\overline{m}, \overline{n})$ reduces to a numeral. But we have to use as hypothesis that *for every n,* $\phi \ (\overline{m}, \overline{n})$ reduces to a numeral. In other words, we cannot simply describe a sequence of steps for the construction of the reduction; rather we argue that if a reduction of $\phi \ (\overline{m}, \cdot)$ can always be constructed, then so can one of $\phi \ (S\overline{m}, \cdot)$.

The same sort of argument would convince us that primitive recursion always defines a function, and this in turn would convince us of the soundness of *PRA*—that is, that if $s = t$ is provable in *PRA*, and $s' = t'$ results from $s = t$ by substituting numerals for the free variables, then s' and t' reduce to the same numeral.[18]

The above argument does show us how to go from a *proof* that $\lambda y\phi xy$ is defined to a proof that $\lambda y\phi \ (Sx,y)$ is defined. Thus it seems that the addition of proofs to our ontology will suffice to justify the introduction of ϕ . This way of putting the matter, however, misleads as to what is really assumed about proofs. We can still

---

[17]See Parsons, "Hierarchies of Primitive Recursive Functions," §5. Similar examples have been known for many years. Cf. the original, slightly more complex example of Ackermann, "Zum Hilbertschen Aufbau der reellen Zahlen."

[18]Indeed, section 5 of my paper cited in note 17 implies that this argument can be formalized in *PRA* with the addition of the defining equations for ϕ.

construe proofs as configurations of symbols, defined in an elementary syntactical way; the fact that we *use* about proofs is that what they prove is true. In other words, what we need is not new entities but a *reflection principle*,

m is a proof of A $\supset$ A

or, more precisely, since A may contain parameters

$(\forall x)(\forall y)(y$ is a proof of $A\bar{x} \supset Ax)$.

Thus, we can say that the expansion of our ontology is only apparent.

On the basis of ideas such as this we could develop the number theory called "finitist" in Kreisel's Edinburgh lecture of 1958,[19] which is proof-theoretically equivalent to classical elementary number theory. I shall not, however, pursue the question how far one can go on a constructive basis with these ideas.

The reflection that even in this context proofs can be construed as configurations of symbols should suggest to us a prospect of going beyond elementary number theory: if proofs can be construed as configurations of symbols, then so surely can predicates. Doesn't this open up for us the prospect of so construing *sets*? In such a way as to yield a strong set theory, perhaps not: we are not assured that all sets are extensions of predicates and, given a formalism, we can by going beyond it define a set which is not the extension of any predicate in the formalism.

Nonetheless, it does seem that in weaker set theories we can identify sets with the predicates of which they are the extensions. In particular, this should be true of some theories of sets of natural numbers. The predicates could then be construed as Gödel numbers in some system of numbering, thus reducing the theories to an ontology of natural numbers.

For definiteness we shall discuss theories of sets of natural numbers which are built on classical first-order (elementary) number theory (*ENT*). This does not accord directly with the constructivistic conceptions of the existence of numbers which we have mentioned above, but a parallel treatment for intuitionistic theories could be given.

Thus we work in a two-sorted language in which we have variables $x, y. . .$ ranging over numbers and $X, Y,. . .$ ranging over sets

[19]See note 12. Gödel ("Über eine bisher noch nicht benützte Erweiterung," p. 281, n. 2) considers Kreisel's construction an extension of the "original finitary point of view" and says that it uses "abstract concepts which relate to nothing but finitary concepts and objects, in a finitary combinatorial way." Since such concepts are "counted as finitary mathematics" it appears that the "finitary concepts and objects" are such in the extended, not the original, sense.

of numbers. We add to the notation of *ENT* the predicate $x \in Y$. Identity of sets is not primitive but introduced by the definition

$$X = Y \text{ for } (\forall z)\, (z \in X \equiv z \in Y).$$

We shall also discuss "ramified" languages in which each set variable has a natural number (or in further developments a transfinite ordinal) as its level. The level of a variable will be indicated by a superscript. RA^n will be a theory in which variables of all levels $< n$ are admitted; RA^ω a theory in which variables of all finite levels are admitted.

Consider first the theory of arithmetic sets. A set X is arithmetic if there is a formula Az of elementary number theory such that $(\forall z)\, (z \in X \equiv Az)$ is true. The theory of arithmetic sets is a formal system in the above two-sorted language with the axioms and rules of *ENT* and the comprehension schema

$$(\exists\, Y)\, (\forall x)\, (x \in Y \equiv Fx)$$

where '*F*' stands for any formula without bound set variables. Induction is allowed for all formulae of the expanded notation.

The conception of an arithmetic set suggests that any such set X can be identified with some formula of which it is the extension. But how are we to interpret the predicate $\in$? '$z \in X$' will now say that X is a formula Ax (with x as its sole free variable) which is true of the number z—that is, $A\bar{z}$ is true. Thus we need the truth predicate for elementary number theory, which, Tarski's theorem shows, is not definable in elementary number theory. Hence, if we take the further step of identifying formulae with their Gödel numbers, we can talk of arithmetic sets without an expansion of our *ontology*, but we need an expansion of our *stock of predicates*.

Let T_0 be the truth predicate of *ENT*, given a suitable Gödel numbering. The theory of arithmetic sets can be reduced to a theory Z_1 which is *ENT* augmented by T_0 (see Appendix). We can carry the process further. Let T_1 be the truth predicate for Z_1. Let Z_2 be Z_1 augmented by T_1. Then we can reduce to Z_2 the theory of sets arithmetic *relative to* T_0—that is, those sets which are extensions of formulae of Z_1.

Iteration of this procedure shows that the systems of ramified analysis can be reduced in a natural way to an ontology of natural numbers. RA^n ($n > 0$) has variables of levels $< n$. An instance of the comprehension schema where the variable Y has level $m < n$ is admitted if 'F' contains bound set variables only of level $< m$ and free set variables only of level $\leq m$.[20] RA^ω (ramified analysis with arbitrary finite levels) is the union of the systems RA^n.

In the intended interpretation of ramified analysis, the variables of level m range over those sets which are extensions of formulae Ax with one free variable x, in which the bound set variables are of level $< m$. Thus we hope to identify these sets with the formulae of which they are extensions. Although this can be done easily enough, it is technically a bit smoother to note that these sets are just the extensions of formulae of Z_m (*ENT* if $m = 0$). Thus we can reduce RA^{n+1} first to the theory of sets arithmetic relative to $T_0 \ldots T_{n-1}$, and then to Z_{n+1}. RA^ω is thus reduced to the theory Z_ω which is the union of the Z_n (see Appendix).

Extension of this method into the transfinite reduces to an ontology of natural numbers any part of the transfinite sequence of systems of ramified analysis which count as predicative according to an analysis due to Feferman and Schütte.[21] The reduction itself is accomplished by means which are predicative according to this conception.

Note that the intended interpretation of ramified analysis also allows construing the quantifiers as substitutional. In some ways this is more natural than the procedure for identifying sets with numbers.[22]

---

[20]This restriction is adopted for the sake of predicativity—that is, so that no set is defined by means of quantification over a totality of which it is itself a member. The first theory developed along these lines was Russell's ramified theory of types. For general discussions of the concept of predicativity, see the first part of Feferman, "Systems of Predicative Analysis," and Gödel, "Russell's Mathematical Logic." A briefer and more elementary treatment is in my "Mathematics, Foundations of."

It should be observed that the description of RA^n is also meaningful if n is a transfinite ordinal. The conception of an extension of the ramified hierarchy into the transfinite is quite clear and simple (see, e.g., Feferman), but the construction of a formalism has technical complications. See Feferman; also Schütte, *Beweistheorie* and "Predicative Well-Orderings."

[21]Feferman, "Systems of Predicative Analysis," Schütte, "Predicative Well-Orderings" and "Eine Grenze für die Beweisbarkeit der transfiniten Induktion."

[22]In particular, in the latter procedure identity of sets is not translated by identity (or could be only by the highly artificial procedure of identifying each set with some *one* of the formulae of which it is the extension). This problem does not arise with the substitutional interpretation. Cf. "A Plea for Substitutional Quantification" (Essay 2 of this volume).

The predicative analyst is not limited to ramified systems, however. More workable formalisms are obtained by using the two-sorted formalism and insuring predicativity by restrictions on the comprehension principle. Given that such a formalism has quantifiers ostensibly ranging over all sets of natural numbers, how are we going to construe the theory so that the ontology will consist only of natural numbers?

Of course this can be done, and done by a natural model. The model will consist of all sets which are extensions of formulae of ramified analysis up to a certain level, and the sets can be construed as numbers in the same way as before.

This has the difficulty, however, that the model in question, though not artificial in the way in which one yielded by the Skolem-Löwenheim theorem would be, is not the standard model of the formalism in question. From a predicative point of view, there is no standard model, and our model is a *minimal* model. We can think of the formalism in question as representing a certain stage in a process of laying down principles for the construction of sets of natural numbers. In the minimal model, the quantifiers range over only those sets constructible by the means so far specified. But this is not the intended meaning of the quantifier in the *unramified* formulation, although it is in the ramified formulation. Rather, in talking about "all sets" we mean something like "the extensions of all predicates which we shall *ever* understand predicatively."

The following consideration might suggest that this difficulty is somewhat spurious. It is possible to describe systems with variables ranging over the natural numbers which are at least as strong as the (only impredicatively justifiable) totality of all systems of predicative analysis. For example, by a so-called "generalized inductive definition" one could define at one stroke the whole transfinite hierarchy of predicative truth predicates. The ontology of the theory resulting from elementary number theory by adding one such generalized inductive definition would certainly comprise only natural numbers, but it would suffice to interpret the whole of predicative analysis. In a more contrived way, we could set up a theory of natural numbers which would be in important respects as powerful as axiomatic set theory.

Thus from a certain nonpredicative point of view we can interpret predicative analysis so that its quantifiers range over natural numbers. But the alternative also exists of interpreting the set quantifiers as ranging over the nondenumerable totality of all sets of natural numbers, since an unramified predicative system is sound on the (impredicative) classical interpretation. Both the ontologically deflationary interpretation and the ontologically inflationary interpretation, however, are made "from outside" in terms of theories which are not predicatively acceptable.

If we confine ourselves to predicative interpretations, then an interpretation of the quantifier over all sets of numbers as really ranging over numbers necessarily falsifies its intent and also imposes a necessary ambiguity on such quantification: at a later stage in the development of predicative analysis, there will be more defining formulae for sets and therefore a more comprehensive range for the quantifier. However, all possible predicative ways of attributing to the quantifier a definite totality as its range have the feature both of yielding a denumerable range and of failing to yield all (predicative) sets of natural numbers.[23] If we think of the quantifier literally, as ranging over all sets of natural numbers, our failure to enumerate them and thus to identify them with numbers might be interpreted as meaning not that there are "more" such sets than there are natural numbers, but that it is not determinate what sets there are.

The fact that the ontology of predicative mathematics can be reduced to natural numbers from a nonpredicative point of view, but not from a predicative point of view, points to a rather obvious limitation of the notion of ontology. The ontology of a theory is, roughly, what the theory says that there is. But this provides only a minimal characterization of what someone commits oneself to who accepts the theory, since he might be able to justify or even understand the propositions of the theory only by relating them to statements which involve additional entities.

Appendix

To see that the formal theory of arithmetic sets is sound on the intended interpretation, we cannot simply take the quantifier $(\exists Y)$ in a comprehension axiom to be satisfied by 'F' itself, since if it contains free set variables, it is not a formula of *ENT*. Rather, if 'F' is a formula $A(x, x_1 \ldots x_m, Y_1 \ldots Y_n)$ with free variables as indicated,

---

[23] Of course, according to Feferman ("Systems of Predicative Analysis," part II) there is a single formal system (called IR) which codifies the whole of predicative analysis, and it has a minimal model which contains *all* predicative sets of natural numbers. But this model is not itself predicatively definable, and the system IR as a whole is not predicatively recognizable as sound.

we must show that if $Y_1 . . . Y_n$ are extensions of arithmetic formulae $B_1(z) . . . B_n(z)$, then for each $x_1 . . . x_m$ a Y can be found such that

$$(\forall x) \ [x \in Y \equiv A(x, x_1 . . . x_m, Y_1 . . . Y_n)].$$

Clearly the required y is the extension of the formula

(1) $\qquad A[x, \bar{x}_1 . . . \bar{x}_m, \hat{z}B_1(z) . . . \hat{z}B_n(z)]$

obtained from $A(x, \bar{x}_1 . . . \bar{x}_m, Y_1 . . . Y_n)$ by replacing each part $t \in Y_i$ by $B_i(t)$.

A similar argument shows that the systems of ramified analysis are sound on the intended interpretation.

The system Z_1 is the result of adding to *ENT* a new predicate T_0 with axioms corresponding to the inductive definition of arithmetic truth.[24] In terms of T_0 we can define the predicate $T_0'(x, y)$.

> y is the number of a formula A (z) of *ENT* with one free variable, and A $(\bar{x})$ is true.

Then to interpret RA^1 in Z_1, take the set variables to range over the numbers of formulae of *ENT* with one free variable, and interpret $x \in Y$ as T_0' (x, y). Then all instances of the comprehension schema are transformed into provable formulae of Z_1.

Consider the comprehension axiom above and let $y_1 . . . y_n$ be the numbers of $B_1(z) . . . B_n(z)$ respectively. The number of (1) is given primitive recursively from $x_1 . . . x_m, y_1 . . . y_n$ by a function ϕ (which depends on A) and we can prove in Z_1

$$(\forall x) \ [T_0'(x, \phi \ (x_1 . . . x_m, y_1 . . . y_n))$$
$$\equiv A(x, x_1 . . . x_m, \hat{z}T_0'(z, y_1) . . . \hat{z}T_0'(z, y_n))]$$

which yields the translation of the comprehension axiom by existential generalization.

It is only because the comprehension schema allows free set var-

---

[24]E.g., Hilbert and Bernays, *Grundlagen der Mathematik*, II, 347-348. In Z_1 induction is again admitted for all formulae.

iables that the notion of truth enters essentially. If the comprehension schema is admitted only for formulae of *ENT* proper, then the resulting system is a conservative extension of *ENT*.

Consider a system Z^* resulting from *ENT* by adding additional predicates $S_1 \ldots S_k$ and perhaps additional axioms. The theory $RA^1(S_1 \ldots S_k)$ of sets arithmetic relative to $S_1 \ldots S_k$ is obtained from Z^* in the same way that RA_1 was obtained from *ENT*.

Then we can form a theory Z_1^* by adding to Z^* a new predicate T^* with axioms for the inductive definition of truth for Z^*. The above method yields an interpretation of $RA^1(S_1 \ldots S_k)$ in Z_1^*.

Let Z_{n+1} be the result of adding in this way the truth predicate T_n of Z_n, so that Z_{n+1} is Z_n^*. Then we have in Z_{n+1} an interpretation of $RA^1(T_0 \ldots T_n)$.

In RA^2 we can eliminate the set variables of level o in favor of T_0. This yields an interpretation of RA^2 in $RA^1(T_0)$. Then the above method yields an interpretation of $RA^1(T_0)$ in Z_2 and hence of RA^2 in Z_2.

Suppose we have constructed by this method an interpretation of RA^n in Z_n. Then we can use it in RA^{n+1} to eliminate the variables of level less than n in favor of $T_0 \ldots T_{n-1}$. This yields an interpretation of RA^{n+1} in $RA^1(T_0 \ldots T_{n-1})$. Then the above method yields an interpretation of $RA^1(T_0 \ldots T_{n-1})$ in Z_{n+1} and hence of RA^{n+1} in Z_{n+1}.

Intuitively, the procedure corresponds to construing sets of level m as extensions of formulae of *ENT* with the additional predicates $T_0 \ldots T_{m-1}$.

Postscript

In the remarks about primitive recursion in section V some important issues are glossed over. The justification of individual primitive recursions is perhaps best viewed as suggesting a picture, which no longer fits when we go to nested double recursion. To formalize the argument, we would have to assume some operations on strings of symbols of the same character as elementary primitive recursive functions. If we give ourselves this much, and allow ourselves induction on predicates of the form $(\exists y)Axy$, then we can show of *each* primitive recursive function that it is everywhere defined. For the double recursion, as the text indicates, we would have to give ourselves induction on a predicate of the form $(\forall y)(\exists z)Axyz$. More needs to be said on why this difference is significant. One reason given in the literature, that the latter does, while the former does not, involve the general notion of function, is not congenial to the general thesis of this section.[25]

Again, one can ask how we can see the truth of the reflection principle for, say, *PRA*. The problem about defining truth which would arise for a formalism with quantifiers does not arise here, since truth for a given substitution for the free variables can be defined by a predicate of the form $(\exists y)Axy$ (with A primitive recursive, or in the language about strings of symbols alluded to above). But in order to carry out a proof of the reflection principle in the form given in the text, we shall have to resort again either to a more powerful recursion or to induction on a predicate of the form $(\forall y)\,(\exists z)Axyz$. Even if it is conceded that an expansion of ontology is unavoidable at this point, however, the point of the text that *intensional* objects are not really required still stands. Moreover, the "general notion of function" that would be involved is that of *constructive* function and would therefore still be a lot less than a notion of function which would be equivalent to the notions of *set* considered in the remainder of the essay.

The reduction of predicative theories of sets of natural numbers requires the concept of truth in the form of truth predicates for given first-order theories. Such predicates can of course be explicitly defined in second-order languages, that is, by quantifying over sets of elements of the domain of the theory. Our reduction is a *reduction* only if we envisage some other way of understanding the truth predicates than by such a definition, as the late Leslie Tharp reminded me. The example of the predicate "natural number" shows that we do understand predicates introduced by inductive conditions in a more direct way. The reduction of inductive to explicit definitions has an inherently impredicative character, in that the range of the second-order quantifier in the explicit *definiens* has to include the extension of the predicate defined, in order for the introduction of the new predicate to have the consequences required. Now we could view this in a typically set-theoretic manner, where what the definition does is to define a predicate by reference to the given totality of subsets of the domain of the theory. However, such cannot be the way in which the inductive generation of the natural numbers themselves is understood, nor is it appropriate to a predicative conception of sets of integers. Here it is not of particular moment whether what is inductively defined is a cer-

---

[25]For example Tait, "Finitism," p. 544.

tain totality of sets (say, those at a particular level of the ramified hierarchy) or a truth predicate. If a second-order quantifier is involved at some point, we do not suppose its range to be definite prior to the introduction of the inductively defined predicate. Thus an inductive definition serves to introduce a predicate which was not previously understood. Where it does have an inherently second-order character is in the fact that constitutive of our understanding of it is the associated induction principle. It is quite true that generalization of predicate places is involved. But we cannot look to any more ultimate source than the inductive definition itself for the justification of treating the new predicate as within the scope of such generalization. If we assume the direct method of such generalization, then the range of the quantifier must be regarded as quite open-ended. If we use the method of semantic ascent, then the implicit notion of satisfaction and truth must have a similar character; otherwise there would be circularity since it is only by inductive definitions that *definite* truth predicates are introduced.

Feferman's analysis of predicativity implies that "generalized inductive definitions" such as that of Kleene's recursive ordinal notations are impredicative, even though they may introduce predicates of natural numbers that can be added to the language of first-order arithmetic. The above remarks indicate that we are not obliged to disagree with him about the predicative character of generalized inductive definitions. There is obviously a technical sense in which they need not increase ontology, since a theory resulting from first-order arithmetic by adding one or more generalized inductive definitions will still have quantifiers ranging only over the natural numbers. The question still remains about the generalization of predicate places that is involved in our understanding of the introduced predicate and of the formal theory.

2

A Plea for Substitutional Quantification

In this note I shall discuss the relevance to ontology of what is called the *substitutional* interpretation of quantifiers. According to this interpretation, a sentence of the form '$(\exists x)Fx$' is true if and only if there is some closed term 't' of the language such that 'Ft' is true. This is opposed to the *objectual* interpretation according to which '$(\exists x)Fx$' is true if and only if there is some object x in the universe of discourse such that 'F' is *true of* that object.[1]

Ontology is not the only connection in which substitutional quan-

From *Journal of Philosophy*, 68 (1971), 231-237; reprinted by permission of the editors.

I owe to Sidney Morgenbesser and W. V. Quine the stimulus to write this paper. I am also indebted to Hao Wang for a valuable discussion and to Quine for illuminating comments on an earlier version.

[1]It is noteworthy that the substitutional interpretation allows truth of, say, first-order quantified sentences to be given a direct inductive definition, while in the objectual interpretation the fundamental notion is truth *of* (satisfaction), and truth is defined in terms of it. Davidson's version of the correspondence theory of truth would not be applicable to a substitutional language. See "True to the Facts."

Quine has pointed out to me that in a language with infinitely many singular terms, a problem arises about defining truth for *atomic* formulae. Objectual quantification can be nontrivial for a language with no singular terms at all, but, if there are such, the problem can be resolved by an auxiliary definition which assigns them denotations, relative to a sequence that codes an assignment to the free variables. In the substitutional case, we need to define the truth of an atomic formula $Pt_1 \ldots t_m$ directly for closed terms $t_1 \ldots t_m$, for example, inductively by reducing the case for more complex terms to that for less complex terms. In the usual language of elementary number theory, closed terms constructed from 'o' by applications of various function symbols which (except for the successor symbol, say 'S') have associated with them defining equations that can be regarded as rules for reducing closed terms to canonical form, namely, as numerals (a numeral is 'S' applied finitely many times to 'o'), so that an atomic formula $Pt_1 \ldots t_m$ is true if and only if $Pn_1 \ldots n_m$ is true, where n_i is the numeral corresponding to t_i. The truth of predicates applied to numerals is defined either trivially (as in the case of ' $=$ ') or in a similar recursive manner.

Quine discusses this issue in section 6 of "Truth and Disquotation."

tification has been discussed in recent years. It has been advocated as a justification for restrictions on the substitutivity of identity in intensional languages[2] or has been found necessary to make sense of restrictions adopted in certain systems.[3] With this matter I shall not be concerned.

It might seem that in discussions of ontology the substitutional interpretation of quantifiers would be advocated in order to make acceptable otherwise questionable ontological commitments. In fact this has not been done widely, although it does seem to be an important part of Wilfrid Sellars's account of abstract entities.[4] The issue of the ontological relevance of substitutional quantification has been raised most explicitly by Quine, essentially to debunk it. He argues that only an interpretation of a theory in terms of *objectual* quantification attributes an ontology to it.

Classical first-order quantification theory on the objectual interpretation, according to Quine, embodies the fundamental concept of existence. He acknowledges the existence of other possible concepts of existence. But he holds that substitutional quantification does not embody a genuine concept of existence at all.

I should like to argue that the existential quantifier substitutionally interpreted has a genuine claim to express a concept of existence which has its own interest and which may offer the best explication of the sense in which "linguistic" abstract entities—propositions, attributes, classes in the sense of extensions of predicates[5]—may be said to exist. I shall then raise a difficulty for the view (which Sellars may hold) that all quantification over abstract entities can be taken to be substitutional.

Quine argues as follows for the view that substitutional quantification does not correspond to a genuine concept of existence:

[2]Marcus, "Modalities and Intensional Languages," and Quine, "Reply to Professor Marcus."

[3]Particularly Hintikka's. See Føllesdal, "Interpretation of Quantifiers," and papers by Hintikka and Føllesdal referred to there.

[4]"Abstract Entities." Sellars's general strategy is to treat attributes and classes as analogous to linguistic types and then to quantify substitutionally over them. This seems to be open to the objection presented below. Sellars's account of classes seems to yield a predicative theory of classes and thus would not justify set theory. See "Classes as Abstract Entities and the Russell Paradox." In studying these two papers I have relied heavily on Gilbert Harman's lucid review of *Philosophical Perspectives*.

[5]This account of classes would suggest distinguishing *classes* in this sense from *sets* as the objects of set theory. One can, I believe, motivate the requirement of predicativity for the former. It is they which are parallel to attributes. It is noteworthy that there is no reason to make impredicative assumptions about the existence of attributes, unless (as in Quine's interpretation of Russell's no-class theory) one seeks to reduce sets to attributes.

Substitutional quantification makes good sense, explicable in terms of truth and substitution, no matter what substitution class we take— even that whose sole member is the left-hand parenthesis. To conclude that entities are being assumed that trivially, and that far out, is simply to drop ontological questions. Nor can we introduce any control by saying that only substitutional quantification in the substitution class of singular terms is to count as a version of existence. We just now saw one reason for this, and there is another: the very notion of singular terms appeals implicitly to classical or objectual quantification.[6]

In answer to the first objection, we should point out two formal features of the category of singular terms that mark substitutional quantification with respect to it as far less trivial than with respect to, say, the left parenthesis. First, it admits identity with the property of substitutivity *salva veritate*. Second, it has infinitely many members that are distinguishable by the identity relation. This has the consequence that '$(\forall x)Fx$' is stronger than any conjunction that can be formed of sentences of the form 'Ft', while '$(\exists x)Fx$' is weaker than any disjunction of such sentences.

With respect to the claim that the very notion of singular terms appeals implicitly to classical or objectual quantification, we might hope for a purely syntactical characterization of singular terms. However, that would not yet yield the distinction between singular terms that genuinely refer and those which do not; in a language in which the latter possibility arises, the substitutional quantifier for singular terms would express not existence but something closer to Meinong's "being an object."

However, we can concede Quine's point here for a certain central core class of singular terms, which we might suppose to denote objects whose existence we do not expect to explicate by substitutional quantification. We might then make certain analogical extensions of the class of singular terms in such a way that they are related to quantifications construed as substitutional. The criterion for "genuine reference" is given in other terms.

For example, the following is a natural way to introduce a predicative theory of classes (extensions of predicates). Let 'F' stand for a one-place predicate of some first-order language. We first rewrite 'Ft' as '$t\epsilon\{x:Fx\}$' and (taking 'α', 'β',. . .as schematic letters for expressions of the form '$\{x:Fx\}$') define '$\alpha = \beta$' as '$(\forall x)(x\epsilon\alpha \equiv x\epsilon\beta)$'. So far we have just made the contextual definitions involved in the

---

[6]"Existence and Quantification," p. 106. Cf. *Ontological Relativity*, pp. 63-64.

theory of virtual classes.[7] We then allow the abstracts to be replaced by quantifiable variables of a new sort. The substitution interpretation gives truth-conditions for formulas in the enlarged notation. The process can be repeated to introduce classes of higher levels.[8]

The advantage of substitutional quantification in this particular case is that it fits the idea that the classes involved are not "real" independently of the expressions for them. More precisely, we know the condition for a predicate to "have an extension" (that it be true or false of each object in the universe) and for two predicates to "have the same extension" without independently identifying the extension. The fact that the substitution interpretation yields truth-conditions for quantified sentences means that everything necessary for speaking of these classes as entities is present, and the request for some more absolute verification of their existence seems senseless.

The obstacle to the introduction of attributes in the same way is, of course, the problem of the criterion of identity. But the procedure goes through, given a suitable intensional equivalence relation. For example, we might introduce "virtual attributes" by rewriting 'Ft' as '$t\,\delta\,[x{:}Fx]$', introducing 'ξ', 'η'. . . as schematic letters, defining '$\xi = \eta$' as '$\Box(\forall x)\,(x\,\delta\,\xi \equiv x\,\delta\,\eta)$', and then introducing attribute variables and substitutional quantification. Then two predicates express the same attribute provided they are necessarily coextensive.

The same procedure could be followed for other intensional equivalences, provided that they can be expressed in the object language.

Consideration of examples such as these leads to the conclusion that in the case where the terms involved have a nontrivial equivalence relation with infinitely many equivalence classes, substitutional quantification gives rise to a genuine "doctrine of being" to be set alongside Quine's and others. It parallels certain idealistic theories of the existence of physical things, such as the account of perception in Husserl's *Ideen*.

It might be thought preferable in our case and perhaps in all cases where the substitution interpretation is workable, to formulate the theory by quantifying over expressions themselves or over Gö-

---

[7]Quine, *Set Theory and Its Logic*, p. 15.
[8]Quine carries out the first stage of this substitutional introduction of classes in *Philosophy of Logic*, pp. 93-94. If classes are introduced in this way in a language that is not extensional, such as modal logic, then the restrictions on substitutivity of identity associated with a substitutional semantics have point.

del numbers that represent them. If one is talking of expressions or numbers already, this has the advantage of ontological economy, and in some cases, such as when one begins with elementary number theory, it makes more explicit the mathematical strength of the theory.[9] The substitutional approach avoids the artificiality involved in introducing an apparatus (be it Gödel numbering or some other) for talking *about* expressions, and it avoids the unnatural feature that identity of expressions does not correspond to identity of extensions or attributes.

In the case of attributes, however, some proposed criteria of identity, such as synonymy, are metalinguistic. Then the above substitutional introduction of attributes does not apply. Here some form of quantification over linguistic types gives the most natural formulation. An example is Sellars's construal of attributes as synonymity-types of expression-tokens.

The manner in which we have introduced classes suggests a rather arbitrary limitation in the case where the universe for the first-order variables contains unnamed objects. For consider a two-place predicate 'F'. In a theory of virtual classes, we would admit the abstract '$\{x:Fxy\}$' with the free variable 'y', but substitutional quantification as we have explained it encompasses only closed terms. If we wish to say that for every y the class $\{x:Fxy\}$ exists, then the notion of substitutional quantification must be generalized. Suppose a language has variables of two sorts, 'X', 'Y', . . .which are substitutionally interpreted, and 'x', 'y'. . .which are objectually interpreted over a universe U. Then the fundamental notion (see fn 1) is satisfaction of a formula by a sequence of elements of U. A sequence s satisfies '$(\exists X)FX$' if and only if for some term 'T' of the upper-case sort, with free variables only of the lower-case sort, some extension of s satisfies 'FT'.[10]

---

[9] See for example the last section of my "Ontology and Mathematics" (Essay 1 of this volume).

[10] The direct construal of classes as expressions fails in this case, and if the universe is of larger cardinality than the set of expressions of the language, classes cannot be so construed even artificially. But of course classes *can* be construed as pairs consisting of an expression and a sequence of substituends for the free variables.

[Here finite sequences were meant. I wrote here "some extension of s" to take care of the possibility that 'T' might contain free variables not already free in 'FX'. Thus this condition already differs from the one used by Quine in *The Roots of Reference*, pp. 106-110, which might be stated thus:

s satisfies '$(\exists X)FX$' if and only if for some term 'T' of the upper-case sort, with free variables only of the lower-case sort and free in 'FX', s satisfies 'FT'.

This difference was missed by some readers (including at this time Quine; see

In this case we can no longer say that classes are not real independently of expressions for them, but each class is a projection of a relation of which this *can* be said. We can say that classes are not real independently of expressions for them and individuals of the universe.

The instances of substitutional quantification we have discussed would suggest that in the process of analogical extension of the category of singular terms, the syntactic characteristics of this category are not all essential. Thus in an extensional language the role of identity can be taken over for one-place predicates by the truth of '$(\forall x)$ $(Fx \equiv Gx)$', and the introduction of "ramified second-order" substitutional quantifiers seems to differ only notationally from the above introduction of classes.[11] However, such a notational step as the introduction of 'ϵ' and abstracts is necessary if one is to take the further step of reducing the two-sorted theory to a one-sorted one. (The resulting quantifiers would have an interpretation which mixes the substitutional and the objectual.)

In connection with attributes one might be inclined to retreat another step from singular terms and forsake identity. In other words, one might regard quantification over attributes as substitutional quantification of predicates, with no equivalence relation with properties corresponding to those of identity. Here I would agree with Quine that these "attributes" would be at best second-class entities. The ability to get at "the same object" from different points of view—different individual minds, different places and times, different characterizations by language—is one of the es-

---

ibid., p. 107). The conception of the text is the same as that of "Sets and Classes" (Essay 8 of this volume) and avoids Quine's difficulty (ibid., pp. 108-109) about the law of unit subclasses.(For detail on this point, see Gottlieb, *Ontological Economy*, pp. 78-79.)

The above formulation had a defect which was corrected in "Sets and Classes": *s* may assign objects to variables free in '*T*' but not in '*FX*' and thus constrain their assignment in an unintended way. But in fact this makes no difference to what formulæ are satisfied, since '*T*' can always be chosen to contain no free variables not free in '*FX*' which are assigned objects by *s*.]

[11]Montgomery Furth in "Two Types of Denotation" reconstructs along these lines Frege's idea of predicates as "denoting" concepts rather than objects. He does not remark that on this reconstruction Frege is unjustified in using full (impredicative) second-order logic for quantifying over concepts (more generally, functions). But replacing it by a ramified second-order logic would require him to abandon his Cantorian conception of cardinal number.

Frege's principle that words have meaning (*Bedeutung*) only in the context of proposition, and his use of it to defend his thesis that numbers are objects, suggests a substitutional account of quantification over "logical objects" such as (on his view) numbers and extensions. Cf. my "Frege's Theory of Number" (Essay 6 of this volume).

sentials of objective knowledge. If this is lacking, then the entities involved should be denied objective existence.

Can all quantification over abstract entities be construed as substitutional? Evidently not if sets as intended by the usual (impredicative) set theories are included. Otherwise we could certainly set up a theory that talked of numbers, of classes in a ramified hierarchy, and even (again in a ramified hierarchy) of propositions and attributes, where the quantifiers over these entities admitted a substitutional interpretation.[12]

Nonetheless there is an obstacle to taking this as implying that nonsubstitutional quantification over abstract entities is avoidable, short of set theory. The difficulty is that the truth-conditions for substitutionally quantified sentences themselves involve quantification over expression-types:

'$(\exists x)Fx$' is true if for some *closed term* 't', 'Ft' is true.

Thus the question arises whether the language in which we give this explanation quantifies substitutionally or objectually over expressions. If we give the latter answer, we have of course given up the claim to rely only on substitutional quantification of abstract entities.

In the former case, the same question arises again about the semantics of the metalanguage. We are embarked on a regress that we shall have to end at some point. Then we shall be using a quantifier over expressions that we shall have to either accept as objectual or show in some other way to be substitutional. One might hope to interpret it so that it does not range over abstract entities at all, for example so that there is quantification only over tokens. The ways of doing this that seem to me at all promising involve introducing modality either explicitly or in the interpretation of the quantifier, so that one does not get an ontology of tokens in a Quinean sense.[13]

The only way I can see of showing the quantifier to be substitutional would be to show that the *given* sense of quantification over expression-types in a natural language, say English, is substitutional. This seems to me very implausible.

However, the most this argument would show is that we could not know or prove that the only quantification over abstract entities

---

[12]Whether in the original sense or in the generalized sense of two paragraphs back would depend on whether the universe of concrete individuals contained unnamed objects. One would suppose the projections referred to above would be needed for some applications.

[13]See section III of "Ontology and Mathematics" (Essay 1 of this volume).

that we relied on was substitutional. It could still be that an outside observer could interpret our talk in this way. To refute the view that abstract entities (short of set theory) exist only in the substitutional sense, one needs to give a more convincing analysis of such entities as expression-types and numbers, as I have attempted to do elsewhere.[14]

[14]Ibid.

3

Informal Axiomatization, Formalization, and the Concept of Truth

Axiomatization is an especially thoroughgoing and rigorous instance of a process in the organization of knowledge that might be called systematization. The objective is to organize a body of knowledge (or of theory that aspires to be knowledge) in such a way as to clarify its structure and strengthen its justification *as a whole*. In particular, one seeks to single out certain concepts and principles as "primitive" or "fundamental" and others as "defined" or "derived." The method of axiomatization, first applied to geometry in ancient times and epitomized by Euclid's *Elements*, presents a theory by singling out certain primitive notions and defining others from them, and singling out certain propositions as *axioms* and deriving all other propositions of the theory by *deduction*.

These notions of definition and deduction are not without ambiguity. It is a mark of an informal axiomatic theory that a general background is used in developing it that is not itself axiomatized in the theory itself. In modern mathematics this background can include logic, arithmetic, and even some analysis and set theory.

It would seem desirable to restrict this background to pure logic, since it should contain the most general rules of definition and deduction which should be applicable to all sciences. Everything else used in developing the theory would have to be axiomatized. Although Euclid may already have had such a procedure in mind,

Revised and expanded version of a lecture given at a conference on the Philosophy of Mathematics in Santa Margherita Ligure in June 1972, sponsored by the Unione Matematica Italiana; published in *Synthese* 27 (1974), 27-47, and also in *Bollettino della Unione Matematica Italiana* (4) 9, 87-107. Copyright © 1974 by D. Reidel Publishing Company, Dordrecht, Holland. Reprinted by permission of D. Reidel Publishing Company and the editors of *Synthese*.

I am grateful to Maria Luisa Dalla Chiara, G.H. Müller, and Dag Prawitz for valuable comments on the lecture on which the paper is based.

one could probably not have achieved it for any serious mathematical theory until an exact characterization of the logical inferences used in mathematics was available: that is, until the time of Frege.

Increasing demands for exactness of axiomatization and for elimination of assumptions not explicitly given as axioms leads to the questioning even of logic as unaxiomatized "background." Even before this point is reached, the validity of the deductions in the theory comes to depend less and less on the intended meaning of the primitives: since whatever is assumed about what the theory is about is to be explicitly stated in axioms, the theory might be interpreted to be about *anything* for which the axioms turn out to be true.

One possible outcome of rigorous axiomatization is the reduction of axiom systems to definitions of general types of structure. Such, for example, is the usual interpretation of the "axioms" of the "theory" of groups, rings, of fields.

Another possible outcome is *formalization*. If all background theory, including logic, and everything given by the meaning of the primitives are to be taken up into the axiomatization, then there seems to be no alternative to describing the axioms and the inferences from them in terms that do not appeal to interpretation: in effect, in terms of syntactical descriptions of the linguistic forms involved. But to do this precisely and simply, we replace the language involved by an artificial syntax. A theory presented in this way has been *formalized*.[1]

Although a formalized theory is described in a way which does not appeal to interpretation, to be of interest it must *have* interpretations. If it has been derived from an informal axiomatic theory, the informal theory survives as a possible interpretation of the formalism. It is this relation that I want to discuss in this paper, particularly in the light of classical metamathematical theorems on the "limitations" of formal systems.

I

A formalized theory is a mathematical structure which has as its base something analogous to a language. Certain combinations of

[1] I have discussed these matters more fully in "Mathematics, Foundations of," pp. 190-192.

symbols are designated by a purely combinational definition as "formulae" or "sentences." The definition endows them with a "grammatical" structure: in effect the formation rules of a formalism constitute a complete grammar. Further definitions specify certain sequences or trees of formulae as "proofs" or "deductions," usually by way of specifying certain formulae as *axioms* and then giving *rules of inference* that enable one to construct a proof of a formula from given proofs of other formulae.

For the present we do not need to answer the question whether a formalism of this sort should be viewed as a real language which might actually be used to present mathematical proofs, or simply as a mathematical model of some aspects of mathematical discourse. These two points of view correspond to two purposes that formalism is intended to serve: rigor and exactitude in the formulation of theories and proofs, and amenability to structural investigation by mathematical methods.

What is important is rather what is omitted both from the formalized language and its description by the formal definitions. A real language is of course not just a mathematical structure, and if a formalism *is* a real language there must be more to it than is given by the formal definitions, and if it is not, there is something essential to what it represents that the mathematical structure does not give us. Of course we want to say that the formulae of a formal system are meaningful, or at least that they can be subjected to various interpretations. But the formal structure is what it is independently of these interpretations.

This obvious remark should remind us of something else, equally obvious. The formalized language is typically interpreted to talk only about mathematical objects and their relations and thus leaves out much that belongs to the actual language of mathematicians— reference to themselves and each other, talk of what is "obvious" and what is not, metalinguistic remarks. These matters are generally not taken to be part of the subject matter of mathematics (at least pure mathematics). Nonetheless particularly the fact of *semantical* talk about mathematical statements—what sentences mean, under what conditions they are true, what interpretations of theories render them true—provides at least heuristic guidance for extending mathematical theories. The existence of such talk leads many to claim that we always know more mathematics than is represented in one formalized theory, with what justice we shall consider.

In part, however, this observation shows an incompleteness in our description so far of the process of formalization. For the possible interpretations of a formalized theory can of course also be characterized mathematically. We can put the matter this way: The dimension of meaning which mathematical language of course as, and which enters very prominently into informal mathematical discourse, can at least in one of its aspects be the object of a mathematical theory, in which the relations of *reference* involved in a theory's talking about a particular mathematical structure can be characterized by mathematical definitions.

"Reference" is here used in the sense of Frege's *Bedeutung*. It thus includes not just denotation as a property of names and other singular terms but also satisfaction and truth as properties of predicates and sentences. For formalisms based on classical first- or second-order logic, we all know how to define interpretations, which can be described as assigning a domain for the quantifiers and a reference for each of the primitive expressions in a number of syntactical categories—names, functors, predicates. Then the reference of syntactically more complex expressions (relative to assignments of values to the free variables) is determined by conditions which can take the form of an inductive definition. In particular, since the reference of a closed sentence in this sense is its truth-value, we have a mathematical definition of what it is for a sentence to be true under an interpretation. (This is, however, relative to a specification of references for the primitives. We do not have a categorical definition in mathematical terms of the relation of reference, but rather a definition of how the reference of the syntactically complex expressions of a formal language are determined by the syntactically simpler expressions. But such a categorical definition can be given if the primitives are themselves mathematical.)

That in the actual use of language a given expression has this or that reference depends in a very complex way on the practices of the users of the language, the intentions of the individual speakers, extralinguistic facts, and perhaps internalized structures in the human mind or brain. Tarski's discovery that one could take certain such facts as given and in terms of them characterize all the relations of reference for a formalized language in mathematical terms is certainly very remarkable.[2] It also opened the way to much larger applications of the method of formalization and to a deeper un-

---

[2]For an interesting recent discussion of its general significance, see Field, "Tarski's Theory of Truth."

derstanding of the relations of informal mathematical theories and their formalized counterparts.

The actual carrying out of Tarski's construction derives from the earlier discovery of Gödel, that the syntax of formalized languages, including the characterization of proofs, can be modeled in elementary arithmetic. In particular, a formalized theory that includes arithmetic, or, equivalently, the theory of finite sets, has the property that it can interpret the theory of its own syntax.

We could go further and actually *identify* formal expressions with numbers—for example the simple symbols of an alphabet with certain particular numbers, and then the sequences and trees which are expressions (e.g., formulae) and proofs with numbers which code them. In that case, syntactical predicates (e.g., 'variable', 'term', 'formula', 'axiom', 'proof') are extensionally equivalent to predicates of elementary arithmetic.

How close they are to being intensionally equivalent depends on how much of the intended application, i.e., the relation to real language, is built into the sense of the syntactical terms. To begin with we remark that only the expressions as *types* are identified with numbers, but the normal concept of a linguistic expression is of a form which can be embodied in instances which are physical (patterns of soundwaves or of ink-on-paper) or perceptual. This interpretation indicates that if we insist on an *identification* of syntax with its arithmetic model, then some generality is lost. Although we can (somewhat unnaturally) identify the types of utterances or inscriptions with numbers, syntax as a structural theory does not at all depend on this construal. It is probably best for looking at mathematical logic as *mathematics* to think of the syntax of a particular formalized language as an abstract structure.

II

With this background, I want to make some comments on a few of the classic theorems on the "limitations" of formal systems: Tarski's theorem on the indefinability of truth and Gödel's incompleteness theorems.

For our general theme, the relation of intuitive theories and formalized theories, Tarski's theorem is more immediately significant and illuminating than Gödel's. For definiteness, I shall suppose we are dealing with a formalized theory F whose logic contains the usual classical first-order logic and such that the fragment Q

of elementary number theory is relatively interpretable in F.[3] Suppose now we have a specific interpretation of the language of F, so that the quantifiers are assigned a domain D, names a denotation in D, n-place functors function from D^n to D, and n-place predicates, functions from D^n to $\{\top, \bot\}$. Suppose we have a specific coding of the expression of F by numbers.

Now consider a formula $T(a)$ of F with one free variable. We suppose that T expresses the predicate of *truth* under our given interpretation. If so we would expect that if A is a closed formula with number p, then

(1) $T(\bar{p}) \equiv A$

would at least be true.[4] But almost trivially we can show that some formula of the above form is *refutable* in F. Hence if our interpretation is actually a *model* of F, then T is not (extensionally) the truth-predicate. The mere consistency of F implies that not all formulae of form (1) are provable in F.

The argument for this is of course a formalization of the liar paradox: For a certain closed formula $\sim T(t)$ with number g, we can prove $t = \bar{g}$, $\sim T(t)$ "says" of itself that it is not true. $T(\bar{g}) \equiv \sim T(t)$ clearly leads immediately to a contradiction.

Let us try to visualize how this theorem works in a concrete situation, where the formal system in question is a formalization of number theory.

As an informal axiomatization of number theory I offer that of Dedekind-Peano. This can be stated in a number of ways. I suppose that 'o', '(natural) number', and 'successor' are primitives and that the axiomatization is stated as follows:

(A1)   o is a number

(A2)   Successor is a one-one mapping from the numbers onto the numbers other than zero.

(A3)   A property which holds for o and for the successor of every number for which it holds, holds for every number.

This formulation evidently envisages that set-theoretic notions such as "mapping" and the perhaps obscure notion of "property" are part of the general background about which informal reasoning

---

[3]Tarski, Mostowski, and Robinson, *Undecidable Theories*, p. 51.
[4]$\bar{p}$ is the formal numeral for the number p.

is possible. But of course in formalizing we cannot leave things at that.

The usual number-theoretic formalisms take the natural numbers as the domain of individuals, and hence the formalism does not contain the predicate 'number'. (A1) is then taken care of simply by admitting 'o' as a name, (A2) is taken care of by admitting a 1-place functor S and the axioms

$$\forall x(Sx \neq o)$$
$$\forall x \forall y(Sx = Sy \supset x = y)$$

Identity is here counted as logical.[5]

(A3) poses problems. A formalization of number theory in effect calls for an axiomatization of the notion of "property." We can suppose that well-defined predicates give us properties, so (A3) yields

(A3′)   A predicate true of o, and such that for any number x, if it is true of x then it is true of the successor of x, is true of all numbers,

which suggests that in the formalism we might take as axioms all formulae of the form

(2) $Ao \wedge \forall x(Ax \supset A(Sx)). \supset \forall x Ax$

But how comprehensive is the language here? We have in fact not obtained from the informal axioms a unique formal system. The reason is that in intuitive number theory, the natural numbers have more structure than is given by o and S.

The most satisfying solution is perhaps to give ourselves *second-order logic*, and then (A3) is simply rendered as

(3) $\forall F(Fo \wedge \forall x(Fx \supset F(Sx)). \supset \forall x Fx$

and it turns out that much of the further structure that arises in

---

[5](A2) would suggest the additional axiom

$$\forall x(x \neq o \supset \exists y(x = Sy)).$$

Given induction, this formula is provable.

ordinary number theory, such as the arithmetic operations, can be explicitly defined.

Nonetheless we shall below cast doubt on the hypothesis that this formalization captures everything intended by the informal axiomatization, so that we cannot neglect weaker systems, in particular the *first-order number theories* obtained by adding additional functors with recursive defining equations, allowing only the logical apparatus of first-order logic, and then rendering (A3) by the schema (2). We can confine our attention to the most usual such system Z, where addition and multiplication are the only additional functions represented by primitive functors.[6]

In the case both of the second-order theory Z^2 and of first-order number theory Z, this transition from the informal axiomatic theory to the formalism contains the material for an *interpretation* of the formal system that results—we simply assign to 'o' the number o, to 'S' the successor function, to recursively defined functors the appropriate functions, and choose as domain of individuals the natural numbers. In the first-order case, a rather simple inductive characterization of the truth-predicate can be given, with the help of which we can prove that all provable sentences of Z are true, and hence that Z is consistent.

Then Tarski's theorem tells us that this truth-predicate T_0 is not coextensive with any formula of Z. That is, for every formula $A(x)$ there is a number n such the either $T_0 n$ and $A(x)$ is not true of n, or $\sim T_0 n$ and $A(x)$ is true of n.

If we add to the formalism a new predicate T_0 and some axioms corresponding to the inductive characterization of truth, then for each closed formula A of Z we can prove in the expanded formalism

$$(4) \qquad T_0(p) \equiv A$$

and then, applying the liar paradox argument, prove for any formula $T(x)$ of Z with one free variable x

$$\sim(T_0(\bar{q}) \equiv T(\bar{q}))$$

for a suitable q. Hence T_0 is a genuine addition to the means of expression of Z.

Tarski offered the truth of all biconditionals (4) for A sentences

---

[6]The first-order theory of o and successor is comparatively trivial. It is shown decidable in Hilbert and Bernays, *Grundlagen der Mathematik*, I, § 6.

of a given formalized language as a criterion for T_0 to be a truth-predicate for that language. But we might remark that if we merely added these formulae as axioms to Z, we should not be able to prove that all closed theorems of Z are true or that Z is consistent. For it is easy to see that Z plus the axioms (4) is a conservative extension of Z.[7]

Hence the extension of Z offered to us by the concept of number-theoretic truth is trivial unless we give ourselves some *laws* involving it and not merely characterizations of individual cases. In particular, we need the relations of truth and the logical connectives, such as that $\forall x A x$ is true if and only if $A(\overline{n})$ is true for every natural number n. It is also essential that we allow instances of the induction schema (2) where the formula A contains the predicate T_0.[8]

Now can we regard this extension of Z as remaining within the framework of our informal axiomatization of arithmetic? Observe what this informal axiomatization would have to be taken to contain. First, either a middlingly strong form of definition by induction or enough set theory or higher-order logic to define T_0 explicitly. Second, a concept of "property" that is broad enough so that T_0 and logical compounds containing it count as expressing properties.

Thus the claim that the means for this extension of Z were already implicitly contained in the informal axiomatic theory is a stronger one than appears at first sight. If we accept it, we must admit that making this implicit content explicit involves developing inductive definitions, set theory, or higher-order logic further than the axioms would easily suggest. Nonetheless, in favor of the claim is the fact that these means were not explicitly excluded from it, and it is hard to see how we could *prove*, as we have done for Z itself, that they must be excluded from it. Nonetheless the "openness" of the informal theory seems at this stage to reduce to its vagueness.

Most prominent is no doubt the vagueness about the means of definition and inference that are allowed, which come as it were from "outside" the system of axioms. Concerning the concept of

---

[7]Let C be a formula of Z which has a proof in Z augmented by T_0 and axioms (4), for formulae A of Z. Then only finitely many such axioms, for formulae $A_1, \ldots, A_n$, are used in the proof of C. Let $g_1, \ldots, g_n$ be the Gödel numbers of these formulae. Then these axioms become provable in Z if we define $T_0 x$ as

$$x = \overline{g}_1 \wedge A_1. \vee \ldots \vee . x = \overline{g}_n \wedge A_n.$$

Hence C is provable in Z.

[8]If only the inductive definition of T_0 is added to Z, but formulae containing T_0 are not allowed as instances of induction, the result is again a conservative extension of Z. See note 15.

"property," the following should be remarked: We can clearly understand the axiom to mean that when a predicate, say Ax, is true or false of every object in a domain including the numbers, then it expresses a property in the relevant sense; if this truth or falsity can be relative to further parameters, it is hard to see what more general concept of property we could have. But nothing is specified concerning the means of expression that might occur in such predicates. However, the axiom is clearly to be understood to impose no restriction whatever. The vagueness resides in the absence of any conception of the limits of the possible development of means of referring to objects and expressing propositions. But the precision of the formalism results from a restriction unintended in the informal theory: in effect, we assume only that induction holds for properties expressible in the language of Z.

It might seem that this limitation is decisively overcome in the second-order formalism Z^2. For one thing induction can be stated as an *axiom*: we have a quantifier that we can read as 'for all properties'. However, this is a misleading appearance: the only properties we can express categorically are those definable by second-order formulae. Tarski's theorem still applies, and we can again construct a truth-predicate that is not coextensive with any predicate of the language of Z^2. This construction is rather more complicated than in the first-order case; the method that involves the minimum of extraneous ideas would be to state the inductive conditions for a satisfaction predicate with two arguments, one to code sequences of *properties* and one to code sequences of numbers. Hence this predicate has to be a "second-level" predicate. To replace the inductive definition by an explicit one requires quantification over such "properties"—i.e., *third*-order logic.

Our treating, in second-order logic, predicates defined in terms of quantification over all properties as true or false of every object expresses a commitment to the effect that it is in some way definite (down to extensional equivalence) what "properties" there are or could be. That commitment indeed does yield a further specification as to what properties there are—they are closed under such quantification. But this falls far short of the specification that would be needed to abolish the open-endedness of the induction axiom, for it places only a lower limit, and no upper limit, on what properties there are.

For both the first- and the second-order cases we observe that although semantical reflection seems to be the key to extending the formalism, the mere possession of semantical concepts such as truth is not sufficient to yield such an extension. The *mathematical representation* of these semantical concepts must be developed to a certain extent. Of course the new principles involved have a mathematical content that is independent of their semantical interpretation, however doubtful it may be that, as a practical matter, they could have been discovered without use of this interpretation. (Of course the discovery that they are new—that certain predicates are not already definable or axioms not already provable—is itself metamathematical.)

III

The interest of Gödel's incompleteness theorem in this connection is that it shows that the methods of proof of number-theoretical formalisms are incomplete even for statements expressed in the formal language. For either Z (called here Z^1) or Z^2, for example, we can construct a formula A and prove (by quite elementary means) that *(for $i = 1, 2$)*

(5)   If Z^i is consistent, A is not provable in Z^i.

and moreover prove *in the formalism Z^i*

(6) $Con_i \supset A$

where Con_i syntactically interpreted, says that Z^i is consistent.

That $\sim A$ is also unprovable follows (for the Gödel example) from the hypothesis that Z^i is ω-consistent (or in fact the weaker hypothesis that Z^i is 1-consistent).[9] But (6) convinces us anyway that A is *true*. But how do we know that Z^i is consistent? (5) and (6) together imply of course that the consistency of Z^i cannot be *proved* by a proof that could be represented in Z^i. What convinces most of us that these systems are consistent is our confidence in the notion of natural number and in the logical and (for Z^2 at least) set-theoretical notions that are formalized in these systems. How are we to articulate this intuition so that we can *prove* that Z^i is consistent? The most direct and straightforward way of doing this is that men-

---

[9] A number-theoretic formalism is 1-consistent if for no formula $A(x)$ which in some canonical way expresses a primitive recursive predicate are $\sim \forall x A(x)$, $A(0)$, $A(1), \ldots$ all provable.

tioned above: one proves that all provable formulae are *true* and that no contradiction is true. This would suggest that to give an intuitively satisfying proof of the Gödel formula *A*, we need the additional development of our *mathematical* conceptual apparatus that we described above.

On reflection, of course, we find this proof of the consistency of Z^i not so satisfying, because we have assumed a stronger theory about numbers than that whose consistency we prove by means of it. In my view this does not make the proof worthless, but rather than go into that I should like to discuss the question how much better we can do by appealing to proof theory: For Z at least, there are constructive proofs of consistency.

All of the usual such proofs can be formalized by adding to Z instances (in the language of Z) of transfinite induction on a certain ordering. This ordering is essentially that of terms which, set-the-oretically interpreted, describe the Cantor normal form of ordinals $<\epsilon_0$; but the terms can be coded by natural numbers, so that the ordering is isomorphic to a primitive recursive ordering. Hence from some additional axioms in the language of Z, the consistency of Z can be derived. However, are these axioms at all evident? Not, it seems, without argument. It would defeat the purpose of the consistency proof to take the notion of ordinal from set theory. The most commonly given proofs that transfinite induction holds for an ordering such as used in proof theory use intuitionistic versions either of the notion of arithmetic truth or the existence of arithmetic sets.[10] In either case, one uses modes of defining predicates that go beyond the means of definition available in Z.

Among such means are certain inductive definitions, of which the definition of arithmetic truth alluded to above is an example. By such means it is possible to introduce the notion of ordinal intuitionistically, although the full strength of this notion is much greater than is needed to prove the consistency of Z.[11] The formally most economical method of deriving transfinite induction for our

---

[10]Hilbert and Bernays, *Grundlagen der Mathematik*, II; Schütte, *Beweistheorie*, §22.

[11]The intuitionistic notion of ordinal can be formulated in various ways, but however this is done it allows the reduction of a large class of inductive definitions, including all those referred to or alluded to so far in this essay. A general intuitionistic theory of ordinals (such as that sketched in §14 of Troelstra, *Principles of Intuitionism*) is as strong as the usual intuitionistic analysis with bar induction (Kleene and Vesley, *The Foundations of Intuitionistic Mathematics*) and permits the proof of transfinite induction on orderings of type up to, but not including, the ordinal $\phi_{\epsilon_{\Omega+1}}$ (o) discussed by Howard, Gerber, and others. See Gerber, "Brouwer's Bar Theorem and a System of Ordinal Notations."

ordering is the following: Let $<$ be the ordering of all natural numbers corresponding to our ordering of "Cantor normal form" terms. We can inductively define a predicate Ac of natural numbers, which is to hold for a number x if it holds for all $y < x$. That is:

(7) $\forall y < x Ac(y) \supset Ac(x)$.

Now the idea is that Ac holds for a given x only if it holds by virtue of this condition. This might be further specified in several ways. One is that Ac must be *minimal* so as to satisfy (7). That is, any predicate satisfying $\forall x[\forall y < x Ay \supset Ax]$ must hold for all x satisfying $Ac(x)$. Thus we have the induction principle

(8) $\forall x[\forall y < x Ay \supset Ax] \supset \forall x[Ac(x) \supset Ax]$.

With the help of second-order logic, $Ac(a)$ can be *defined* as $\forall F[\forall x(\forall y < x Fy \supset Fx) \supset Fa]$; in effect, $Ac(a)$ holds if transfinite induction holds up to a.

If (7) and (8) are added to Z, we can prove that $Ac(x)$ holds for every natural number x. The antecedent of (8) then implies $\forall x Ax$, so that we have transfinite induction on $<$ for any predicate Ax (in the language of Z augmented by Ac).

Since Ac is thus coextensive with a predicate (namely $x = x$) of Z, we have avoided the use of extensionally nonarithmetical predicates. But to explain Ac in such a way that (7) and (8) are plausible, we need to use concepts that go beyond first-order arithmetic: the explanation of $Ac(a)$ as "transfinite induction on $<$ holds up to a" appeals either to arithmetic truth or to second-order logic, which also yield the more usual deductions of transfinite induction on $<$. Otherwise, it seems that the explanation of Ac requires a notion of indefinite (transfinite) *iteration*: $Ac(a)$ holds if it results from iterated application of the inference from $\forall y < x\ Ac(y)$ to $Ac(x)$. Such an explanation is on the same plane as an appeal to the intuitionistic notion of ordinal.

Let us return to the more usual procedure of deriving transfinite induction on $<$. Suppose that $<$ encodes terms for Cantor normal forms to the base 2, so that we can code basic terms o and ω and, for any a and b, $a + b$ and 2^a. Then the key step is the following: We want to show that induction on a predicate A holds up to 2^a, i.e.

(9) $\forall x[\forall y < x Ay \supset Ax] \supset \forall x(x < 2^a \supset Ax)$.

This is proved by using induction up to a on the predicate $C(x)$:

$$\forall z[\forall y < zAy \supset \forall y < z + 2^x Ay].$$

Thus informally we would reason thus: if induction up to a holds for *all* predicates, then induction up to 2^a holds for all predicates. Thus if $a_0 = \omega + 1$, $a_{n+1} = 2^{a_n}$, since it is easy to show for any A that induction on A holds up to $\omega + 1$, by induction we have that induction holds for all predicates up to a_n for any n. But $<$ is so defined that any number precedes a_n for some n.

To formalize the statement 'induction up to a holds for all predicates' we need either arithmetic truth or a use of set-theory or second-order logic equivalent to the theory of arithmetic sets. Nonetheless the application of this apparatus observes a restriction that parallels the fact that the argument by Ac uses no extensionally nonarithmetical predicates: Suppose A satisfies $\forall x[\forall y < xAy \supset Ax]$, and we want to show by means of the general argument that for any x, Ax holds. Consider the argument by second-order logic. Then the predicates instantiated for the second-order variables all satisfy the condition $\forall x[\forall y < x Fy \supset Fx]$ and are therefore true of all numbers. Moreover all the reasoning with quantifiers beyond those possibly in A itself is in effect reasoning with iteration of hypotheses that subsequently are revealed to be *true*.

Thus it appears that although *concepts* going beyond those of elementary number theory are needed to prove the consistency of Z, the proof-theoretic proofs accomplish this with a very restricted use of number-theoretic truth or other abstract concepts.

IV

I want now to consider some of the same questions in relation to set theory. Zermelo originally axiomatized set theory informally. One of his axioms, the axiom of separation (*Aussonderung*), involved a notion of property like that occurring in the induction axiom of arithmetic. The stronger axiom of replacement, introduced in the 1920's by Fraenkel and Skolem, can be interpreted in the same way. They would then be stated as follows:

(AS)   For any property P and any set z, there is a set x whose members are just those members of z which have the property P.

(R)   For any relation H and any set z, if for every member x of z there is exactly one y such that x bears H to y, then there is a set w consisting of those y to which some x in z bears H.

Again, systems based on these axioms can be readily formalized in two ways: on the one hand using second-order logic, so that (AS) and (R) become straightforward axioms, and on the other hand using first-order logic and replacing (AS) and (R) by schemata whose instances are formulae built up from ϵ as sole predicate (or ϵ and $=$). The first-order versions with replacement include the familiar Zermelo-Fraenkel set theory (ZF). The corresponding second-order theory will be called ZF^2. It is essentially the theory called Kelley-Morse set theory or the extension of ZF by impredicative classes.

It seems that in these cases we can reason for Tarski's theorem in the same way as we did for number theory. That is, we obtain from the informal theory an interpretation of the formal theory and then, by making this interpretation mathematically precise, we add a truth-predicate to the formal theory. By Tarski's theorem this predicate is not definable in the formalism; hence we have a new formal theory that is stronger.

Such a formal procedure can be carried out in more than one way, for ZF even more simply than for number theory, since the prime formulae of ZF are all of the form $x \epsilon y$ or $x = y$. However, I do not have the same confidence as in the number theoretic case that we know what we are doing here.

One reason for questioning is that we understand the axioms of number theory as describing a definite structure, so that in particular the range of the (individual) quantifiers in the formal theory is a totality that we conceive rather clearly. The temptation is very strong to justify the inductive definitions of truth and satisfaction by appealing to set theory; in particular, we do not hesitate to regard the natural numbers as a set. Then the interpretation of formal number theory is subsumed under a general theory of models which belongs to set theory.

This point of view is also available in set theory, but as we shall see there are obvious objections to regarding it as simply defining mathematically the original intended interpretation of ZF. What we would do is to treat the axioms of ZF as describing a structure, i.e., a domain for the variables which is a *set*, and a two-place relation which for our purposes can be taken to be membership. Such a structure we call an ϵ-structure. (We have informal set theory at our disposal.)

However, in ZF truth of first-order formulae in a structure whose domain is a set is definable. In particular, there is a predicate $Sat(x,y,z)$ which can naturally be read to mean:

> x is a sequence of members of z, y is the Gödel number of a formula A of ZF, and x satisfies A when the quantifiers of A are interpreted to range over z.

We can now prove in ZF the expected laws of satisfaction and of the truth predicate $TS(y, z)$ definable from it. But of course we do not obtain the Tarski biconditionals. Let $A^*(z)$ be the formula A with its quantifiers restricted to z; i.e., each part of the form $\forall xB$ is replaced by $\forall x(x \in z \supset B)$, and $\exists xB$ by $\exists x(x \in z \wedge B)$. Then we can prove in ZF for any closed formula A with number p

$$(10) \qquad TS(\bar{p},z) \equiv A^*(z).^{12}$$

Suppose now we pick out a definite $\in$-structure and interpret ZF in relation to it. That is, z is a certain definable set $\{x:Px\}$. Then we have predicates of satisfaction and truth without really extending the *language* of ZF, if 'Px' is expressible in the language of ZF. If we can show all theorems of ZF true under this interpretation, it must be because we are presupposing stronger *set existence assumptions*. Such interpretations are well known in work on set theory, for example where 'Px' is 'x is a set of rank less than the first strongly inaccessible cardinal.'.

That the informal axiomatization of Zermelo set theory or its expansion by the axiom of replacement actually yields the existence of a set such that the $\in$-structure on it is a model of ZF is dubious to say the least. This is a different question from whether the informal explication of the intuitive notion of set would justify an axiom implying the existence of such a set: that is much more plausible. The simplest such axiom is no doubt the existence of a strongly inaccessible cardinal. That already yields a model of ZF^2, where the second-order quantifiers range over *all* subclasses of the

---

[12]If we replace z by a term t denoting some set, then the liar paradox argument shows that for a certain formula A with number g, $TS(g,t) \equiv A$ is refutable. Then $\sim(A \equiv A^*(t))$ follows by (10): A is not equivalent to its relativization to t. Since A is $\sim TS(s,t)$ for some term s, this is unsurprising: A contains a reference to t, and t is not in the domain of the interpretation, since $t \notin t$. But this shows that our understanding of the truth-predicate involves a different reading of the language of set theory from the one it formulates.

domain. If informal ZF allowed the proof of the existence of a strongly inaccessible cardinal, it is hard to see how a standard model of ZF^2 could fail to contain one. Thus it seems pretty clear that the existence of a strong inaccessible goes beyond informal ZF. Thus if the notion of interpretation requires that the domain of an interpretation be a specific set, it is not plausible that informal ZF yields an interpretation that can be used to strengthen the theory.[13]

But of course there are more direct objections to taking an interpretation by an ϵ-structure as giving the intended interpretation of the informal ZF. Set theory as an informal theory is usually thought of as talking about *all* sets (or possibly about all sets built up by iteration of the power set operation from a (possibly empty) set of individuals). But ZF excludes a universal set; hence any interpretation which makes the universe a set makes it fall short of containing all sets. It seems that it must fall short of being the intended interpretation.

Now we can introduce truth so as to avoid this consequence by a straightforward inductive definition of satisfaction, where the quantifier clause is something like:

s satisfies $\forall x_i A$ if and only if for every x (or for every x in R_α for some ordinal α),[14] the sequence $s^{i,x}$ such that $s_i^{i,x} = x$ and, for $j \neq i$, $s_j^{i,x} = s_j$, satisfies A.

If we take the first choice in this clause, the quantifiers range over everything we talk about in our informal language. The truth-definition then embodies the identity or "homophonic" translation of ZF into the metalanguage. Tarski's argument then shows that the truth-predicate is not expressible in the language of ZF. The argument can easily be extended to the case where the quantifiers range over the sets built up from a particular set of individuals.

It seems that the claim that this extension of ZF by a satisfaction predicate is implicit in the informal theory can be criticized or defended on much the same grounds as the corresponding claim about number theory. It should be remarked that the extension is conservative unless formulae containing the satisfaction predicate are allowed as instances of the axiom of separation (ΛS). This would

---

[13]Of course the statement that there is a set z such that all theorems of ZF are true when relativized to z is much weaker. But as a proposed axiom it has no stronger motivation from the informal axiomatization.

[14]Where R_0 = the set of "individuals" (perhaps ϕ); $R_{\alpha+1}$ is the power set of R_α; $R_\lambda = \cup_{\beta<\lambda} R_\beta$ for any limit ordinal λ.

be an application of the open-ended character of the concept of "property" in the informal axiom. Such instances of (AS) are needed, for example, to carry out the proof by ordinary induction that all theorems of ZF are true. If ZF is consistent, the necessary instances of (AS) are independent in ZF plus the inductive clauses of the satisfaction-definition. For a predicate provably satisfying the inductive clauses can be explicitly defined in von Neumann-Bernays set theory NB, which is a conservative extension of ZF. If the instances of (AS) required to prove all theorems of ZF true were provable in NB, one could prove in NB that ZF is consistent. But then one could prove it in ZF, contrary to Gödel's second incompleteness theorem.[15]

The fact that inductive definitions such as that of a satisfaction predicate for first-order set theory are often justified on set-theoretic grounds seems to imply that the claim that the extension of ZF by such a predicate is implicit in the informal theory is on stronger ground in this case than in the corresponding cases for number theory. But the clearest such justification would presuppose that the bound variables in the inductive definition range over a *set*. In that situation, full second-order logic easily reduces to set theory, since the individual variables range over a set D and the second-order variables range over subsets of D.

However, we introduced our inductive definition because it seemed that the intended interpretation could not allow the universe to be a set. If second-order logic is to be used to justify the inductive definition, it must stand more on its own feet. One line that could be taken would relate the use of second-order logic to the informal theory, by interpreting the second-order variables to range over the properties the informal theory speaks about, in effect attributing to the theory the explicit assumption that these

---

[15]A similar situation obtains for arithmetic. One can add to Z second-order variables with axioms implying $\exists F \forall x (Fx \equiv A)$, for any formula A not containing F or any bound second-order variables. If no new instances of induction are added, the resulting theory is a conservative extension of Z, for example if the induction *axiom* (3) is used. The usual 'arithmetic analysis' admits induction on all formulae, even with bound class variables. In the first theory one can define arithmetic truth and prove the Tarski biconditionals, but only in the second can one prove that all theorems of Z are true.

Thus the theory related to ZF as arithmetic analysis is related to Z is not NB but an extension NB$^+$ of ZF with axioms of class existence equivalent to those of NB, but with *Aussonderung* and replacement allowed for all formulae, not just those without bound class variables.

NB$^+$ has the same theorems in the language of ZF as the theory obtained from ZF by adding the *satisfaction* predicate, its inductive definition, and (AS) and (R) for all formulae.

properties obey the laws of second-order logic, in particular that properties are closed under the operations of second-order logic. But such assumptions are not present in the original formulations of the informal theory. And of course this suggestion does not lead at all to a justification for a truth-predicate for ZF^2.

This procedure is not essentially different from the procedure of von Neumann, Bernays, Gödel, and Quine in postulating in addition to sets another kind of entity, classes, such that the universe of sets is a class but not a set. To be sure, the standard formulation of such a theory, the system NB, assumes only that classes are closed under *first*-order operations over sets. As we remarked, NB suffices for the definition of satisfaction but not for the proof that all theorems of ZF are true.

It might be remarked that class quantification satisfying NB can be defined in terms of satisfaction: to say there is a class X such that $A(X)$ is just to say there is a sequence s and a formula $B(x_i)$ of ZF, perhaps with other free variables, such that $A[\{x: s^{ix}$ satisfies $B(x_i)\}]$.[16] Alternatively and more or less equivalently, we can view class quantification as substitutional quantification of predicates relative to parameters.[17] These considerations might suggest seeking an explanation of the satisfaction-concept independently of the notion of class or of second-order logic.

What seems to be required here is that definitions by induction on the natural numbers should be understood and accepted directly or explained by an argument that is not of a second-order character.[18] The first course is certainly conceivable and seems a reasonable course in dealing with a single inductive definition such as that of satisfaction. However, it renounces the attempt to state the principles involved, and it is hard to see how to do that without quantifying over properties or classes or related entities such as propositions and proofs. It seems to me that there might be some alternative that would use the notions of meaningfulness and truth in a way different from the usual uses in formal semantics, in that their extensions would be gradually constructed rather than being definite for a given context. It should also be pointed out that an

---

[16]$\{x: C(x)\}$ is a virtual class abstract in the sense of Quine, *Set Theory and Its Logic*, chap. 1. $t \in \{x: C(x)\}$ is an 'abbreviation' for $C(t)$; $\{x: C(x)\} \in t$ for $\exists y [\forall x (x \in y \equiv C(x)) \land y \in t]$.

[17]See "A Plea for Substitutional Quantification" (Essay 2 of this volume).

[18]This requirement also obtains for the epistemological interest of the reduction of predicative analysis to an ontology of natural numbers that is discussed in section V of "Ontology and Mathematics" (Essay 1 of this volume). Criticism by Leslie Tharp made me aware that I had not been sufficiently explicit about this or concerned about its implications. [On this matter see the Postscript to Essay 1.]

explicitly second-order argument for the admissibility of an inductive definition like that of satisfaction can be given in which the second-order reasoning is in a sense constructive, in that the reasoning with second-order quantifiers is intuitionistic.[19]

To conclude our discussion, it seems to me that one ought not to reject as totally irrelevant the formulation of our interpretation of formalized set theory that regards the universe as a set and therefore permits the reduction of second-order logic to set theory. To begin with, ZF and other standard set theories apply classical logic to statements involving quantification over the whole universe, so that for any definable property P of sets either there is a set satisfying P or there is not, so that statements about the universe of sets at least behave as if it were quite definite what sets there are. The process by which one moves from this principle to the introduction of classes, then to impredicative reasoning about classes, then perhaps to classes of classes, and so on, is the gradual introduction of modes of reasoning that would be set-theoretically appropriate if the "universe" were a set. This would suggest that the

---

[19]Consider a formulation of NB with different styles of variables for sets and classes but with *intuitionistic* logic. To say that a class X is a set is to say $\exists y \forall x (x \in X \equiv x \in y)$. Now add the law of excluded middle for atomic formulae and the schema

(A) $\forall x(Ax \lor {\sim}Ax) \supset \forall xAx \lor {\sim}\forall xAx.$

Then evidently the law of excluded middle is provable for all statements not containing bound class variables, and therefore the system contains the classical ZF.

Now we add *ordinary induction* on all formulae. Then we can prove that for every n there is a class X that is a satisfaction relation for formulae of ZF of quantifier depth $\leq n$, say $\exists X SatR(X,n)$. $SatR(X,n)$ contains no bound class variables. Then we define $Sat\,(x,y)$ as

$$\exists X[SatR(X, \text{depth } y) \land \langle x, y, z \rangle \in X]$$

and using the extensional uniqueness of X such that $SatR(X,n)$ for formulae of depth $\leq n$, we can derive

$$\forall x \forall y[Sat(x,y) \lor {\sim}Sat(x,y)]$$

Then by (A) we can derive the law of excluded middle for all formulae containing *Sat* but otherwise no bound class variables. We can then prove that all closed theorems of ZF are true.

If we add an axiom schema of replacement in the form

$$\forall x \forall y(Axy \lor {\sim}Axy) \land \forall x \exists! yAxy. \supset \forall u \exists z \forall y[y \in z \equiv \exists x(x \in u \land Axy)]$$

then we can construct an inner model of NB$^+$ (note 15), construing classes by means of satisfaction.

line between sets and classes is somewhat arbitrary. It might be an argument for a point of view that is advocated on other grounds: that the "totality" of sets is irreducibly potential and that therefore classical existential quantification is not applicable over it;[20] hence a theory about all sets that uses classical quantification is either wrong or, more plausibly, falls so far short of characterizing the possibilities of set formation that those it describes are already realized by a set, which the theory therefore does not distinguish from the "universe."

It would be the subject of another paper to deal more adequately with these suggestions. To return to our earlier theme, we seem to be confronted with an ambiguity in the notion of the intended interpretation of formalized set theory: the theory seems to be about a definite "universe" that we are tempted to conceive as the analogy of a set, and the formulae of the theory can be given a sense so that this universe is a set, but to take it in this way falsifies the intent, and at the very least involves a use of set-theoretic language that would not be amenable to an interpretation with the *same* set as universe. Whether statements about all sets can avoid this ambiguity by an irreducibly modal interpretation of their quantifiers is unclear: ordinary statements of necessity and possibility in classical mathematics are equivalent to general and existential statements. Perhaps a certain "dialectical" character is essential to the axioms of set theory.

---

[20]That on the grounds of the potential nature of the totality of sets, reasoning involving quantification over the universe should be intuitionistic has been suggested by Lawrence Poszgay and Saul Kripke. A formulation of ZF on this basis proved to be a rather weak theory. See Leslie H. Tharp, "A Quasi-Intuitionistic Set Theory."

However, Harvey Friedman, "The Consistency of Classical Set Theory Relative to a Set Theory with Intuitionistic Logic," proves the consistency of ZF relative to another formulation of "intuitionistic" set theory. [See also Powell, "Extending Gödel's Negative Interpretation to ZF."]

PART TWO

Interpretations

4

Infinity and Kant's Conception of the "Possibility of Experience"

I. Introduction

In this paper I intend to discuss Kant's theory that space is the "form" of our "outer intuition." This theory is intended, among other things, to explain what Kant took to be a fact, namely that we have synthetic a priori knowledge of certain basic properties of space and of the objects in space. In particular, Kant thought we knew a priori that space is in some respects infinite, for example infinitely divisible.

I shall not challenge Kant's claim that we have such synthetic a priori knowledge, but I shall attempt to show that Kant's theory of space must be taken in such a way that it does not explain this putative knowledge. For the intent of the whole of Kant's epistemology is to prove that our synthetic knowledge is limited to objects of "possible experience." Now when we try to give the notion of "possible experience" a concrete intuitive meaning, we shall find that the limits of possible experience must be narrower than what, according to Kant, is the extent of our geometrical knowledge of space.

The alternative to so limiting the possibility of experience is to define it by what, on mathematical grounds, we take to be the form of our intuition. But then the content of mathematics is not de-

From *Philosophical Review*, 73 (1964), 182-197; reprinted in Robert Paul Wolff, ed., *Kant: A Collection of Critical Essays* (Garden City: Doubleday, 1967); reprinted here by permission of the editors of the *Philosophical Review*.

I am greatly indebted to Burton Dreben for his criticism of the penultimate draft of this paper.

termined by any concrete knowledge of the form of intuition and the limits of possible experience associated with it. In this setting, the notion of "form of intuition" loses much of its force, and as an explanation becomes ad hoc.

The difficulty seems to me deeper than that which gives rise to the most common objections to Kant's theory of space. These rest on a comparatively accidental feature of Kant's view, namely his belief that we can know a priori that space is Euclidean. It is then pointed out that there is no sufficient reason for believing this, and that there are physical reasons for preferring a theory in which space (more strictly, space-time) is not Euclidean. The Kantian might simply concede this point and reply that the form of our outer intuition might indeed not determine the answer to such a question as whether the parallel postulate is true, but it does determine more primitive properties of space, so that these can be known a priori. The infinite divisibility in particular is, so far as I know, not denied in any serious application of geometry to physics. The claim that we know properties of this order a priori is not absurd even in the contemporary context—and indeed I am not denying it, but only the adequacy of the Kantian explanation of it, if it is the case.

In order to carry our discussion further, we must say something by way of elucidation of the terms which Kant uses, and which we have used to state our problem. Our elucidation will not be completely thorough and satisfactory; to give such elucidation would be a larger and more difficult undertaking than the present paper, and there are probably some irresolvable obscurities. I hope, however, that my elucidation will be sufficient to make my own argument clear.

We shall begin with the notion of "form of intuition." According to Kant, intuitions, like anything "in the mind," are representations. A vital feature of representations is that they have what, after Brentano, is called intentionality. That is, they at least purport to refer to an object; moreover, they have a certain content which they represent as in some way belonging to the object. Kant defines intuition as a species of representation which is distinguished by being in immediate relation to objects, and by being in relation to, purporting to refer to, individual objects (A 19, B 33; lectures on logic §1.)[1] The implications of the word "immediate" will be con-

---

[1] I.e., *Critique of Pure Reason*, 1st ed., p. 19; 2d ed., p. 33. With a few minor modifications, all quotations will be from Kemp Smith's translation. The passage from the logic lectures is in *Ak.* IX, 91.

sidered below. Kant assumes that our faculty of intuition is *sensible*; that is, we have intuitions only as a result of being *affected* by objects. The primary instance of this is *sense perception*—seeing, hearing, and so forth.

Our intuitions have certain characteristics which belong to them by virtue of the nature of our capacity to be affected by objects, rather than by virtue of some characteristics of the specific occasions of affection which give rise to them. These characteristics are said to be the form of our intuition in general. Since spatiotemporality is among them, space and time are spoken of as *forms of intuition*.

This must be understood to mean that the nature of the mind determines that the object we intuit should be spatial, and indeed intuited as spatial. Outer intuitions represent objects *as* in space. "By means of outer sense, a property of our mind, we represent to ourselves objects as outside us, and all without exception in space" (A 22, B 37). That its objects are in space is perhaps the definition of "outer" intuition, so that space is the form of outer intuition. Since inner intuition is characterized as of "ourselves and our inner state" (A 33, B 49), outer intuition is also distinguished by representing its objects as in some way outside our minds.

It might be remarked that from the fact "we represent to ourselves objects as...in space" it does not immediately follow that this is not an illusion. Since Kant characterizes phenomenal objects as "things to be met with in space," what the claim of nonillusoriness amounts to is that phenomenal objects really exist, such that perception puts us into immediate relation to them. That this is so is the claim of the Refutation of Idealism, which in turn (in the second edition at least) rests on the Transcendental Deduction.

Kant also supposed that space has certain mathematical properties which are reflected as properties and relations of the objects in space. It is of course the fact that they describe the *form* of our intuition which makes mathematical propositions a priori. Kant, of course, supposed that we know a priori that space is Euclidean. This means, in particular, that it is both infinite in extent and infinitely divisible. It is this which is the source of the difficulties which we shall develop.

A final preliminary remark concerns the concept of "possible experience." A main purpose of the *Critique* is to deduce the principles which describe the general nature of the objects given in experience by showing that they describe "necessary conditions of the possibility of experience." It follows, however, that these principles apply only to objects of possible experience. "The conditions of the *possibility of experience* in general are at the same time conditions of the *possibility of the objects of experience*, and...for this reason they have objective validity in a synthetic a priori judgment" (A 158, B 197).

If this analysis is to yield its result, the limitation of our knowledge to objects of possible experience must mean more than that the objects should be such as might present themselves in some way or other in a possible experience. For Kant allowed the possibility that the objects of experience should have an existence in themselves, apart from their relation to us in our perception and even apart from the general conditions of this relation. But of this we can know nothing; everything about the object which we can know must be able to show itself in experience and must therefore be limited by the general conditions of possible experience. We shall see that applying this dictum to the infinite properties of space produces difficulties for Kant.

II. The "Antinomy" of Intuition

It follows from the fact that the empirical objects of perception are in an infinitely divisible space that they are *indefinitely complex*. For the spatial region which an object occupies can be divided into subregions, which again can be so divided, and so on. This is not to say that the object can be physically separated into parts indefinitely, although Kant does refer to what occupies a subregion of the region occupied by an object as a "part" of the object. Given two disjoint subregions of the region occupied by the object, the "parts" of the object occupying these subregions are distinguishable. One could know a great deal about the state of one "part" while knowing little or nothing about the state of the other.

We shall now develop some apparent implications of the definition of intuition as immediate representation, in such a way as to lead to an absurd conclusion. From the view that space is the form of our outer intuition, it seems to follow that the objects we perceive are represented in intuition as having the structure which objects in space have. Then it seems that they are perceived to be indefinitely complex. Indeed, Kant speaks constantly of a *manifold* of intuition, and says that every intuition contains a manifold.

We shall now take an extreme interpretation of this, and make the following argument. In intuition we have an immediate representation of a spatial extension. Such an extension contains subregions. It follows that we have an immediate representation of these subregions. Consider a particular one, say, one whose surface area is no more than half that of the original one. It follows that if we have an intuition of a region of surface area x, we have at the same time an intuition of a region of surface area $\leq \frac{1}{2}x$. By iterating this argument, we can show that we have an intuition of a region of surface area $\leq \dfrac{x}{2^n}$ for each n. In other words, we have *at the same time* intuitions of all the members of an infinite sequence of regions converging on a point. We must suppose that this is something Kant is denying when he says, in the solution to the Second Antinomy, "For although all parts are contained in the intuition of the whole, the *whole division* is not so contained, but consists only in the continuous decomposition" (A 524, B 552). But we do not have an interpretation of what it means for the parts to be "contained in the intuition of the whole."

We shall now make another deliberate misinterpretation of Kant. Kant also says that we can perceive the manifoldness of something given in intuition only by picking out its parts or aspects one by one:

> Every intuition contains in itself a manifold which can be represented as a manifold only in so far as the mind distinguishes the time in the sequence of one impression upon another; for each representation, *in so far as it is contained in a single moment*, can never be anything but absolute unity [A 99].

Now what is suggested by the last part of this sentence is that what "representing as a manifold" means is apprehending a succession of simple parts one by one at different times. Then by the Threefold Synthesis which it is the purpose of the whole passage (A 97-104) to describe, the mind will impose certain relations on these simple entities so that the system of objects thus related will be a spatial whole. Some such interpretation as this is suggested by a number of statements by Kant to the effect that apprehension of a manifold is a successive act. "We cannot think a line without drawing it in thought" (B 154). It is hard to see what the simple entities might be in cases like this if not the points of the line. But then a "single moment" in the above passage would have to be an instant. Absurdities follow immediately. First, it contradicts Kant's repeated statements that the parts of space are not points but spaces, and that the successive synthesis in the apprehension of a space is a synthesis of these parts. Moreover, it is hard to see how the doctrine could be carried over to a two- or three-dimensional space. Of course the points of such a space can be placed in one-to-one correspondence with the instants of time in a time interval. But it would be fantastic to suppose that in thinking of, say, a square, we run through its points in the order deriving from such a correspondence. Indeed, since there is a one-to-one correspondence between the points of the whole of infinite space and the instants of a finite interval of time, it is hard to see why we should not be able to run through *it* in a finite time, contrary to the position of the First Antinomy.

I should not mention these absurd consequences if so distinguished a commentator as Vleeschauwer did not say that, according to this passage, apprehension involves absorbing a succession of simple instantaneous "impressions" which are *not* intuitions; this is the reason it requires a synthesis, so that these simple elements will be connected to form an "intuition."[2] Kemp Smith also reproaches Kant for failing to say what the simple elements are which are synthesized in the synthesis of apprehension.[3]

III. Solution of the "Antinomy"

We have considered two interpretations which resemble the antithesis and the thesis of the Second Antinomy. The first was a naïve reading of the doctrine that space is a form of sensible intuition, the second a naïve reading of Kant's view that perception presupposes a synthesis. I think that neither of them is plausible either in itself or as an interpretation of Kant. Mentioning them ought to make clear that it is not so obvious what is meant by saying that space is a form of sensible intuition. The measures we take to reconcile the two tendencies expressed by the two sides of our antinomy will reveal the difficulties I mentioned at the beginnning.

In order to interpret the passage from the synthesis of appre-

[2]*La déduction transcendentale dans l'oeuvre de Kant,* II, 242-244.

[3]*Commentary,* p. 87. [The interpretation of Kant criticized here seems to be held by the authors of some of the most important recent works on the Deduction, notably Dieter Henrich (for example, *Identität und Objektivität,* p. 17).I have to defer to their scholarship. However, the line I pursue in the text is of philosophical interest on its own account. Moreover, we do not have to suppose Kant to have been entirely consistent on this point. I am persuaded that in some of the passages discussed here, Kant's phenomenology was better than would fit the kind of theoretical presupposition attributed to him by Kemp Smith, de Vleeschauwer, and Henrich.]

hension, we have to make a distinction which Kant does not explicitly make and which has some consequences which I am not sure that Kant would have accepted.

The most likely interpretation of the passage is that the manifoldness of an intuition can be apprehended *explicitly* only by going over the details one by one, in such a way that the times at which the different details are taken in are distinguished from one another. Although Kant may not be disputing that we can *in some sense* take at least a limited complexity at a glance, it still appears that Kant is claiming something false. For in such cases as the perception of at least short written words, it seems that our perception of what the letters are can hardly be made more explicit by going over them one by one. The same is true of the divisions of a region of space provided that they are marked and the number of them is sufficiently small.

This, however, allows us to save something of Kant's point. The amount of complexity and division which we thus take in is finite and in fact has an upper limit, even if this limit is indeterminate. We can take in the letters of a four-letter word at a glance, but not the letters of a printed page. If we consider a ten-letter word, we might find that some could, some could not, and others would not be sure whether to call what they did "taking the word in at a glance."

More generally, we observe that those details or parts of something which we perceive which we can apprehend are themselves given in perception, and therefore have the same indefinite complexity which the whole has. But this complexity is not as explicit to us as what we might call the "first-order" complexity. If it were in general, the antithesis of the above "antinomy" would follow, that is that *all* the complexity of the object is given to us at once. On Kant's grounds, this would imply the decidability of any mathematical question about the continuum, and in general it would imply that our senses are infinitely acute. It is not really possible to imagine what the world would be like if this were the case. A number of examples will show that it is not. Many of us can improve our perception of detail by putting on glasses, and it is likely that some (for example, Ted Williams) see better than we do even then. Our ability to perceive details decreases continuously with an increase in distance, from which it follows that it was not unlimited at optimum distance. The problems about submicroscopic entities arise because our senses cannot take in anything which occupies less than a certain amount of space.

It seems that Kant is not asserting that in momentary perception the complexity beyond what we take in explicitly (on his view *all* the complexity) is not given to us at all. Indeed, it is hard to see what this would mean. We could hardly say that the letters of a word which we look at look simple and undifferentiated. We might take the perception to be of an aggregate of "sense data" which are themselves simple, but then it could be only by virtue of external relations that they could be appearances of objects which are themselves complex (for example, the letters). This is simply the thesis of our antinomy.

The complexity which is given in a nonexplicit fashion can be perceived explicitly by taking a closer look directed specifically at some aspect of it. This seems to be what Kant calls, in the section on the synthesis of apprehension, running through the manifold.

We claim to have shown that the distinction between explicit and implicit givenness is necessary in order to save the doctrine of continuous space as a form of outer intuition from a contradiction with the fact of the limited acuteness of our senses, as is presented in the "antinomy" above sketched. In fact, the distinction is that which is made in Gestalt psychology between "figure" and "ground." The data of the senses at any given time are divided into that toward which the attention is primarily directed, called "figure," and that to which it is not, called "ground." In the simplest case of looking at a physical object the latter includes the "background." The figure appears more clearly and definitely. What needs to be noted is that within the spatial boundary of the figure, many aspects of the object appear as ground—subtle differentiations of its color, irregularities of its shape, all but a few of its spatial divisions. There is here, however, no definite line between appearing as ground and not appearing at all.

In the appearance of objects to us we can thus distinguish three levels: primary complexity or figure, which appears explicitly; secondary complexity or ground, which appears in a nonexplicit way which is difficult to describe; and tertiary complexity, which does not appear at all but which might appear in some other perception of the same object.

IV. Relation to Kant's Actual Views

That Kant was not fully in possession of this distinction can, I think, be shown by analysis of the Threefold Synthesis and the Mathematical Antinomies.

In the Threefold Synthesis Kant argues that an intuition containing a manifold can be represented as manifold only by attending to its different aspects individually at different times (running through) and yet keeping them in mind as aspects of *one* intuition (holding together). This, however, presupposes a *reproduction* of the previous elements of the series and their *recognition* as "the same as what we thought a moment before" (A 103). In terms of the analysis I have given, there seems in the description of the threefold synthesis to be a striking omission: what is given *explicitly* to closer attention must be identified as what was given *implicitly* at an earlier stage of perception.

In the Second Antinomy, Kant speaks in what seems to me a not at all clear way about an object as a "spatial whole given in intuition" in such a way as to insure that a division of it (in the conceptual sense mentioned above) can continue without end. He is not clear as to why he asserts this and yet denies that "it is made up of infinitely many parts," on the grounds that "although all parts are contained in the whole, *the whole division* is not so contained" (A 524, B 552).

If Kant saw the matter clearly in the way we suggest, then he might very well not have made the distinction he makes between the solution of the First Antinomy and the solution of the Second. In the case of the Second, his solution might be taken to mean that if a part is given implicitly, and appears explicitly in a later perception, then this perception is of the same type as the preceding, so that the process can be repeated. There seems, however, to be no reason not to make the symmetrical assumption at least with respect to the infinity of the world in space. For the figure in our perceptual field is surrounded by ground. It is just as natural to assume that for any given direction, some element of the ground immediately in that direction can be made figure by a shift of attention, with a result of the same type as the previous perception, so that the result can always be repeated. This would mean that there would always be something outside any spatial region which we clearly perceived.

There is still a difference, but it is not the one in the text between a *regressus in infinitum*, where we know at the outset that at every point we shall be able to find some further term, and a *regressus in indefinitum*, where we can never know that we shall *not* be able to continue, but need not know that we always *can*. The difference is that the regions in the outward progression may be of objective size decreasing fast enough so that they can all be included in one bounded region. I am not quite convinced that this can be true while in the case of the division the objective size of the parts *must* tend to zero. It seems that in either case we might have some liberty in devising the measuring system so that the objective size of the whole does not become infinite in the first case, or so that the objective size of the parts does or does not tend to zero in the second.

V. Intuition and Infinity

We can now return to the question we raised at the beginning, of the sense in which space and its properties are conditions of the possibility of experience, and in which the objects in space are objects of possible experience. Let us consider the assumption that justifies the continuation at every stage of the regress of the Mathematical Antinomies. This assumption is that whatever appears to us as ground can become figure, in such a way that it will have the same structural properties as the figure of the previous perception. We shall call this the Continuability Principle.

The expression "structural properties" is somewhat vague. By using it we require that the new figure should have a nonnull spatial extension (area) and that the new perception should have primary, secondary, and tertiary complexity. Of course, some of what belonged to the tertiary complexity will now be secondary.

That the objects in arbitrarily small or arbitrarily distant regions of space will be objects of possible perception, and therefore of possible experience, follows from iterated application of the Continuability Principle. The question arises what sort of possibility is in question. I shall argue that it cannot be practical possibility, and that it must be a sort of possibility which implies some circularity in Kant's explanations.

There are a number of circumstances which might prevent us from following out a part of the ground and making it figure. The object might change or be destroyed; *we* might die or be unable to make some necessary motion; we might have reached the limit of the acuteness of our senses.

The problem of change in the object is very complicated. In order to make a transition from ground to figure, it seems there must be some element of stability. For example, one could not look more closely at some part of a surface, and see clearly what one previously saw less clearly, if the surface changed in some wild way. How *much* continuity there has to be, however, depends on what one wants. In particular, I do not believe change gives rise on the macroscopic level to difficulties in principle about identifying subregions of the region occupied by the previous figure, or in identifying regions just outside it.

The limit of the acuteness of our senses seems to me more serious. Something which is too small or too distant cannot be seen as clearly as the things we see in everyday life. For this reason it is doubtful that the Continuability Principle applies in the sense of practical possibility to things which are so small or so distant that they appear as "dots." The range of application can be greatly extended with the aid of optical devices. But even with this latitude, physics would place a certain definite limit on our microscopic perception.[4]

In the case of great distance, we have to bring in the third limitation, that something might happen to *us*. In view of the fact that, according to present-day physics, our speed of motion must be less than that of light, if the object is very distant it will be a very long time before we can get close enough to see it much more clearly. But biology (or, perhaps, common sense) sets a definite limit on the amount of time we shall have to get there. If the question is not what one individual can perceive but what the species can perceive, then it seems that the second law of thermodynamics places an upper limit on the longevity of the species.

There is reason to believe, therefore, that (speaking vaguely) every aspect of our perceptual powers is limited in a definite way by natural conditions. We could call a being of which this is true a *thoroughly finite* being. For such a being, there is an n such that he can have no perception of what is within any sphere of radius $1/n$ or outside a certain fixed sphere of radius n. I believe that in this statement we can replace "perception" by "empirical knowledge," if we limit empirical knowledge to what can be inferred from observations by induction. For by these means, we could not verify that the laws by which we might extrapolate to very distant or very small regions would not fail at these magnitudes.

Thus it appears that the "possibility of experience" for Kant must extend beyond what is practically possible for the sort of being we have reason to think we are. It would be possible to describe the

---

[4] In the case of the most powerful devices, a great deal of physical theory intervenes between our perception and our interpretation of it as of a certain very small object. And it is not clear that this interpretation will be true more than schematically; it may be that, as with a photograph, the fine structure of what we see no longer belongs to the object.

perceptual powers of a being for whom the Continuability Principle would be true, but who would still be finite, in the sense that he would not need to have infinitely acute senses or be immortal, or be able to perform infinitely many acts in a finite time, or anything similar. Such a creature (call him *U*) would be more bizarre than appears at first sight; for example, he would have to be able to increase the acuteness of his senses beyond any limit. The difficulty for Kant, however, is not so much one of forming an abstract conception of such a being, but of explaining how it can be that the *form of our sensibility* leads us to represent "appearances" as having the structure and relations which *U*'s experience would reveal them to have. In the case of the Continuability Principle, this does not seem so unnatural, but if it is spelled out it seems unavoidably dogmatic: namely, the structure of ordinary microscopic perception, in which the figure appears with its internal and external ground, is certainly compatible with the notion that the ground appears as something which "can become figure" in a sense in which this "possibility" is in some sense an aspect of the intuition and not, for example, something which would have to be inferred from general rules. Kant must, however, suppose that "intuition" must represent its objects as having the structure which is revealed in this most favorable situation, even those which are in fact presented only on the margins of our sense experience. As an extrapolation, this is quite natural. But when it is carried to infinity, far beyond the actual limits of our experience, it is not so clear that it is the only possible one. Even if it is, however, it is also not clear why it is to be regarded as a form in which objects must be "given" rather than as an imaginative or intellectual construction. Perhaps Kant came close to saying that it *was* imaginative.[5] The reason he did not regard it as intellectual seems to be that he thought that, apart from something nonconceptual, there was no source of any representation of infinity. That only the form of intuition can allow us to represent indefinite continuation in space seems to me to be the sense of the following dark saying: "If there were no limitlessness in the progression of intuitions, no concept of relations could yield a principle of their infinitude" (A 25).

The upshot of all this is that the "possibility of experience" must be a quite abstract kind of possibility, defined by the form of intuition. It is interesting to consider how it is related to contrary-to-fact possibility. If "object of possible experience" means "object of

[5]Cf. Heidegger, *Kant und das Problem der Metaphysik*, esp. §28.

possible perception," then there must be objects which I *might have* perceived but did not. This is obvious without our earlier argument, if we bring in time and change. Since one cannot be in two places at once, everything about the state of the world *now* which I cannot perceive from my present vantage point must, on our assumption, be something I might have perceived but did not.

There is, however, an important difference between this sort of case and those of very distant and submicroscopic regions of space. Consider these three cases:

(1) I might have seen the Red Sox play yesterday if...
(2) I might have seen the explosion (on a planet 10^{10} light-years away) if...
(3) I might have seen (something happening in a certain region of diameter 10^{-100} cm.) if...

In (1) what might fill the blank would be something which could easily have been the case; for example, if the idea had occurred to me, if I had felt like it, if I had not had to see my tutees, and so forth. What would fill the blank in (2) would be something which could be the case only if my origin and that of the whole human race were quite different from what I take it to be, transposed to a quite different part of the universe. What would fill the blank in (3) would be something which could be the case only if our units of measure, and therefore we ourselves, were of a totally different order of magnitude in relation to such physical quantities as the size of atoms and electrons and the wave length of light. It is not clear that we could even describe the circumstances which, if they obtained, would be such that we could identify some spatial region of diameter 10^{-100} cm., let alone find out what is going on in it.

In any case, the possibilities (2) and (3) are of such a sort as Kant would have regarded as idle and unverifiable, for a reason which makes it difficult for him to make any use of contrary-to-fact possibility. That reason is the view expressed in the Postulates of Empirical Thought (A 230-232, B 282-285) that we cannot know that there is more than one possible experience, and that all our empirical knowledge must be brought under the unity of a single experience. For this reason, he asserts that we cannot know that there is anything really possible which is not actual.

It ought to be remarked that the extension of the possibility of perception to infinity depends on the *iterated* application of the Continuability Principle. Even with the assurance that the Continuability Principle is true, we should not obtain the indefinite complexity of objects and the infinity and infinite divisibility of space without this iteration; we should not be able to deduce them without mathematical induction. It appears that, for Kant, mathematical induction is in some way founded on the form of our intuition, but Kant gives no explicit statements to make clear how. I do not see that the case is much better with the modern Kantians in mathematics (for example, Brouwer and Hilbert).[6]

VI. Conclusion

Kant's theory that space is a form of intuition has two conflicting pulls. On the one hand, it implies that the objects of outer intuition have certain characteristics, in particular that of being in some form of mathematical space. But in order to be an explanation of this fact, and in view of the fact that intuition is representation, nothing ought to be attributed to the form of intuition which is not revealed in the way objects present themselves to us in perception.

The first pull leads Kant to assume an aspect of the form of our intuition, namely the potential infinity of certain kinds of series of perceptions, which is beyond our actual powers of perception. He must hold that we represent objects as being in a space and time having parts which are beyond the experience of a thoroughly finite being, and that this arises from the form of our sensibility. But this cannot be justified phenomenologically. I should like to say that in this situation the term "form of sensibility" has no explanatory force. I am inclined to think that we cannot be justified in saying that our having a concept of the infinite is explained by the fact that the form of our sensibility contains the possibility of indefinite continuation and division.

Thus, if he is to maintain the view that the form of intuition is the *ratio essendi* of the mathematical infinite—in particular the infinite divisibility of space—Kant must take the uncomfortable position of saying that mathematical considerations are the *ratio*

---

[6]Poincaré seems to me to shed more light on induction (*La science et l'hypothèse*, chap. 1), but although he is called an "intuitionist," I do not count him as a Kantian. He seems quite uninfluenced by Kant's special conception of "pure intuition."

[On the relation of induction to Kantian intuition, see also "Kant's Philosophy of Arithmetic" (Essay 5 of this volume) p. 141, and my "Mathematical Intuition," pp. 164-165. However, these remarks do little more than repeat the difficulty raised here. I discuss the matter further in my work in progress.]

cognoscendi of the form of intuition, which in turn defines the possibility of experience. The content of mathematics is thus not determined by concrete knowledge of the form of intuition and the limits of possible experience associated with it. In addition to the difficulty about explanation, there are two further limitations of this position. First, it is hard to see how on this basis the notion of form of intuition can be used for critical purposes, to determine the boundaries of mathematical evidence. Second, this position has for the argument of the *Critique of Pure Reason* as a whole the implication that at some point the "synthetic" method which it purports to follow, of arguing directly from the nature of experience to the conclusions of transcendental philosophy, must be supplemented by the "analytic" method of the *Prolegomena*, of starting with the given content of science and arguing hypothetically to its a priori conditions, a method which Kant regarded as acceptable only as an expository device in a semi-popular work. (See *Prolegomena*, §§4-5).

The situation we have described provides some explanation and justification of Heidegger's claim that Kant took the finiteness of our understanding more seriously than did his predecessors, but even he did not follow this through to the end. One reason he did not is that he also took seriously the fact that mathematics requires that there be some source in our cognitive apparatus for some sort of representation of infinity. And as he said, the possibility of mathematics is proved by the fact that it exists. But he did not fully succeed in reconciling this so often cited support of rationalism with the empiricism implicit in restricting the content of our knowledge to "possible experience."

5

Kant's Philosophy of Arithmetic

The interest and influence of Kant's philosophy as a whole have certainly been great enough so that this by itself would be enough to make Kant's philosophy of arithmetic of interest to historical scholars. It is also possible to show the influence of Kant on a number of important later writers on the foundations of mathematics, so that Kant has importance specifically as a figure in the history of the philosophy of mathematics. However, my own interest in this subject has been animated by the conviction that even today what Kant has to say about mathematics, and arithmetic in particular, is of interest to the philosopher and not merely to the historian of philosophy. However, I do not know how much of an argument the following will be for this.

Kant does not discuss the philosophy of arithmetic at any great length, so that it is virtually impossible to understand him without making use of other material. What I have used consists mainly of two considerations: the integration of Kant's theoretical philosophy as a whole, and modern knowledge on the foundations of logic and mathematics. The justification for using the second is twofold; first, I think experience shows that one does not get far in understanding

From Sidney Morgenbesser, Patrick Suppes, and Morton White, eds., *Philosophy, Science, and Method: Essays in Honor of Ernest Nagel* (New York: St. Martin's Press, 1969), pp. 568-594. Reprinted by permission of St. Martin's Press, Inc., and the editors. The Postscript has been added for this volume.

An earlier version of this paper was written while I was George Santayana Fellow in Philosophy, Harvard University, and presented in lectures in 1964 to the University of Amsterdam and the Netherlands Society for Logic and the Philosophy of Science. I am indebted to the Instituut voor Grondslagenonderzoek en Filosofie der exacten Wetenschappen of the University of Amsterdam for its hospitality during that period, and to J. J. de Iongh, J. F. Staal, and G. A. van der Wal for helpful comments. I am also grateful to Jaakko Hintikka for sending me two then unpublished papers on the subject of this essay.

a philosopher unless one tries to think through the problems on their own merits, and in this one must use what one knows; second, if one is today to take Kant seriously as a philosopher of mathematics, one must confront him with this modern knowledge, which after all in major respects shows immense progress from the situation in his lifetime.

I shall be concentrating mainly on one question, which I think must be answered before one goes farther with the subject: Why did Kant hold that arithmetic depends on sensible intuition, indeed that arithmetical propositions in some way refer to sensible intuition? This is, of course, closely related to the question of why he regarded such propositions as synthetic rather than analytic. In considering this question, one must very soon consider Kant's views on logic and its relation to arithmetic. Also since the answer to the above question is much clearer if "arithmetic" is replaced by "geometry," we shall also give some consideration to Kant's views on geometry.

I

In order to clarify our problem, let us first briefly consider Kant's concept of intuition. Intuition is a species of representation (*Vorstellung*) or, in the language of Descartes and Locke, idea. Having intuitions is one of the primary ways in which the mind can relate to or be conscious of objects. The nearest thing to a definition in the *Critique of Pure Reason* occurs in a classification of representations:

This [knowledge] is either *intuition* or *concept* (*intuitus vel conceptus*). The former relates immediately to the object and is singular [*einzeln*], the latter refers to it mediately by means of a feature which several things may have in common.   [A 320 = B 376 − 7][1]

In the opening sentence of the Transcendental Aesthetic, Kant says:

In whatever manner and by whatever means a mode of knowledge may relate to objects, *intuition* is that through which it is in immediate

---

[1] I.e., 1st ed., p. 320, 2d ed., pp. 376-377. All passages are quoted in the translation of Norman Kemp Smith, with slight modifications. Other translations from German are my own. Translations of Kant's Inaugural Dissertation are by John Handyside, *Inaugural Dissertation and Early Writings on Space.*

relation to them, and to which all thought as a means is directed.   [A 16 = B 33]

A passage in §1 of Jäsche's edition of Kant's lectures on logic reads:

> All modes of knowledge, that is, all representations related to an object with consciousness are either *intuitions* or *concepts*. The intuition is a singular representation (*repraesentatio singularis*), the concept a *general* (*representatio per notas communes*) or *reflected* representation (*representatio discursiva*).[2]

Intuitions are thus contrasted with concepts, which relate to objects only mediately, by way of certain properties and by way of intuitions which instantiate them and which relate indifferently to all the objects which possess the required properties.

What is meant by calling an intuition a singular representation seems quite clear. It can have only one individual object. The objects to which a concept "relates" are evidently those which fall under it, and these can be any which have the property which the concept represents, so that a concept will only in exceptional cases have a single object. Thus far, the distinction corresponds to that between singular and general terms.

One might think that the criterion of "immediate relation to objects" for being an intuition is just an obscure formulation of the singularity condition. But it evidently means that the object of an intuition is in some way directly present to the mind, as in perception, and that intuition is thus a source, ultimately the only source, of immediate knowledge of objects. Thus the fact that mathematics is based on intuition implies that it is immediate knowledge and thus, even though synthetic a priori, does not require the elaborate justificatory argument which the Principles do (A 87 = B 120). By the immediacy criterion Kant's conception of intuition resembles Descartes's, while by the singularity criterion and his insistence on a nonintuitive conceptual factor in all knowledge, Kant's theory of intuition differs from that of Descartes.

That what is immediately present to the mind are individual objects seems to be an axiom of Kant's epistemology, or one might also say metaphysics, since it goes with the conviction that objects, the primary existences, are in the first instance individual objects. Thus what satisfies the immediacy criterion of intuition will also satisfy the singularity criterion.

[2]*Ak.*, IX, 91.

It does not seem that the converse must be true. The idea of a singular representation formed from concepts seems quite natural to us. Such a representation would relate to a single object if to any at all, but it hardly seems immediately. By associating it with a definite description rather than with a general term, we would distinguish it from a concept under which exactly one object falls (even if necessarily). For Kant, however, the passage from A 320 = B 376 − 7 seems to allow such a representation to be a concept; this might also be suggested by the fact that the idea of God is called a concept; it is nowhere suggested that it is an intuition. However, Kant never remarks, so far as I know, on the implications of the possibility of nonimmediate singular representations for the concept of intuition.

This omission may give support to a theory which has been advanced by Jaakko Hintikka according to which the singularity criterion is the sole defining criterion: An intuition is simply an individual representation.

> In Kant and his immediate predecessors, the term "intuition" did not necessarily have anything to do with appeal to imagination or to direct perceptual evidence. In the form of a paradox, we may perhaps say that the "intuitions" Kant contemplated were not necessarily very intuitive. For Kant, an intuition is simply anything which represents or stands for an individual object as distinguished from general concepts.[3]

Many of the passages Hintikka cites also mention the immediacy criterion, and it is not clear why Hintikka thinks it nonessential. The main reason, which we shall consider later, is that this assumption supports a theory of Beth and Hintikka to explain Kant's notion of "construction of concepts in intuition" and the resulting analysis of mathematical demonstration. Another seems to be the absence of the immediacy criterion in the *Logic* and the fact that Kant makes remarks on concepts which seem to exclude essentially singular concepts and thus to imply that all singular representations are intuitions.[4]

Hintikka also points out that the part of the Transcendental

---

[3]"Kant's 'New Method of Thought' and His Theory of Mathematics," p. 130. Hintikka argues in detail for this thesis in "On Kant's Notion of Intuition (*Anschauung*)." The same idea seems to underlie the analysis of Kant's theory of mathematical proof in Beth, "Über Lockes 'allgemeines Dreieck.' "

[4]"It is a mere tautology to speak of general or common concepts" (*Logic*, §1, *Ak.*, IX, 91).

Aesthetic where Kant argues that space is an intuition argues essentially that the representation of space is singular. However, he has opened the Aesthetic by stating the immediacy criterion (A 16 = B 33, cited above) and in the proof of intuitivity he does say that space is *given* (B 39, also A 25). Moreover, in arguments for the same thesis in the Inaugural Dissertation of 1770, Kant does appeal to immediacy: Immediately after arguing that space is a pure intuition because it is a "singular concept," Kant says of geometrical propositions that they "cannot be derived from any universal notion of space but only as it were *seen* in space itself as if in something concrete" (§15c, emphasis Kant's). Later he says, "Geometry makes use of principles which fall under the gaze of the mind."

It seems to me that the textual evidence for Hintikka's view is not sufficient to outweigh the clear statements and emphases on the immediacy criterion, even though the alternative view must assume that Kant in discussing these matters did not keep in mind the possibility of nonimmediate singular representations.[5] But Hintikka's theory really stands or falls on the interpretation of the role of intuition in mathematics.

A thesis about intuition which is of great importance for Kant is that our mind can acquire intuitions of actual objects only by being *affected* by them. Just what this "affection" is I shall not venture to say, but it involves for the subject a certain passivity, so that our perceptions are not on the face of it brought about by our own mental activity, and also a certain exposure to contingency in our relations with objects. Thus we do not perceive objects unless they physically affect our sense organs.

A particular and highly important twist of Kant's philosophy is that the nature of our capacity to be affected by objects, our *sensibility*, already determines certain characteristics of our intuitions. These are said to be the form of our intuition in general. Among them is *spatiotemporality*. This must be understood to mean that the

---

[5]One might attribute to Kant the view that there are no such representations. The classification Kant makes in A 320 = B 376-377 and *Logik* §1 is of *Erkenntnisse*, which Kemp Smith translates as "modes of knowledge" but which in many contexts would be more accurately though inelegantly translated as "pieces of knowledge." Then the relation of a representation to its object is that through which one can *know* its object, and it might be held that intuition in the full sense is the only singular representation which can provide such knowledge. This view would have the perhaps embarrassing consequence that an object which is not in some way perceived is not really known as an individual.

[On the discussion of Kant's conception of intuition since the first publication of this paper, see the Postscript.]

nature of the mind determines that the objects we intuit should be spatial and temporal, and indeed intuited as such. The intuition which plays a role in mathematics, which is not the direct result of the affection of our mind by objects, expresses an intuitive insight which we have into our forms of intuition and is in that sense still an intuition of sensibility. It is apparently also sensible intuition in the sense of being intuition of *inner* sense.

As Hintikka rightly emphasizes, this intrinsic connection between intuition and sensibility does not come directly from the concept of intuition but represents a characteristic of man, or more generally of finite intelligences. Such an intelligence derives the content of its consciousness from outside with the resulting exposure to contingency and the necessity of concepts in order to represent objects not present. Thus not only sensibility but also thinking, or consciousness through concepts (knowledge through concepts, B 94), are characteristics of finite intelligences. The alternative is an "intuitive understanding" whose activity would *create* the objects of its awareness. Its awareness would be *only* intuition; it is called "intellectual intuition" because it has the spontaneity which for us is characteristic of thought and because the unity which with us is the result of synthesis of the *given* is for it already present in the intuition. It seems clear that intellectual intuition would satisfy the immediacy criterion.

II

Let us now turn to Kant's views on logic. What must strike a person with modern training most forcefully in considering Kant's outlook on logic is the limitation of his knowledge of and conception of it. Kant learned and taught the established logical lore at a very uncreative time in the history of the subject. Thus the formal logical analysis he undertakes is pretty well limited to the categorical proposition-forms of the theory of the syllogism, with gestures toward hypothetical and disjunctive propositions. The inferences which are covered are the syllogisms and immediate inferences of the Aristotelian theory and a few propositional inferences such as *modus ponens*. Of propositional logic as an additional developed theory, or of the additional possibilities of quantification theory, Kant had no idea.

Kant not only had very limited technical resources at his command; what is more striking and more damaging to his standing as a philosopher, he was largely satisfied with logic as he found it. Technically he could hardly in any case have gone very far beyond the state of the science in his own time, and he was not a creative mathematician. But what would have been needed for Kant to be dissatisfied with "traditional logic" might only have been more insight into his own discoveries.

As is well known, Kant attributed the lack of progress in logic to the absence of any need for it. He held that logic was established as a science and then finished off once and for all by Aristotle. This is a false view not only of the possibilities of discovery in logic but also of the history of the subject, which, far from not being "able to advance a single step" nor "required to retrace a single step" since Aristotle, had done both more than once. Kant's opinion was also influential and served to create resistance to more reasonable views both of logic itself and of its history.

Why Kant should have thought the science of logic both completable and completed is a question which I shall not attempt to answer here. I do not know whether a serious effort to answer it would uncover interesting ideas of Kant, which as it is we do not understand. In general, it can be said that the view harmonized extremely well with the more rationalistic side of Kant's way of thinking and with the belief, which he was not the only great philosopher to hold, that his own work finished off an important part of philosophy. Kant certainly thought that there were inexhaustible sources of problems, even philosophical problems, for the human intellect to wrestle with. But he held that this inexhaustibility lay within limits fixed by a form, the basic properties of which could be exhaustively described. This form would belong to the human faculty of thought itself, which so long as it was dealing with "itself and its form" and not with objects given from outside or with the manner in which they might be given from outside, was bound to be capable not only of being on sure ground but of uncovering and analyzing every relevant factor. Reason and the understanding are "perfect unities" (A xiii, A 67 = B 92). We also find an echo of the Cartesian idea that the self is better known than objects: "I have to deal with nothing save reason itself and its pure thinking; and to obtain complete knowledge of these, there is no need to go far afield, since I come upon them in my own self" (A xiv).

Logic is, according to Kant, the most general of all divisions of knowledge. It applies to all objects of our thought in general, and all true statements and sound inferences must conform to it. In particular and especially important, logical possibility is the most inclusive kind of possibility. If something is possible in any respect whatsoever, it is logically possible; its concept does not involve a contradiction. In particular, at least as far as Kant's explicit statements are concerned, the applicability of logic is not limited by the forms of our sensibility.

The relation between logic and the forms of intuition can best be seen by contrast with geometry: The forms of intuition provide the basis for certain necessary truths, in particular those of geometry, in the sense that if the forms of intuition were not as they are the truths in question would not hold, and if we did not have a certain insight into our forms of intuition, we would not know them. The application of these truths, however, is limited to the objects which affect our senses. Moreover, the principles are true of these objects only as they appear and not as they are in themselves.

These limitations do not obtain for logic. In particular, there are states of affairs which are logically possible but which are excluded by the forms of intuition, such as the existence of spatial configurations contrary to the theorems of Euclidean geometry; so that geometry is a more special theory than logic, not only in the sense that it deals with a more restricted type of object but also in the sense that it makes statements about these objects which are not logically necessary, although they are necessary in another way.

Logic is also not subject to the great limitation of knowledge based on intuition, that of applying only to appearances. When Kant says that it must be possible to *think* of things in themselves, he implies first that such a conception does not contradict the laws of logic, and second that in the statements we make about them, the logical laws are still a negative criterion of truth. If he could not trust logic in this realm, Kant's metaphysics of morals would not be able to get off the ground.

Already on this level, it is possible to see quite clearly some reasons why Kant should have regarded geometry as synthetic a priori and used an idea such as that of a form of intuition in order to explain how such a science was possible. Geometry is a more special theory than logic first in the sense that it contains nonlogical primitives, second in that its theorems cannot in general be proved merely by means of definitions and logic, as Leibniz apparently thought. Indeed this is much more obvious to us than in Kant's time, given that we have non-Euclidean geometry and are in general less tempted to overestimate the power of logic, especially traditional logic. It is worth pointing out that Euclid's postulates are what are in effect existence assumptions, so that here Kant's general views about existence would imply that they could not be analytic.

That Kant should then found geometry on the form of our sensible intuition is not difficult to understand. On the one hand spatiotemporality is a characteristic property of the objects given to the senses. Moreover, Kant emphasized that space was an individual the notion of which was understood in a way analogous to ostension, and the same ostensive understanding would be necessary for the particular primitives of geometry. On the other hand, Kant started from the idea that geometry was a body of necessary truths with evident foundations. That the axioms of geometry should be empirically verified directly is contrary to their necessity; that they should be some sort of high-level hypotheses is contrary to their evidence.

The second observation to make about Kant's views on logic is that he never suggests a conventionalist account of logical validity. It is true that the very general character of logical and analytical truths goes with their *uninformativeness*. They reflect the nature of the mind and of certain particular concepts, and apparently not at all how the world is otherwise. But this nature and the manner in which particular concepts give rise to the analytic truths that they do seem to be something given, which will be in fact the same for all discursive intelligences, even if their forms of intuition are quite different from ours.

Kant does not give much explanation of how this is, and perhaps he felt some doubt as to the possibility of giving such an explanation. If we try to apply the insight which we might get from the Transcendental Deduction to this question, we get into a very difficult dilemma. Namely the essential activity of the understanding seems to be in relation to material given in intuition, to bring it to the unity expressed in an objective judgment. In other words, the notions of object, concept, judgment get their whole sense from their application to experience. Nonetheless the understanding has a greater generality than intuition: The forms of intuition are not logically necessary; and in operating logically with a given notion, it is not necessary to appeal to intuition or even to suppose that the notion has an intuition corresponding to it. It is possible in some way for us to recognize that what can be given in experience is not the whole of possible reality, and even to recognize that with the help of intuition we can know objects only in a relative way, as they appear. All this points, even apart from the requirements of Kant's moral philosophy, to the presence in us of more general conceptions of object, concept, judgment, and *a fortiori* inference.

This dilemma will occupy us again later, since it has an application to the problems of arithmetic.

III

With respect to Kant's philosophy of geometry, the difficulties do not concern why Kant thought geometry to be a priori intuitive knowledge, but rather whether this is true and what precisely the theory was by which he proposed to explain how it could be true. When we turn to Kant's philosophy of arithmetic, there is even less difficulty as to why he should have thought arithmetical propositions a priori. But it is already by no means easy to see why Kant regarded them as synthetic, as based in some way on our forms of intuition, in particular on the form of inner intuition, time, and as limited in their application to appearances.

It will become clearer why Kant regarded arithmetical propositions as *synthetic* if we observe that Kant's concept of analytic proposition most likely had a much narrower extension than the corresponding concept in more recent philosophy, e.g., in Frege and logical positivism. Kant does not formulate his concept with enough precision so that we can be altogether sure about this. But it seems rather clear from the examples that when Kant speaks of the concept of the predicate of an analytic judgment as *contained* in that of the subject, the situation is analogous to that in which the subject concept is defined by the conjunction of the predicate concept with perhaps certain others. This would be a paradigm case where the connection of subject and predicate is "thought through identity" (A 7 = B 11). An idealized version of an analytic judgment would be one of the form 'All *AB* are *A*', or 'All *C* are *A*', where '*C*' is defined as '*A* and *B*'. This is idealized because, according to Kant, outside mathematics concepts do not in general have definitions in the proper sense.

It seems certain that a number of other forms would have to be admitted as analytic, e.g,. 'No *AB* are not *A*' or the propositional 'If *p* and *q*, then *p*'.[6] But there is no particular reason why '7 + 5 = 12' should be. Kant says (B 15) that for '7 + 5 = 12' to be analytic, it would have to follow from the concept of a sum of 7 and 5 by the law of contradiction. This would be as if it were

---

[6]Cf. the examples of "truths of reason" given by Leibniz, *Nouveaux essais*, Book IV, ii, §1.

provable from definitions by a very restricted logic, probably included in the limited traditional apparatus at Kant's command, and it is hard to see how it could be true otherwise.

However, it is one thing to say there is no reason to expect this and another to understand Kant's specific reason for thinking it false. Kant indicates that the way you find out that $7 + 5 = 12$ is by a process like counting, of progressing from 7 to 12 by successive additions of 1, in which one must operate with a *particular instance* of a group of objects, which can only be given in intuition.

> We have to go outside these concepts, and call in the aid of the intuition which corresponds to one of them, our five fingers, for instance, or, as Segner does in his *Arithmetic*, five points, adding to the concept of 7, unit by unit, the five given in intuition. For starting with the number 7, and for the concept of 5 calling in the aid of the fingers of my hand as intuition, I now add one by one to the number 7 the units which I previously took together to form the number 5, and with the aid of that figure [the hand] see the number 12 come into being. [B 15 - 16]

It is, however, still not clear why that process cannot be either itself put in the form of a purely logical argument or replaced by something quite different which can.

There was an attempt to do just this with which Kant was in a position to be familiar, by Leibniz in the *Nouveaux Essais*.[7] Leibniz worked with '$2 + 2 = 4$', but the type of argument suffices for any addition formula. He assumed as an axiom the substitutivity of identity, which Kant would in all probability have regarded as analytic. Leibniz took as definitions

$$2 = 1 + 1,$$
$$3 = 2 + 1,$$
$$4 = 3 + 1,$$

which is approximately what is done in modern formalizations. Then the proof goes as follows:

$$
\begin{aligned}
2 &= 2 + 1 + 1 && \text{(def. of ``2'')} \\
&= 3 + 1 && \text{(def. of ``3'')} \\
&= 4 && \text{(def. of ``4'')}
\end{aligned}
$$

The standard modern objection to this argument is that Leibniz should have inserted brackets, so that it goes

$$2 + 2 = 2 + (1 + 1) = (2 + 1) + 1 = 3 + 1$$

[7]Ibid., IV, vii, §10.

and therefore assumes an instance of associativity. We cannot exclude the possibility that this was known to Kant when he was working on the *Critique of Pure Reason*, since it occurs in effect in the book *Prüfung der kantischen Kritik der reinen Vernunft*, vol. I (Königsberg, 1789), by Kant's pupil Johann Schultz, professor of mathematics in Königsberg.

Putting great weight on the evidence of writings by Schultz and other disciples of Kant, Gottfried Martin has put forth the hypothesis that Kant envisaged an axiomatic foundation of arithmetic similar to the classical axiomatizations of geometry.[8] He sees the claim that arithmetic is synthetic as resting on the first instance on the logical point that arithmetical propositions such as '7 + 5 = 12' cannot be proved by mere logic from definitions such as those Leibniz uses. An axiomatic foundation of the sort which would answer to Martin's ideas is given in Schultz's *Prüfung*. Without explicitly mentioning Leibniz, Schultz points out that the sort of proof of an arithmetical identity that Leibniz gives rests on the assumption of associativity. He gives, for '7 + 5 = 12', an argument which also rests on commutativity, and seems, wrongly, to think this assumption unavoidable. But of course commutativity has to be used sooner or later in arithmetic.[9]

Schultz gives two axioms, the commutativity and associativity of addition. He neither asserts nor denies the independence of the corresponding laws of multiplication and of the distributive law. He also gives two "postulates" which are worth quoting in full:

1. From several given homogenous quanta, to generate the concept of one quantum *by their successive connection*, i.e., to transform them into *one whole*.

2. To *increase* and to *diminish* any given quantum *by as much as one wants*, that is, *to infinity*.[10]

The second postulate implies that Schultz is not thinking specifically of the arithmetic of integers but also of continuous quantities. In any case, the first postulate is the basis for the supposition that

---

[8]*Arithmetik und Kombinatorik bei Kant; Kant's Metaphysics and Theory of Science*, chap. 1; *Klassische Ontologie der Zahl*, §12.

[9]Neither Leibniz nor Schultz seems to mention the fact that in order to prove formulae involving *multiplication*, such as '2 · 3 = 6', one also needs instances of the distributive law.

[10]1. Aus mehrern gegebenen gleichartigen Quantis *durch ihre successive Verknüpfung* den Begriff von einem Quanto zu erzeugen, d. i. sie in *ein Ganzes* zu verwandeln.
2. Ein jedes gegebenes Quantum, *um so viel, als man will*, d. i. *sie ins Unendliche zu vergrössern, und zu vermindern* (*Prüfung*, I, 221).

the function of addition is *defined*; i.e., given numbers *m, n*, there actually exists a number *m + n*.

If we accepted this as actually giving Kant's conception, there would still remain the question how intuition enters into the foundation of these axioms and postulates. About this Schultz has in fact something to say. But in transferring the conception to Kant we are faced immediately with the difficulty that he explicitly says that arithmetic does not have axioms.

> As regards magnitudes (*quantitas*), that is, as regards the answer to be given to the question, "What is the magnitude of a thing?" there are no axioms in the strict meaning of the term, although there are a number of propositions which are synthetic and immediately certain (*indemonstrabilia*). [A 163 – 4 = B 204]

He considers two possibilities, rules of equality, which he asserts to be analytic (a proper axiom must be synthetic), and the elementary arithmetical identities, such as '7 + 5 = 12', which are what he seems to be referring to at the end of our quotation, which are indeed synthetic and indemonstrable, but which he declines to call axioms because they are singular.

This position is reaffirmed in a letter from Kant to Schultz dated November 25, 1788,[11] in which he comments on the manuscript of volume I of the *Prüfung*. There he gives a reason, which I shall mention later, why arithmetic should not have axioms. He does say that arithmetic has *postulates*, "immediately certain practical judgments." The general tone of his discussion suggests that he might regard the general directive to carry out addition, on the presupposition that this can always be done, as a postulate, i.e., that he might accept Schultz's first postulate. But what he seems to have specifically in mind in what he elsewhere calls numerical formulae, i.e., 7 + 5 = 12.

We cannot be certain, however, that the mathematical material of the published version of the *Prüfung* was present in the manuscript that Kant was commenting on. For it seems from the letter, as Martin points out,[12] that the manuscript maintained that arithmetical propositions were analytic, and thus it is clear that it was considerably revised *after* Schultz received Kant's letter. The fact that in the published version the axiomatic analysis is used to support the conclusion that arithmetic is synthetic does not prove that it was not present in the manuscript, although the supposition that

---

[11]*Ak.*, X, 554-558.
[12]*Arithmetik und Kombinatorik bei Kant*, pp. 64-65.

the *postulates* were there is a bit strained. But that Schultz might have argued that the commutative and associative laws were analytic is not at all impossible. (Leibniz argued this at least for commutativity.)[13]

Even so, unless one accepts Martin's rather unlikely idea that the axiomatic analysis was contributed by Kant to Schultz *after* the letter, it is hard to escape the conclusion that Schultz understood the mathematical issue in at least one respect better than Kant himself: Kant does not seem to have had an alternative view of the status of such propositions as the commutative and associative laws of addition. He can hardly have denied their truth, and it seems that if they are indemonstrable, they must be axioms; if they are demonstrable, they must have a proof of which he gives no indication.

If when speaking of the axiomatic character of arithmetic, Martin means that according to Kant arithmetic must make use of propositions which cannot be deduced by logic and definitions, then there can be no disagreeing with him. But if he means that Kant had in mind setting up arithmetic as an axiomatic system of which Schultz's is a very primitive instance and that it is in the verification of such laws as the commutative and associative that the primary application of intuition in arithmetic is to be found, then Kant's actual words go against him.

Even if Martin's view of the matter is quite correct as far as it goes, it cannot satisfy us. In the first place, it does not answer the question why arithmetic should depend on intuition, except in the sense, entirely bound to the primitive level of axiomatics in Kant's time, that so far as one can see the obvious alternative is insufficient. In the second place, it carries over to arithmetic the considerations which were at work in geometry while our original sense of difficulty arose from the difference between the two. And there are many indications, in particular some remarks in the letter to Schultz, which I shall discuss, that he saw some of this difference and did not intend to give an entirely symmetrical account.

IV

The problem of the asymmetry of arithmetic and geometry could be solved by an interpretation suggested by E.W. Beth[14] and de

---

[13]*Leibnizens mathematische Schriften*, ed. Gerhardt, VII, 78. Leibniz gives a definition of addition from which he claims commutativity follows immediately. One could read his argument as deriving the commutativity of addition from the commutativity of set-theoretic union.

[14]Über Lockes 'allgemeines Dreieck'."

veloped by Hintikka.[15] From their interpretation it seems to follow that if a proposition B of geometry is proved by a proof which appeals to axioms $A_1 \ldots A_n$ (I here include postulates),[16] then in general the conditional $A_1 \& \ldots \& A_n. \supset B$ is synthetic; at any rate an appeal to intuition is made over and above any which is made in verifying the axioms. One could then argue that since arithmetic according to Kant does not have axioms, only the first type of appeal to pure intuition occurs in arithmetic.

Beth's and Hintikka's hypothesis is that for Kant certain arguments which can nowadays be formulated in first-order predicate logic involve an appeal to intuition. In view of the singularity criterion for intuition, the natural candidates for such arguments are arguments involving singular terms. For Beth the form of argument involved is illustrated by the proof that the base angles of an isosceles triangle are equal:

> We proceed, as is well known, as a rule as follows: first we consider a particular triangle, say ABC, and suppose that AB = AC; then we show that $\angle$ ABC = $\angle$ ACB and have thus proved that the assertion holds in the particular case in question. Then one observes that the proof is correct for an arbitrary triangle, and therefore that the assertion must hold in general.[17]

The general form of the argument is as follows: We want to prove '$(x) (Fx \supset Gx)$'. We assume a particular a such that Fa. We then deduce 'Ga'. We then have '$Fa \supset Ga$' independently of the hypothesis. But since a was arbitrary, '$(x) (Fx \supset Gx)$' follows.

This form of argument, as for example in Beth's case, is the characteristic form of a proof in Euclid. In the "Discipline of Pure Reason in Its Dogmatic Employment," the section of the *Critique* where Kant sets forth his view of mathematical proof as proceeding by "construction of concepts in pure intuition," this form of ar-

---

[15]"Kant's 'New Method of Thought'," "On Kant's Notion of Intuition," also "Are Logical Truths Analytic?" and "Kant on the Mathematical Method."

[16]It ought to be remarked that while no doubt the distinction which Kant makes between axioms and postulates derives historically from that of "common notions" and postulates in Euclid, Kant's distinction does not correspond exactly to Euclid's. Euclid's division is between more general principles and specifically geometrical ones. For Kant postulates are "immediately certain practical judgments," the action involved is construction, and their purport is that a construction of a certain kind can be carried out. The role they play is thus that of existence axioms. Euclid's common notions are all of a type which Kant asserted to be analytic propositions (A 164 = B 204, B 17), while axioms proper must be synthetic.

[17]"Über Lockes 'allgemeines Dreieck'," p. 365.

gument appears clearly in the geometrical example (A 716 − 7 = B 744 − 5). The geometer "at once begins by constructing a triangle." By a series of constructions on *this triangle* and applications of general theorems to it "through a chain of inferences guided throughout by intuition he arrives at a fully evident and universally valid solution of the problem."

Hintikka concentrates attention rather on the rule of existential instantiation, that is on arguments of the form

$$(\exists x)Fx$$
$$Fa$$
$$.$$
$$.$$
$$.$$
$$p$$

where *a* is introduced to indicate an *F*, in view of the fact that the previous line affirms that there are *F*'s.[18]

Both of these arguments have in common that they turn on the use of a free variable which indicates *any* one of a given class of objects, so that an argument concerning it is valid for *all* objects of the class. They thus have a formal analogy with the appeal to pure intuition, in that a *singular* term is used in such a way that what is proved of it can be presumed generally valid. Moreover, the manner in which this generality is assured, namely by not allowing anything to be assumed about *a* except what is explicitly stated in premises, is reminiscent of a statement of Kant about the role of a constructed figure in a proof: "If he is to know anything with *a priori* certainty he must not ascribe to the figure anything save what necessarily follows from what he has himself set into it in accordance with his concept." (Bxii)

It is noteworthy that in traditional algebra calculations are carried out on terms and formulae with free variables, where the derivation of such an equation serves to prove a general proposition. Hintikka interprets the rather obscure remarks about "symbolic construction" in algebra in this sense.[19]

It would naturally follow from the conception of intuition as simply individual representation that the mere form of these arguments is such that they involve intuition. Of course, it would not give any plausibility to Kant's more far-reaching philosophical theses which turn on the connection of mathematics with the form of

---

[18]Cf. Quine, *Methods of Logic*, §28 of 2d ed.

[19]"Kant's 'New Method of Thought'," p. 130, also "Kant on the Mathematical Method." The texts are A 717 = B 745, A 734 = B 762.

sensibility. Thus the philosophically interesting aspects of the concept of pure intuition seem to lose their point when it is pointed out that these arguments can be formalized in pure quantification theory. This is exactly the conclusion which Beth draws.

One might object that this seems to presuppose that logic itself does not pose philosophical problems which the notion of pure intuition might be needed to answer, but on this at least Beth is in agreement with Kant in most of his utterances. But anyway it seems unlikely that the break between arguments which turn on the generality interpretation of free variables and logical arguments which do not is the philosophically most significant break within mathematical proof.[20]

One could wish more clear-cut evidence for the attribution of such a view to Kant or even for the most modest thesis that he started with this idea in developing his philosophy of mathematics. If it was his mature view, Kant's mathematically astute pupil Schultz seems not to have suspected it since there is no suggestion of it in the *Prüfung*. Schultz took for granted that an adequate axiomatization would be such that if the axioms were analytic so would be all the theorems. Mathematics fails to be analytic just because in its deductive development synthetic *premises* must be used. The same view is expressed by Kant when he says:

> For as it was found that all mathematical inferences proceed in accordance with [*nach*] the principle of contradiction (which the nature of all apodictic certainty requires), it was supposed that the fundamental propositions of the science can themselves be known to be true through that principle. This is an erroneous view. For though a synthetic proposition can indeed be discerned in accordance with the principle of contradiction, this can only be if another synthetic proposition is presupposed, and if it can then be apprehended as following from this other proposition.   [B 14]

---

[20]In "Are Logical Truths Analytic?" Hintikka develops a distinction between analytic and synthetic according to which some logical truths are synthetic. He suggests that the logical truths which are analytic according to this criterion are roughly those which Kant would have regarded as analytic. It follows, however, that in some of the arguments which according to Beth and Hintikka involve for Kant an appeal to intuition, the conditional of their premises and conclusion is analytic. In particular, this is true of the example that Beth works out in detail in "Über Lockes 'allgemeines Dreieck'," §7. In order to be applied to mathematical examples like Kant's, Hintikka's criterion would have to be extended to languages containing function symbols. The way of doing this which seems to me most in the spirit of Hintikka's definition has some anomalous consequences.

See also *Logic, Language-Games, and Information*, chaps. 6-9. (These chapters first appeared in 1966; see pp. ix-x.)

Against this, it is pointed out[21] that Kant says of a geometric proof that it proceeds "through a chain of inferences guided throughout by intuition" (A 716 – 7 = B 744 – 5). In view of the description Kant gives of the proof, this could easily mean that in the course of the proof one is constantly appealing to the evidences formulated in the axioms and postulates. It would obviously be anachronistic to attribute to Kant a picture of proof modeled on a formal deduction where the axioms are stated at the beginning and everything else is logic and where the purpose is not to show the *truth* of the proposition proved but merely that it follows from the axioms. On the contrary, for Kant a Euclidean proof is convincing because on each particular application of an axiom or postulate the correctness of what it claims in this particular case is evident.

It must be conceded that it might be true that inference by certain rules from analytic premises might yield analytic conclusions while inference according to the same rules from synthetic premises could lead to conclusions which are not only themselves synthetic but such that the conditional of premises and conclusions is also synthetic. In particular, the rule of existential instantiation can only come into play in the presence of an existential quantifier, and it is not clear that, for Kant, a statement in which an existential quantifier occurs essentially can be analytic. I can only say that in such cases the text of Kant does not clearly indicate that the necessity of an appeal to intuition arises for the *inference* and not merely for the verification of the premise.

If Hintikka were right, one could expect that in the passages on algebra the role of variables would be emphasized. It is possible to find this emphasis in the passage on A 717 = B 745, but it is not really explicit. The emphasis of A 734 = B 762 seems different, where Kant says, "The concepts attached to the symbols, especially concerning the *relations* of magnitudes, are presented in intuition" (emphasis mine). The relations would seem to be expressed by algebraic function signs. Although the passages on algebra offer some support for Hintikka's theory, it is less than decisive. I shall show that there are other possible ways of looking at these passages.

The direct evidence thus seems to me on the whole opposed to the Beth-Hintikka theory. However, it would have strong indirect support if there were not other ways to explain how *arithmetic* can require pure intuition and to interpret the notion of "construction

---

[21]Beth, "Über Lockes 'allegmeies Dreick'," p. 363.

of concepts," especially in algebra. To this end we now return to the problem of the difference between arithmetic and geometry.

V

The difficulty can be put in this way: The synthetic and intuitive character of geometry gets a considerable plausibility from the fact that geometry can naturally be viewed as a theory about actual space and figures constructed in it. This space is related to the senses by being a field in which the objects given to the senses appear, and geometry seems to give quite substantial information about this space which from the point of view of abstract thought might be false.

The content of arithmetic does not immediately suggest such a special character or such a connection with sensibility. Of course in the first instance it speaks of numbers and purely abstract operations and relations—equality, addition, subtraction, etc. Then the question is—what is the field of application of numbers? That is, what sorts of things can be counted, assigned cardinal or ordinal numbers, or measured and thus assigned continuous quantities? On the face of it, there is no reason to believe that the application of arithmetic need be to objects in space and time. Although this has certainly become more evident since the rise of abstract mathematics, that mathematical objects themselves could be numbered was something which Kant was certainly in a position to be aware of. If the application of arithmetic is to be limited to appearances, this limitation has to be understood rather broadly in order to reconcile it with obvious facts.

In the case of geometry, it was possible to mention logical possibilities which the concepts allowed but which did not exist according to the mathematical theory; Kant gives the example of a two-sided plane figure, and many more such possibilities were opened up in the development of non-Euclidean geometry. It was probably impossible in Kant's time to be clear about whether such a possibility exists in arithmetic. If it did, it would give rise to a clear separation of arithmetical from logical truth. This sort of argument was not available to Kant. The difficulty is made more acute, some would say insoluble, by subsequent developments in logic, particularly the efforts of Frege and others to do just what Kant thought impossible—to reduce arithmetic to logic, to deduce arithmetical propositions from definitions and propositions of pure logic.

Of course the extent of what counts as "logic" here is considerably wider than what Kant regarded as such. At the very least, we need for this type of construction to incorporate some of the theory of classes into logic; not just the notion of class and some elementary operations concerning them, but also at least some modest axioms of class existence—how modest depending on how much arithmetic one wants to deduce.

Both to set forth what we need of this construction for our purposes and to indicate how far one can go without using set-theoretic devices, I shall discuss a logical truth which is closely related in meaning to '2 + 2 = 4' and provides the key to the proof of '2 + 2 = 4' in more extended formalisms. This example will help to indicate how far the cases of arithmetic and geometry are symmetrical.

Consider the following schema of the first-order predicate calculus with identity:

$$(1) \qquad (\exists x)Fx \cdot (\exists x)Gx \cdot (x){-}(Fx \cdot Gx) \cdot \supset (\exists x)(Fx \text{ v } Gx)$$
$$ 2 2 4$$

where '$(\exists x)\ Fx$' is an abbreviation for '$-(\exists x)Fx$' and '$(\exists x)Fx$' for
$$0 n+1$$

$$(\exists x)[Fx \cdot (\exists y)(Fy \cdot y{\neq}x)].$$
$$ n$$

so that '$(\exists x)Fx$' can be expanded as
$$2$$

$$(2) \qquad (\exists x)(\exists y)[Fx \cdot Fy \cdot x \neq y \cdot (u)(Fu \supset \cdot u = x \text{ v } u = y)]$$

and '$(\exists x)(Fx \text{ v } Gx)$' as
$$4$$

$$(3) \qquad (\exists x)(\exists y)(\exists z)(\exists w)[Fx \text{ v } Gx \cdot Fy \text{ v } Gy \cdot Fz \text{ v } Gz \cdot Fw \text{ v } Gw \cdot$$
$$x \neq y \cdot x \neq z \cdot x \neq w \cdot y \neq z \cdot y \neq w \cdot z \neq w.$$
$$(u)(Fu \text{ v } Gu \cdot \supset \cdot u = x \text{ v } u = y \text{ v } u = z \text{ v } u = w)].$$

Interpretations

Intuitively, the proof of this schema goes like this: Suppose

$$(\exists x)Fx\ (\exists x)\ Gx.$$
$$\quad\ _2\qquad\quad _2$$

Then in view of (2) and its counterpart for

'$(\exists x)Gx$' there are x, y, z, *and* w such that
$_2$

$$Fx \cdot Fy \cdot x \neq y \cdot (u)\ (Fu \supset \cdot u = x \vee u = y)$$
$$Gz \cdot Gw \cdot z \neq w \cdot (u)\ (Gu \supset \cdot u = z \vee u = w).$$

We then go out to argue, with the help of '$(x) - (Fx \cdot Gx)$', that x, y, z, w satisfy the condition in the scope of (3), and so we infer that there *are* x, y, z, w such that this condition holds.

This schema requires for its formulation only predicate letters, variables, quantifiers, identity, and logical connectives. The only notion involved which could possibly be different in principle from what Kant regarded as general logic is identity, and since that is used in application to quite arbitrary objects, it does not immediately suggest a restriction as to application as the geometrical concepts do. Moreover, the schema is proved without the application of existence axioms: The range of values of the variables can be any universe whatsoever, even the empty one.[22]

Frege and his twentieth-century followers certainly thought that by their construction they had refuted the view that arithmetic depends in any way on "pure intuition," sensibility, or time. Thus the temporal notion of the successive addition of units, or the even more concrete one of combining groups of objects, is replaced in Frege's construction by the *timeless* relation of one class being the union of two others, which can be defined in terms of the logical connective alternation as it occurs in (1). Moreover, the construction provides a framework for the application of the concept of number

---

[22]In fact, (1) is analytic according to the criterion of "Are Logical Truths Analytic?" However, according to another criterion which might be more in the spirit of Kant, to consider as synthetic a conditional whose proof involves formulae of degree higher than its antecedent, (1) is synthetic. Hintikka takes account of this in chapter 7 of *Logic, Language-Games, and Information* by making an additional distinction between analytic and synthetic *arguments*, such that in the relevant sense the argument from the conjuncts of the antecedent of (1) as premises to its consequent as conclusion is synthetic.

far beyond the scope of concrete appearances, in particular, in the elaboration of set theory.

Analogous to a non-Euclidean space would be a possible world in which the arithmetical identities turned out differently, for example, in which $2 + 2 = 5$. But would that not be a world in which there was a counterinstance to our schema, and therefore in conflict with logic? Only, of course, if the connection of meaning between '$2 + 2 = 4$' and the schema (or '$2 + 2 = 5$' and a similar schema) is preserved. I am inclined to regard the breaking of this connection as a change in the meaning of addition.

There is, however, one way out of this dilemma. With '$2 + 2 = 5$' we would associate the schema

$$(4) \qquad (\exists x)Fx \cdot (\exists x)Gx \cdot (x) - (Fx) \cdot Gx) \cdot \supset (\exists x)(Fx \ v \ Gx)$$
$$ 2 2 5$$

Now suppose we had a universe U in which for any choice of extensions of 'F' and 'G' this schema came out true. Even according to our notions of logic, there is a possible case in which this happens, and in which (since (1) is valid) there is also no conflict with (1), namely in which U contains fewer than four elements. In that event the antecedent of the above would always be false.[23]

If one considers the minimal existence axioms which would be needed to prove the categorical '$2 + 2 = 4$' in modern set theory, we find that again they require the universe to contain at least four elements, which can be identified with the numbers 1, 2, 3, 4.

If we accept first order quantification theory with identity as a logical framework, then it seems that we can maintain the symmetry of arithmetic and geometry in a weak sense, that such propositions as '$2 + 2 = 4$' imply or presuppose existence assumptions which it is logically possible to deny. To draw the line at this point and to declare thus that set theory is not logic seems to me eminently reasonable; but I shall not argue for this now, particularly since I have done so elsewhere.[24]

I think the presence of existence propositions in mathematics is one of the considerations at stake in Kant's views on mathematics, but it is not clearly differentiated from others. His general views on existence imply that existential propositions are synthetic, but he never applies this doctrine directly to the existence of abstract

[23]Cf. Wang, "Process and Existence in Mathematics," p. 335.
[24]"Frege's Theory of Number" (Essay 6 of this volume).

entities. In the letter to Schultz cited above, Kant says that arithmetic, although it does not have axioms, does have postulates. Postulates as to the possibility of certain constructions, for Kant constructions in intuition, played the role of existence assumptions in Euclidean geometry. Schultz states as a postulate in the *Prüfung* essentially that addition is defined.

This factor is also present in Kant's remarks about "construction of concepts in pure intuition," which he regards as the distinguishing feature of mathematical method. If the geometer wants to prove that the sum of the angles of a triangle is two right angles, he begins by *constructing a triangle* (A 716 = B 744). This triangle, as we indicated above, can serve as a paradigm of all triangles; although it is itself an individual triangle, nothing is used about it in the proof which is not also true of all triangles. The proof consists of a sequence of constructions and operations on the triangle.

Kant's view was that it is by this construction that the concepts involved are developed and the existence of mathematical objects falling under them is shown. Although we need not regard this theorem as implying or presupposing that there are triangles, Kant regarded a general proposition as empty, as not genuine knowledge, if there are no objects to which it applies. In this instance only the construction of a triangle can assure us of this. Apart from that, further existence assumptions are used in the course of the proof, in the example of A 716 = B 744 of extensions of lines and of parallels.

The same factor is also suggested in the rather puzzling passage in which Kant says that the operation with variables, function symbols, and identity in traditional algebraic calculation involves "exhibiting in intuition" the operations involved, which he calls "symbolic construction." In fact, such operation presupposes that the functions involved are *defined* for the arguments we permit ourselves to substitute for the variables. Moreover, the construction of an algebraic expression for an object to satisfy a certain condition is the very paradigm of a constructive proof of the existence of such an object. However, I think there is something else at stake in this passage, which I shall come to.

VI

It is by no means obvious that the existence assumptions which must be made in the deductive development of mathematics have any connection with sensibility and its alleged form. Frege for one was quite convinced that they did not. What Kant says that bears on this point is not completely clear, partly because in the nature of the case it is bound up with some difficult notions in his philosophy, partly because again he did not disengage this issue from some others.

As a preliminary remark, we must observe that Kant certainly did not regard arithmetic as a special theory of, say, time, in the sense in which he regarded geometry as a special theory of space. It does not turn up in this connection in the proofs of the apriority of time in either the Aesthetic or the corresponding discussion in the Inaugural Dissertation (§12, §14 no. 5).

Nevertheless it is clear that according to Kant, the dependence of arithmetic on the forms of our intuition is in the first instance only on time. I should venture to say that space enters the picture only through the general manner in which inner sense, and thus time, depends on outer sense, and thus space. We shall be clear about the intuitive character of arithmetic when we are clear about the manner in which it depends upon time.

Whenever Kant speaks about this subject, he claims that number, and therefore arithmetic, involves *succession* in a crucial way. Thus in arguing that intuition is necessary to see that $7 + 5 = 12$:

> For starting with the number 7, and for the number 5 calling in the aid of the fingers of my hand as intuition, I now add *one by one* to the number 7 the units which I previously took together to form the number 5, and with the aid of that figure [the hand] see the number 12 *come into being*. [B 15-16; emphasis mine]

When he gives a general characterization of number in the Schematism, the reference to succession occurs essentially:

> The pure image of all magnitudes (*quantorum*) for outer sense in space; that of all objects of the senses in general is time. But the pure *schema* of magnitude (*quantitatis*), as a concept of the understanding, is *number*, a representation which comprises the successive addition of homogeneous units. [A 142 = B 182]

As I said, this seems to conflict not only with the interpretation which number and addition acquire in such constructions as Frege's, in which instead of the *successive addition* of "units" we have a *timeless* relation, for example, that one set is the union of two others; but also with the application of these notions within modern mathematics, in which arithmetical statements can be made about structures which are entirely timeless, and in reference to which any talk of "successive addition" is in on the face of it entirely metaphorical.

In the letter to Schultz, Kant qualifies his position in a way which does more justice to this more general character of arithmetic:

> Time, as you quite rightly remark, has no influence on the properties of numbers (as pure determinations of magnitude), as it does on the property of any alteration (as a quantum), which itself is possible only relative to a specific condition of inner sense and its form (time); and the science of number, in spite of the succession, which every construction of magnitude [*Grösse*] requires, is a pure intellectual synthesis which we represent to ourselves in our thoughts.

Earlier in the letter he writes:

> Arithmetic, to be sure, has no axioms, because it actually does not have a *quantum*, i.e., an object of intuition as magnitude, for its object, but merely *quantity*, i.e., a concept of a thing in general by determination of magnitude.

Kant is here in fact reaffirming a position affirmed in the Dissertation:

> To these there is added a certain concept which, though itself indeed intellectual, yet demands for its actualization in the concrete the auxiliary notions of time and space (in the successive addition and simultaneous juxtaposition of a plurality), namely, the concept of number, treated of by arithmetic. [§12]

These remarks place arithmetic less on the intuitive and more on the conceptual side of our knowledge. If arithmetic had for its object "an object of intuition as magnitude," i.e., forms such as the points, lines, and figures of geometry, then it would refer quite directly to a form of intuition. But instead it refers to "a concept of a thing in general"; the science of number is a "pure intellectual synthesis." This latter phrase especially suggests that arithmetical notions might be definable in terms of the pure categories and thus be associated with logical forms which do not refer at all to conditions of sensibility. Such a view would seem to conflict with the statement of the Schematism that number is a schema.

The reference to "a concept of a thing in general" is no doubt to be meant in the same sense as that in which the categories are said to specify the concept of an object in general, and the pure intellectual synthesis is no doubt that of the second edition transcendental deduction, which is the synthesis of a manifold of intuition in general, which is for us realized so as to yield knowledge only in application to intuitions according to *our* forms of intuition. Thus the "concept of an object in general" could give rise to actual knowledge *of objects* only if these objects can be given according to our forms of intuition.

But does this merely mean that objects in space and time provide the only concrete application of these concepts which we can know to exist, as one might expect from the absence of special reference to intuition? Whether it means this or something more drastic is, I think, a special case of the general dilemma about the understanding which I mentioned in the beginning. In either case, however, it would be a plausible interpretation of Kant to say that the forms of intuition must be appealed to in order to verify the existence assumptions of mathematics.

However, it is not very clear how to apply the general conceptions derived from the Aesthetic and the Transcendental Deduction to the case at hand. The direct existence propositions in pure mathematics are of *abstract* entities, and it is only in the geometric case that they can be said to be in space and time. I do think that the objects considered in arithmetic and predicative set theory can be construed as *forms* of spatiotemporal objects. Full set theory would of course not be accommodated in this way, but it is not reasonable to expect that from a Kantian point of view impredicative set theory should be intuitive knowledge or indeed genuine knowledge at all. It could legitimately be said to postulate entities beyond the field of possible experience.[25]

VII

It is natural to think of the natural numbers as represented to the senses (and of course in space and time) by numerals. This

---

[25]An interesting intermediate case is how constructive proofs as the object of intuitionist mathematics could be interpreted from a Kantian point of view. According to Kant as I interpret him, certain empirical constructions can function as paradigms so as to establish necessary truths because of the intention or meaning associated with them. Intuitionism would require that our insight into these meanings be sufficient not only to establish laws directly relating to objects in space and time but also to establish laws concerning the *intentions* as "mental constructions." I leave open the question of whether this is possible from Kant's point of view or not.

does not mean mainly that numerals function as names of numbers, although of course they do, but that they provide instances of the structure of the natural numbers. In the algebraic sense, the set of numerals generated by some procedure is isomorphic to the natural numbers in that it has an initial element (e.g., 'o') and a successor relation which the notion of natural number requires. In this sense, of course, the numerals are abstract mathematical objects; they can be taken as geometric figures. But of course concrete tokens of the first n numerals are likewise a model of the numbers from 1 to n or from o to $n - 1$. A set of objects has n elements if it can be brought into one-to-one correspondence with the numbers from 1 to n; a standard way of doing this is by bringing them in some order into correspondence with certain numerals representing these numbers, that is by counting. (The numerals used in work in formal logic, for example where the initial element is 'o' and the $(n + 1)$st numeral is obtained by prefixing 'S' to the nth numeral, have the further property that each numeral contains within itself all the previous ones so that the nth numeral is itself a model of the numbers from o to n).

The basis for the use of a concrete perception of a sequence of n terms in verifying general propositions is that, since it serves as a representative of a structure, the same purpose could be served by any other instance of the same structure, that in any other perceptible sequence which can be placed in a one-one correspondence with the given one so as to preserve the successor relation. This might justify us in calling such a perception a "formal intuition." We might note that the physical existence of the objects is not directly necessary, so that we can abstract also from that "material" factor.

An empirical intuition functions, we might say, as a pure intuition if it is taken as a representative of an abstract structure. Such a perception provides the fullest possible realization before the mind of an abstract concept. One of the important questions about Kant's philosophy of arithmetic is whether a comparable realization exists beyond the limits of scale of concrete perception.

Before we can enter into this question, let me point out another closely related reason in Kant's mind for regarding mathematics as dependent on intuition. This comes out in particular in the concept of "symbolic construction." The algebraist, according to Kant, is getting results by manipulating *symbols* according to certain rules, which he would not be able to get without an analogous intuitive representation of his concepts. The "symbolic construction" is essentially a construction with *symbols* as objects of intuition:

> Once it [mathematics] has adopted a notation for the general concept of magnitudes so far as their different relations are concerned, it exhibits in intuition, in accordance with certain universal rules, all the various operations through which the magnitudes are produced and modified. When, for instance, one magnitude is to be divided by another, their symbols are placed together, in accordance with the sign for division, and similarly in the other processes; and thus in algebra by means of a symbolic construction, just as in geometry by means of an ostensive construction (the geometrical construction of the objects themselves) we succeed in arriving at results which discursive knowledge could never have reached by means of mere concepts. [A 717 = B 745]

That this is a source of the clarity and evidence of mathematics and provides a connection of mathematics with sensibility is indicated by the following remark: "This method, in addition to its heuristic advantages, secures all inferences against error by setting each one before our eyes" (A 734 = B 762).

A connection of mathematics and the senses by way of symbolic operations is already claimed in Kant's prize essay of 1764, *Untersuchung über die Deutlichkeit der Grundsätze der natürlichen Theologie und der Moral*, [26] which presents a prototype of the theory of mathematical and philosophical method of the Discipline of Pure Reason in its Dogmatic Employment. For example, consider the statement of the latter:

> Thus philosophical knowledge considers the particular only in the universal, mathematical knowledge the universal in the particular, or even in the single instance, although still always *a priori* and by means of reason. [A 714 = B 742]

This distinction corresponds in the prize essay to the following, where the distinctive role of signs in mathematics is explicitly emphasized: "Mathematics considers in its solutions proofs, and inferences the universal in [unter] the signs in concreto, philosophy the universal through [durch] the signs in abstracto."[27] The certainty of mathematics is connected with the fact that the signs are *sensible*:

---

[26]*Ak.*, II, 272-301.
[27]Die Mathematik betrachted in ihren Auflösungen, Beweisen, und Folgerungen das Allgemeine unter den Zeichen *in concreto*, die Weltweisheit das Allgemeine durch die Zeichen *in abstracto* (Erste Betrachtung, §2, heading, *Ak.*, II, 278).

Since the signs of mathematics are sensible means of knowledge, one can know with the same confidence with which one is assured of what one sees with one's own eyes that one has not left any concept out of account, that every equation has been derived by easy rules, etc.; thereby attention is made much easier in that it must take account only of the signs as they are known individually, not the things as they are represented generally.[28]

The prize essay suggests a position incompatible with the *Critique of Pure Reason*, namely that since in mathematics signs are manipulated according to rules which we have laid down (in contrast to philosophy, where the value of any definition turns on its having a certain degree of faithfulness to preanalytic usage), operation with signs according to the rules, without attention to what they signify, is in itself a sufficient guarantee of correctness.[29]

These passages show that a connection between *sensibility* and the intuitive character of mathematics existed in Kant's mind before he developed the theory of space and time of the Aesthetic. However, unlike in the later work, no inference is drawn at this stage from this connection to a limitation of the application of mathematics to sensible *objects*.

The general point behind the observations on symbolic construction can be put in the following way: In general, a mathematical proposition can be verified only on the basis of a proof or calculation, which is itself, a construction in intuition. But in view of the remarks about '$7 + 5 = 12$', a more special fact may have influenced Kant. Certain "symbolic constructions" associated with propositions about number actually involve constructions isomorphic to the numbers themselves and their relations, or at least an aspect of them. Thus in Leibniz' proof that $2 + 2 = 4$, '$2 + 2$' must be written out as '$2 + (1 + 1)$' and the two 1's as it were added on to the '2'. A corresponding proof of '$7 + 5 = 12$' would involve *five* such steps instead of two.

---

[28]Denn da die Zeichen der Mathematik sinnliche Erkenntnismittel sind, so kann man mit derselben Zuversicht, wie man dessen, was man mit Augen sieht, versichert ist, auch wissen, dass man keinen Begriff aus der Acht gelassen, dass eine jede einzelne Vergleichung nach leichten Regeln geschehen sei u.s.w. Wobei die Aufmerksamkeit dadurch sehr erleichtert wird, dass sie nicht die Sachen selbst in ihrer allgemeine Vorstellung, sondern die Zeichen in ihrer einzelnen Erkenntnis, die da sinnlich ist, zu gedenken hat (Dritte Betrachtung, §1, *Ak.*, II, 291).

[29]But cf. the following: in der Geometrie, wo die Zeichen mit den bezeicheten Sachen überdem eine Ähnlichkeit haben, ist daher deese Evidenz noch grösser, obgleich in der Buchstabenrechnung die Gewissheit ebenso zuverlässig ist (ibid., p. 292).

A similar observation concerning the schema (1) has been made by a number of writers. Although the schema does not *imply* that the universe contains any elements or that any construction can be carried out, the *proof* of it involves writing down a group of two symbols representing the F's, another such group representing the G's, and putting them together to get four symbols. So that it is not at all clear that '$2 + 2 = 4$' interpreted as a proposition about the combinations of symbols is not more elementary than the logically valid schema (1).

I have already suggested that the "symbolic" construction in generating numerals is already enough to settle the question of their reference. In the same way the actual carrying out of the calculations shows the well-defined character for individual arguments of recursively defined functions. However, induction, which I have wanted to leave out of account here, is involved in seeing that they are defined for all arguments. Maybe Kant ought to have said that apart from intuition I do not even know that there is such a number as '$7 + 5$'. And it seems that one could not see by a particular construction that there is such a number without also seeing it to be 12. This is in agreement with Hintikka's statement that the sense of Kant's statement that numerical formulae are indemonstrable is that the construction required for their proof is already sufficient.

The considerations about the role of symbolic operations apply equally to logic and therefore undermine Kant's apparent wish to distinguish them on this basis. This appears more forcefully in modern logic, where instead of a short list of forms of valid inference one has an infinite list which must be specified by some inductive condition. In my opinion this is a consequence to be accepted and is even in general accord with Kant's statements that synthesis underlies even the possibility of analytic judgments.[30]

The special connection of arithmetic and time can, I think, be explained as follows: If one constructs in some way, such as on paper or in one's head, such a sequence of symbols as the first n

---

[30][This "general accord" now seems to me quite tenuous, and Manley Thompson is probably right in saying that the synthesis required for analytic judgments is clearly distinguishable from that in mathematical judgments ("Singular Terms and Intuitions in Kant's Epistemology," p. 342, n. 23).

Nonetheless, a reply to the main point, that logic is not entirely independent of intuitive construction, would demand a lot of Kant's distinction between intuitive and discursive proofs, as is clear from Thompson's interesting discussion of this distinction (ibid., pp. 340-342). His interpretation implies rather extreme limits on the role of logic in mathematics. This raises a doubt whether Kant's distinction is in the end tenable.]

numerals, the structure is already represented in the sequence of operations and more generally in the succession of mental acts of running through a group of *n* objects, as in counting. Thus time enters in through the succession of acts involved in construction or in successive apprehension. This connects with Kant's remark about number in the Schematism. In the operations involved in representing a number to the senses, we also generate a structure in *time* which represents the number. Time provides a universal source of models for the numbers. In particular, Kant held that it is only by way of successively perceiving different aspects of a manifold and yet keeping them in mind as aspects of one intuition that we can have a clear conception of a plurality. For quite small numbers this seems doubtful although not for larger ones. Nonetheless the element of succession appears even for the smaller ones in the comparison involved in generating or perceiving them in *order*, and the order is certainly part of our concept of number. What would give time a special role in our concept of *number* which it does not have in general is not its necessity, since time is in some way or other necessary for all concepts, nor an explicit reference to time in numerical statements, which does not exist, but its sufficiency, because the temporal order provides a representative of the number which is present to our consciousness if any is present at all.

Of course it is one thing to speak of representation in space and time and another to speak of representation to *the senses*. What is represented to the senses is presumably represented in space and time, but maybe not vice versa. To establish a link of these two Kant would appeal to his theory of space and time as forms of sensibility. The relevant part of this theory is that the structures which can be represented in space and time are structures of *possible* objects of perception. The kind of possibility at stake here must be essentially mathematical and go beyond "practical" or physical possibility.[31]

Consider once again a procedure for generating numerals, say by starting with 'o' and prefixing occurrences of 'S'. The actual use of these as symbols requires that they be perceptible objects. Nonetheless we say it is *possible* to iterate the procedure indefinitely and

---

[31]One might say that it is possible to construct tokens. The sense of possibility in which this is possible is, however, derivative from the mathematical possibility of constructing types (or mathematical *existence* of the types). For we declare that the tokens are possible either directly on the basis of the mathematical construction, or physically on the basis of a theory in which a mathematical space which is in some way infinite is an ingredient.

therefore to construct indefinitely many numerals. Thus it is clear that the numerals (numeral types)[32] which it is in this sense possible to construct extend far beyond the numeral-tokens which have ever been produced in history or which could in any concrete sense actually be used as symbols.[33] This possibility of iteration is necessary for the constructibility of indefinitely many numerals and therefore for the infinity of natural numbers to be given by intuitive construction. Moreover, some insight into such iteration seems necessary for mathematical induction.

Insofar as the appeal to pure intuition for the evidence of mathematical statements is supposed to be an analogy of mathematical and perceptual knowledge, it holds less well for propositions involving the concept of indefinite iteration, such as these proved by induction, than for propositions such as $2 + 2 = 4$. There seem to be two independent types of insight into our forms of intuition which a Kantian view requires us to have, that which allows a particular perception to function as a "formal intuition" and that which we have into the *possible progression* of the generation of intuitions according to a rule. To speak of a peculiar *kind* of intuition in the second case seems quite misleading. The mathematical knowledge involved has a highly complex relation to "intuition" in the more specifically Kantian case.

The complexity must be in some way present in the "intuitions" of space and time since space is an individual which is *given*, but its structure also determines the limits of *possible* experience and contains various infinite aspects. No doubt the plausibility of the idea that space is present in immediate experience made it more difficult for Kant to appreciate the differences of the kinds of evidences covered by his notions of pure intuition. I am sure that more could be done to explicate the Kantian view of their connection.

In our discussion of intuition, we have somewhat lost sight of the view of logic which at the start we attributed to Kant, which except for the question of existence resembles the modern views called Platonist. Although Kant's view of intuition fits better with

---

[32]Cf. "Infinity and Kant's Conception of the 'Possibility of Experience' " (Essay 4 of this volume).

[33]This does not imply that there is an upper limit on the numbers which can be individually represented, once we admit notations for faster-growing functions than the successor function. This happens already in Arabic numeral notation. The number 1,000,000,000,000, if written in 'o' and 'S' notation with four symbols per centimeter, would extend from the earth to the moon. That there is such an upper limit follows, of course, from the assumption that human history must come to an end after a finite time.

the modern tendencies called constructivist or intuitionist, it seems certain that the concept of pure intuition was meant to go with this view of logic and not to replace it. Without using notions like "concepts" and "object" in a quite general way, it is probably not possible to describe it. It would be hasty for that reason to identify Kant's conception of intuition with that of Dutch intuitionists, although Brouwer's undoubtedly shows some affinity. It would also be hasty to regard Brouwer's critique of classical mathematics as altogether in accord with Kantianism.

Postscript

The remarks in this paper about Kant's conception of intuition provoked some controversy. Hintikka's reply to my criticisms of his views concentrated on this issue.[34] It is agreed that Kant's basic conception of an intuition is of a representation of an individual object that relates to its object immediately. The dispute concerned what I called the "immediacy criterion" and how it is related to the individuality or singularity criterion.[35]

According to Hintikka, an intuition is simply a singular representation, the analogue among Kantian *Vorstellungen* of a singular term. In reply to my question why he thought the immediacy criterion nonessential, Hintikka says that it is "simply a corollary of the individuality criterion" (p. 342). He cited the following well-known passage: "This knowledge is either *intuition* or *concept* (*intuitus vel conceptus*). The former relates immediately to the object and is singular [*einzeln*],[36] the latter refers to it mediately by means of a feature which several things may have in common" (A 320 = B 376-7). What is not immediate about concepts, as Hintikka reads this passage, is that they refer to their objects "only through the

---

[34]"Kantian Intuitions." In this Postscript this paper is cited merely by page number.

[35]Particularly since the singularity criterion itself is not in dispute, I should emphasize that my own understanding of its importance owes much to Hintikka's writings and conversation. I should belatedly thank him for explaining his ideas to me some years before they were published in English.

[36]Kemp Smith translates *einzeln* in this passage as "single." The translation "singular" fits its use in *Logic*, §1 (see above, p. 112), where it is paired with the Latin *singularis*. Thompson suggests that in the *Critique* passage it may mean that an intuition is a *single* occurrence. ("Singular Terms and Intuitions," p. 328, n. 13.) If intuitions are thus in effect *events*, that would rule out Hintikka's interpretation (though not the view of Robert Howell discussed below). Though this seems to me to agree with Kant's characteristic way of speaking about intuitions, the point is not so clear as to be a serious argument in the present dispute.

mediation of a characteristic which several objects may share" (p. 342). Thus the immediacy of intuitions consists in their not representing their objects by way of properties that they may share with other objects.

This is, so far, a defensible reading of the passage. To support Hintikka's thesis, however, it would have to be stretched to say that no representation that is mediate in this sense could be singular. Hintikka does not explicitly argue for this consequence, and strong textual arguments against it were subsequently offered by Robert Howell.[37] An important difficulty is that it would make singular judgments an exception to Kant's view that a judgment is a combination of two concepts. This would make nonsense of Kant's assimilation of singular judgments to universal.[38] Kant does not, to be sure, say that there are singular concepts. It is not concepts themselves but only their *use* that can be classified as universal, particular, and singular.[39] But there is no indication that the singular use of a concept makes it an intuition.

A more principled difficulty with Hintikka's interpretation is how it could be, on his account, that all our intuitions are sensible. A definite description is a singular term that refers to its object by means of concepts; quite apart from the passage where he seems pretty clearly to say the contrary,[40] I do not see how even after Kant's logical analysis of mathematics (however that is to be interpreted) and the argument of the Aesthetic, that can be taken to be a representation of sensibility.

Rejection of Hintikka's view of the immediacy criterion does not of itself imply acceptance of my own. A *via media* is proposed by Howell. He agrees with Hintikka that the above-cited passage gives the strict definition of immediacy; it is simply the absence of "mediation" by marks or characteristics. So, far from agreeing with Hintikka that it is merely a corollary of singularity, he goes on to understand it as direct reference in something like the sense of modern theories of names and demonstratives.[41] Indeed, his view

[37] "Intuition, Synthesis, and Individuation in the *Critique of Pure Reason*," p. 210.
[38] A 71 = B 96; *Logic*, §21, n. 1 (*Ak.*, IX, 102).
[39] *Logic* §1, Note 2 (*Ak.*, IX, 91). This point is discussed at some length by Thompson, "Singular Terms and Intuitions," pp. 316-318.
[40] In the discussion of "the black man" in a letter to J. S. Beck, July 3, 1792 (*Ak.*, XI, 347); cf. Howell, "Intuition, Synthesis, and Individuation," p. 210. Other examples that can be given, such as the Idea of God, involve either Ideas of Reason or mathematics and might therefore be regarded as exceptional.
[41] "Intuition, Synthesis, and Individuation," pp. 210-211. The distinction between what he takes to be the definition and his further interpretation is not explicit in the paper; here I rely on his clarification of his views in a recent letter.

is that empirical intuitions at least are analogues of demonstratives.[42] Hintikka himself reads immediacy as "direct reference to objects" (p. 342) and may have adopted a view like Howell's, although then he surely should not continue to hold that immediacy is a consequence of singularity.[43]

Howell's view of the strict definition of immediacy has the great advantage of putting into relief the somewhat hypothetical character of all three of the developed views we are considering, Hintikka's, Howell's, and mine. In my view it also shows the necessity of some further assumption, since it makes the bare-bones notion of immediacy very uninformative; indeed, to one trained in modern logic, it appears circular, for what could mediation by marks or characteristics be but some *predicative*, in Kantian terminology conceptual, content of the representation.[44] Nonetheless, the controversy about the immediacy criterion convinced me that my interpretation of its meaning was not so evident as I had thought. Howell draws a reasonable line between what is plain from the text and what has the character of a reconstruction.

Howell's own further interpretation is certainly an interesting line of reconstruction and may be fruitful. As an interpretation of Kant's intentions, it has the difficulty of relying on ideas developed much later in response to problems Kant did not consider. Howell's view, like Hintikka's, attempts to make the distinction between intuitions and concepts entirely within general logic. Kant followed a logical tradition that neglected the distinction between singular and general terms. Without breaking with this tradition more clearly than he did, Kant could not give a clear account of how singular representations could enter into propositions and inferences. My strategy was not even to try to do this on Kant's behalf. In understanding immediacy as some kind of direct presence, I was treating the concept of intuition as from the beginning epistemic. Even if Howell is right about the strict definition of the immediacy of intuition, there is undoubtedly an epistemic sense of immediacy in

[42]Ibid., p. 215. A somewhat similar picture is presented in Thompson, "Singular Terms and Intuitions." However, Thompson rejects Howell's view that certain demonstratives can be the linguistic expression of intuitions; see Thompson, pp. 332-333, and Howell's reply, p. 232. Thompson holds that a Kantian "canonical language" would be virtually without singular terms (ibid., p. 334).

[43]In a discussion in March 1983, after this Postscript had been written, Hintikka stated that Howell's interpretation of immediacy was the view Hintikka had maintained all along.

[44]But obviously one needs to look carefully at how Kant and his contemporaries actually viewed the relation of concepts and their "marks." This matter was explored by my student Alan Shamoon in his dissertation, "Kant's Logic."

Kant's writings, at stake when he says that mathematical axioms are immediately certain (A 732 = B 760).[45] I do not see how to get around regarding some link between the immediacy of intuition and this epistemic sense as an assumption of Kant's, whether or not it was directly embodied in the way he understood the word "immediate" in the definition of intuition. Without some such interpretative hypothesis, I did not see how to make sense of Kant's theory of geometry. It gives a straightforward explanation of how Kant could think that mathematics depends on sensibility, and extends it from the easier case of geometry to the harder case of arithmetic.

I should point out that I intended "present" in a phenomenological sense; in imagination as well as perception an object is "present" in the relevant sense. It follows that intuition does not necessarily involve the *existence* of the object intuited. It is not clear to me how the direct-reference view achieves the same result, which seems necessary for an account of nonveridical perception.[46]

At this point I should mention a misunderstanding to which my view can give rise, which is expressed most clearly in the following remarks by Gordon Brittan:

> If we were to accept Parsons' interpretation—that it is part of the *meaning* of "intuition" that intuitions are quasi-perceptual (and thus that the "immediacy criterion" is independent of the singularity criterion and has epistemological import)—then how would we be able to understand Kant's claim (at B 146) that "as the Aesthetic has shown, the only intuition possible to us is sensible"? On Parsons' interpretation, that (human) intuition is sensible follows as a trivial consequence of the definition and should not require the extended argument of the Aesthetic.[47]

It cannot be my view (or anyone else's) that it is a trivial consequence of the definition that *intuition* is sensible; as I made clear

---

[45]Cf. the contrast on the following page (A 733 = B 761) between discursive and intuitive principles.

[46]I owe this last observation to Manley Thompson. However, the direct-reference view might itself suggest an assumption such as I attribute to Kant; compare the connection between direct reference and sense-perception in the philosophy of Bertrand Russell.

[47]*Kant's Theory of Science*, p. 50, n. 15. Chapter 2 of this book is a very clear and instructive discussion of Kant's philosophy of mathematics, with expositions both of Hintikka's and my own views.

Howell seems not to be free of the same misunderstanding; see "Intuition, Synthesis, and Individuation," pp. 210-211.

above (section I), an intellectual intuition would be a counterexample. I do not see why Brittan should think that the more restricted thesis that *human* intuition is sensible should be immediate from the definition as I read it, since it makes no explicit mention of distinctively human (or even finite) intellectual capacities. It may be that he takes it to be obvious from Kant's point of view that it is only by sensibility that *individuals* are immediately present to us. However, this could be true only if "individual" is taken to mean concrete object. Since clearly only logical singularity is at issue in the definition of intuition, one cannot derive the sensible character of all human intuition in this way.

There is an underlying reason why Brittan's misunderstanding should be natural. I did say that in intuition the object "is in some way directly present to the mind" (p. 112), and that word "present" does suggest that the object is *given* in the sense that Kant regards as characteristic of the intuition of *finite* intelligences, that is, that the mind is to some degree passive and is apprehending an object that is "there" prior to its apprehension. Perhaps we do have to cancel such a suggestion in understanding the idea of an intellectual intuition (to the extent that we can understand it). If so, this reflects the intrinsic difficulty of forming a conception of a mode of intuition different from our own.

Howell's view of Kant's conception of intuition would serve to save Hintikka's analysis of Kant's philosophy of mathematics as a whole. But my own criticisms in section IV above of the rest of Hintikka's analysis would not be affected by the assumption that Kant's basic conception of intuition is as Howell claims. However, Hintikka directly criticizes my statement that "intuition is thus a source, ultimately the only source, of immediate knowledge of objects" (p. 112 above). Appealing to remarks of Kant in §§8-9 of the *Prolegomena*, he argues (p. 343) that my view would lead to a perverse conception of Kant's problem concerning mathematics. His view appears to be that my reading of the immediacy condition would make a priori intuitions "misnomers." He relies here on the puzzle expressed in §8 of the *Prolegomena*:

> For the question now is, "How is it possible to intuit anything *a priori*?" An intuition is such a representation as would immediately depend upon the presence of the object. Hence it seems impossible to intuit spontaneously *a priori*, because intuition would in that event have to take place without either a former or a present object to refer to, and in consequence could not be intuition.[48]

[48]*Ak.*, IV, 281-282, translation from Beck's edition.

I should remark that the claim about intuition made in the second sentence fits Hintikka's conception even less well than mine. But it is clear that Kant is here expressing a common conception of intuition, not necessarily his own, and that in the end his view is that it fits only empirical intuition. In fact it is quite clear (especially from §9) that when Kant in this passage talks of "objects," he means *actual* concrete objects. A priori intuition of *such* objects has to be prior to their being given. This, he says in §9, is possible only "*if my intuition contains nothing but the form of sensibility, antedating in my mind all the actual impressions through which I am affected by objects.*"[49] Kant's problem in this passage is thus how intuition can be "of" an object not yet given. I am not sure why, in Hintikka's view, there should be a *problem*. Hintikka grants that on my view there is but holds it to be insoluble. I hold that the claim that a priori intuition "contains nothing but the form of sensibility" is the main idea of Kant's solution. In §9 of the *Prolegomena*, as in many other passages, his stress is on the thesis that this form's being a condition of my mind on the intuition of objects makes a priori knowledge of such objects possible. The question how knowledge resting on the form of sensibility is *intuitive* is prior. My own writing on mathematical intuition undertakes to offer a model of how *forms* may be given in intuition which are yet the forms of objects not yet given. Kant, in giving geometrical examples, must have thought this tolerably clear.

An obstacle to complete clarity is the absence in Kant's philosophy of a theory of mathematical objects. Of course Kant's writing on mathematics abounds in what would ordinarily be read as references to mathematical objects. Sometimes he seems to commit himself to them in a more philosophical way, as when he says that '7 + 5 = 12' is a singular proposition (A 164 = B 205), and that we can "give it [the concept of a triangle] an object wholly *a priori*, that is, construct it" (A 223 = B 271). In sections V and VI of this essay I assumed that Kant was concerned with mathematical objects of the usual kind.

This was an incautious assumption. The concept of object in terms of which construction gives the concept of triangle an object is not Kant's primary one, and indeed in that passage Kant partly takes away what he has just given in saying that the triangle is "only the form of an object." Kant never talks explicitly of the *existence* of mathematical objects; existence for him seems to be concrete

---

[49]*Ak.*, IV, 282, emphasis Kant's.

existence; this is quite explicit in its schematization as actuality. He seems to decline to attribute existence to mathematical objects at all.[50]

But what, then, are a priori intuitions, as singular representations, intuitions *of*? Mathematics contains a priori knowledge, which is knowledge of objects in the full-blooded sense, that is of the objects given in *empirical* intuition. Sometimes, as in *Prolegomena* §8, Kant talks as if these objects were the objects of a priori intuition. But how can a priori representations have such reference and still be singular? This is a difficulty common to my own and to Hintikka's view of intuition. A picture common to us is of pure intuitions as analogous to *free variables*, with predicates attached to them representing the concepts they "construct." If we are not to import into Kant the "mathematical-objects picture," then it seems we have to take the *range* of these variables to be *empirical* objects. Then a mathematical argument cannot, strictly speaking, establish existence. What plays the role of mathematical existence in Kant is constructibility. The most plausible reconstruction of Kant would be, in my view, to take constructibility of a concept to be a kind of possible existence of a (nonabstract) object falling under the concept. Kant's view would then be in line with the modal interpretation of quantifiers discussed elsewhere in this volume, in particular, in connection with intuition, in Essay 1, section III.

However, I say "a kind of possible existence" because it cannot be possible existence in the precise sense Kant gives to those words. The difficulty is not with existence but with possibility, which for Kant is what we might call real possibility. The fact that the concept of triangle can be constructed makes it appear that we can see the possibility of a triangle "from its concept in itself" (A 223 = B 271).

---

[50] "But in mathematical problems there is no question of this [the conditions under which the perception of a thing can belong to possible experience], nor indeed of existence at all, but only of the properties of the objects in themselves, solely in so far as these properties are connected with the concept of the objects" (A 719 = B 747). This passage is instructively discussed by Thompson, "Singular Terms and Intuitions," pp. 338-339. It is largely through this paper that I became aware of the difficulties faced by my own views about mathematical objects and existence in Kant.

Brittan comments on the same passage (*Kant's Theory of Science*, p. 66), but he seems to me to misread it in saying that "mathematical problems" have to do with real possibility. That seems to me to neglect the clear statement of A 223−4 = B 271 that construction does not establish such possibility, and even the remark in the present passage that in mathematical problems there is "no question" of the conditions under which perception of a thing can belong to possible experience. The point is subtle because Kant holds that mathematics is about really possible objects and that this can be established. But it is not *mathematics* that establishes it.

What makes the construction itself fall short of showing possibility is that possibility involves agreement with the formal conditions of experience with respect to intuition *and concepts*, that is, not only the forms of intuition but the categories.[51] Thus in order to see the possibility of a triangle, we have to observe that space is a formal condition of outer experience and that "the formative synthesis through which we construct a triangle in imagination is precisely the same as that which we exercise in the apprehension of an appearance, in making for ourselves an empirical concept of it" (A 224 = B 271). This, of course, repeats the considerations advanced in the Axioms of Intuition. The "objective reality" in the full sense of mathematical concepts seems to be a proposition not of mathematics but of philosophy.

To return to the original issue about the immediacy criterion for being an intuition: I would say that Kant as I interpreted him is of interest to the philosopher today. My line of interpretation parallels Kant's most important influence on twentieth-century foundational research, through Brouwer and Hilbert. A concept of intuition like that I attributed to Kant is of interest in its own right.[52]

Hintikka, too, could defend his interpretations partly on the grounds of philosophical interest. I must also grant a certain justice to the closing remarks of his comment on my paper (pp. 344-345). Indeed, a central problem for Kant was "why the knowledge so obtained [in mathematics] can be applied to all experience a priori and with certainty." There is an important aspect of Kant's answer to that question that I hardly touched on, namely the argument in the Analytic for the claim that mathematics necessarily applies to the objects of *empirical* intuition. However, I do not find an analysis of that argument in Hintikka's writings either. I do not think that either of us has undertaken the task of constructing a truly Kantian explanation of the a priori character of mathematics. In my own case, I doubt that such an explanation could be given without appealing to Kantian transcendental psychology.[53]

[51]A 218 − B 265. Cf. the whole discussion of possibility this statement introduces.
Thompson (p. 341, n. 21) sees a difficulty with a modalist interpretation of how Kant might deal with reference to mathematical objects in Kant's distinction between demonstrations and discursive proofs. I am not sure I understand what the difficulty is, but it is not evident that the attenuated version of modalism suggested for Kant in the text is more exempt from it than the direct version. But cf. note 30 above.
[52]See above, Essay 1, section III; also "Mathematical Intuition."
[53]I wish to thank Robert Howell and Manley Thompson for their helpful comments on an earlier version of this Postscript.

6

Frege's Theory of Number

It is impossible to compare Frege's *Foundations of Arithmetic*[1] with the writings on the philosophy of mathematics of Frege's predecessors—even with such great philosophers as Kant—without concluding that Frege's work represents an enormous advance in clarity and rigor. It is also hard to avoid the conclusion that Frege's analysis increases our understanding of the elementary ideas of arithmetic and that there are fundamental philosophical points that his predecessors grasped very dimly, if at all, which Frege is clear about.

I mention this impression which Frege's book makes because it is often forgotten in critical discussion of his ideas, and still more forgotten in discussion of "the Frege-Russell view," "the reduction of mathematics to logic," or "logicism." Frege's main thesis, that arithmetic is a part of logic, is not fashionable now. It seems to me that this is justified, and the accumulated force of the criticisms of this thesis is overwhelming. But even though Frege is more studied now than at the time when his thesis was regarded by many as having been conclusively proved, I find that we still lack a clear

From Max Black, ed., *Philosophy in America*, pp. 180-203. Copyright under the Berne Convention by George Allen & Unwin Ltd. First published in the United States of America, 1965. Used by permission of Professor Black and the publishers, Cornell University Press and George Allen & Unwin, Ltd.

This paper benefited from the comments of hearers of earlier versions at Cornell and Columbia in 1962 and 1963 and from the comments of Professor Black and especially Burton Dreben on the penultimate version.

[1]*Die Grundlagen der Arithmetik*. Quotations are in Austin's translation; page references will serve equally for that translation and any of the German editions in existence.

[I have verified Austin's assertion in his Translator's Note that the pagination of the German text printed with his translation is the same as that of the first edition; it is exactly correct.]

view of what is true and what is false in his account of arithmetic. What follows is intended as a contribution toward such a view.

I

It will help with this task not to focus our attention too exclusively on the thesis that arithmetic is a part of logic. An examination of the argument of the *Foundations* shows that this thesis is introduced only *after* some of the confusions of his predecessors have been cleared up by other analyses. It seems to me that we might best divide Frege's view into three theses, which I shall discuss in turn.

(1) Having a certain cardinal number is a property of a *concept* in what we may take to be Frege's technical sense. It appears that the basic type of singular term referring to numbers is of the form 'the number of objects falling under the concept F', or, more briefly, 'the number of F's', or in symbolic notation '$N_x Fx$'.

(2) Numbers are *objects*—again in Frege's technical sense.

(3) Arithmetic is a part of logic. This may be divided into two:
   (a) The concepts of arithmetic can be defined in terms of the concepts of logic.
   (b) The theorems of arithmetic can be proved by means of purely logical laws.

The first thesis does not require much discussion. The appeal to Frege's special sense of 'concept' is of course something which would give rise to difficulty and controversy, but it is not essential to the main point. Everyone will agree that we cannot get far in talking about cardinal numbers without introducing singular terms of the form mentioned, or others in which the general term is replaced by a term referring to a class or similar entity.

There is a futher question whether, in elementary examples involving perceptual objects, we could attribute the number to something more concrete, or more in accord with the demands of nominalism, than a class or a concept. Frege himself was apparently not interested in this question, and it does not seem to me very important for the foundations of mathematics.

II

I shall now discuss the thesis that numbers are objects. It will prove to be closely connected with the third thesis, that arithmetic is a part of logic. The first thing to note is that for Frege the notion of an object is a *logical* one. He held that linguistic expressions satisfying certain syntactical conditions at least purport to refer to objects. I do not have a precise general account of what these conditions are. Being a possible subject of a proposition is the primary one, but it must be so in a *logical* sense; otherwise 'every man' in 'Every man is mortal' would refer to an object. The occurrence of the definite article is an important criterion. Such examples as 'The number 7 is a prime number' seem to show that numerical expressions satisfy the syntactical criteria.

Another criterion of great importance in the *Foundations* is that *identity* must have sense for every kind of object.

Frege took this in a very strong sense: if we think of '———' and '. . .' as object-expressions, then '——— = . . .' must have sense even if the objects to which they purport to refer are of quite different categories, for example, if '———' is 'the Moon' and '. . .' is 'the square root of 2'.[2] Moreover, the principle of substitutivity of identity must be satisfied.

Now for every object there is one type of proposition which must have a sense, namely the recognition-statement, which in the case of numbers is called an identity. . . . When are we entitled to regard a content as that of a recognition-judgement? For this a certain condition has to be satisfied, namely that it must be possible in every judgement to substitute without loss of truth the right-hand side of our putative identity for its left-hand side.[3]

This is a view which Quine expresses succinctly by the maxim, "No entity without identity." One of the main efforts of the positive part of the *Foundations* is to explain the sense of identities involving numbers.

That terms satisfying these conditions, and perhaps others besides, occur in places accessible to quantification, and that we make such inferences as existential generalization (e.g., from '2 is an even prime number' to 'There is an even prime number') might be taken to show that numerical terms purport to refer to objects. From Frege's point of view, however, I should think that this shows only

---

[2]This follows from Frege's general doctrine that a function must be defined for every object as argument, since identity is for Frege a function of two arguments. See *Funktion und Begriff*, pp. 19-20; *Grundgesetze*, §§56-65 (Geach and Black, 1st ed., pp. 159-170); *Foundations*, pp. 78, 116-117.

[3]*Foundations*, pp. 116-117.

that they purport to *refer*, for quantification could also occur over functions, including concepts.

Shall we accept Frege's criteria for expressions to purport to refer to objects? We might, I think, separate those explicitly stated by Frege from the criterion, added from Quine, of accessibility to quantification. The latter has some complications having to do with mathematical constructivity and predicativity. These do not make it an unacceptable criterion, but might lead us to distinguish "grades of referential involvement." With this reservation, I do not know any better criteria than the ones I have mentioned. I am still not very clear about their significance, what it is to be an object. In particular, the central role of identity is something which I do not know how to explain. Perhaps it has to do with the fact that the cognitive activities of human beings are spread out over space and time.

III

I have been very careful to speak of criteria for expressions to *purport* to refer to objects. Indeed, it would seem that if we explain number-words in such a way that they will be shown in at least some of their occurrences to satisfy these criteria, then we shall at most have shown that they purport to refer to objects but not that they actually *do* refer to them, i.e., that in these occurrences they actually *have* reference. I shall now consider how this matter stands in Frege's analysis of number.

The simplest account of it is as follows: Frege finds[4] that a necessary and sufficient condition for the number of F's to be *the same* as the number of G's is that the concept F and the concept G stand in a relation he called *Gleichzahligkeit*, which may be translated as 'numerical equivalence'. The concept F is numerically equivalent to the concept G if there is a one-to-one correspondence of the objects satisfying F and the objects satisfying G. If we express this by '$Glz_x(Fx, Gx)$', we may express the result of this stage of Frege's analysis as the principle

(A) $$N_x Fx = N_x Gx \,.\, \equiv Glz_x(Fx, Gx).$$

He then gives an *explicit definition* whose initial justification is that (A) follows from it. The number of F's is defined as the extension of the concept *numerically equivalent to the concept F*, in other words the class of all concepts numerically equivalent to the concept F.

---

[4]Ibid., pp. 72-81.

Then it seems that the problem of the existence of numbers is merely reduced to the problem of the existence of extensions. If this is so, Frege is in two difficulties.

The first is that the paradoxes make it not very clear what assumptions as to the existence of extensions of concepts are permissible. Frege sought a general logical law by which one could pass from a concept F to its extension $\hat{x}Fx$, but his axiom (V):

$$\hat{x}Fx = \hat{x}Gx \; . \; \equiv \; (x)(Fx \equiv Gx)^5$$

led directly to Russell's paradox. What stands in its place in later set theory is a variety of possible existence assumptions of varying degrees of strength.

Frege may have seen no alternative to the recourse to extensions in order to secure a reference for numerical terms. It may be for that reason that he saw Russell's paradox as a blow not just to his attempt to prove that arithmetic is a part of logic, but also to his thesis that numbers are objects:

> And even now I do not see how arithmetic can be scientifically established; how numbers can be apprehended as logical objects, and brought under review, unless we are permitted—at least conditionally—to pass from a concept to its extension.[6]

He never mentions the possibility of apprehending numbers as objects of a kind other than logical.

It is possible to identify at least finite numbers with quite unproblematic extensions. A natural way of doing this is to identify each number with some particular class having that number of members, as is done in von Neumann's construction of ordinal numbers.[7]

The second difficulty is that if we admit enough extensions, on some grounds or other, then there are too many possible ways of

---

[5]My symbolism, essentially that of Quine's *Methods of Logic*, follows the *Foundations* rather than the ideas of the *Grundgesetze*. Theroughout I largely neglect the fact that in Frege's later writings concepts are a special kind of function; the axiom stated is more special than the actual axiom V of the *Grundgesetze*, which relates arbitrary functions and what Frege calls their *Werthverläufe*. This expository convenience is not meant to exclude the *Grundgesetze* from the scope of my discussion.

[6]*Grundgesetze* II, 253; translation from Geach and Black, 1st ed., p. 234. [The Appendix in which this passage occurs is translated in full by Furth in Frege, *Basic Laws*.]

[7]See for example Quine, *Set Theory and Its Logic*, chap. 7.

identifying them with numbers. In fact, *any* reasonably well-behaved sequence of classes can be chosen to represent the natural numbers.

Frege must have thought that his own choice was more natural than any alternative. The relation of numerical equivalence is reflexive, symmetric, and transitive; and he thought of a number as an equivalence class of this relation. He motivates it by discussing the notion of the direction of a line as an equivalence class of the relation of parallelism.[8] But this will not do. Ordinary equivalence classes are subclasses of some given class. But the application of numbers must be so wide that, if *all* concepts (or extensions of concepts) numerically equivalent to a concept F are members of N_xFx, then it is by no means certain that N_xFx is not the sort of "unconditioned totality" that leads to the paradoxes. The difference is reflected in the fact that in the most natural systems of set theory, such as those based on Zermelo's axioms, the existence of ordinary equivalence classes is easily proved, while, if anything at all falls under F, the *nonexistence* of Frege's N_xFx follows.

It is odd that we should have to identify numbers with extensions in order to insure that number terms have a reference, but that we should then be able to choose this reference in almost any way we like. We might entertain the fantasy of a tribe of mathematicians who use the ordinary language of number theory and who also all accept the same set theory. In their public life, the question whether the numbers are to be identified with classes never arises. However, each one *for himself* identifies the natural numbers with a certain sequence of classes but does not tell the others which it is. If one says that two terms of number theory refer to the same number, whether another assents or dissents in no way depends on whether *his* natural numbers are the same as the speaker's.

The reader will be reminded of a well-known passage in the *Philosophical Investigations* (I, 293). If it makes no difference in mathematics to which class a term of number theory refers, what is the relevance to the thesis that numbers are objects of the *possibility* of an identification of numbers and classes?

IV

Dummett, in the paper cited in note 8, suggests another way of looking at this problem. He appeals to Frege's principle that "only

---

[8]Cf. Dummett, "Nominalism," esp. pp. 46-47.

in a proposition have words really a meaning,"[9] that "we must never try to define the meaning of a word in isolation, but only as it is used in the context of a proposition."[10] Dummett takes this to mean that in order to determine the sense of a word, it is sufficient to determine the sense of the sentences in which the word is used. This is a matter of determining their truth-conditions. If a word functions syntactically as a proper name, then the sense of the sentences in which it occurs will determine its sense; and which sentences, as a matter of extralinguistic fact, express true propositions will determine its reference.

Although Dummett says that when this position is stated it is a "banality," it seems to me to have a serious ambiguity which makes it doubtful that it can get around the difficulties we have mentioned. Dummett apparently takes the principle to mean that if the specification of truth-conditions of the contexts of a name can be done at all, then the name *has* a reference. But this comes into apparent conflict with Frege's principles. For a sufficient explanation of the sense of a name by no means guarantees that it has a reference, and therefore that the sentences in which it occurs express propositions that are true or false. So it seems that there will be a further question, once the sense has been specified, whether the reference actually exists, which must be answered before one can begin to answer questions about the truth-values of the propositions. Or perhaps it will prove impossible to specify the sense of all the expressions required without in the process specifying the reference or presupposing that it exists. In either case, the usefulness of Frege's principle for repudiating philosophical questions about existence would be less than Dummett thinks.

However, I do not think that these considerations show that Dummett's interpretation of Frege is wrong. Another possible way of taking the principle is that if we can show that sentences containing a certain name have well-determined truth-values, then we are sure this name has a reference, and we do not have to discover a reference for the name antecedently to this. Frege indicates that the contextual principle is the guiding idea of his analysis of number in the *Foundations*; I shall try taking it this way. I shall also suppose that the same principle underlies Frege's attempted proof in the *Grundgesetze der Arithmetik*[11] that every well-formed name in the formal system there set forth has a reference.

[9]*Foundations*, p. 71.
[10]Ibid., p. 116.
[11]I, 45-51. I am indebted to M. D. Resnik for pointing out to me the relevance

There is a general difficulty in the application of the principle which will turn up in both these cases: it is hard to see how it can be applied unless it provides for the elimination of the names by contextual or explicit definitions. We shall see that, for a quite simple reason, the kind of contextual definition Frege's procedure might suggest is impossible. And in the case of the *Grundgesetze* argument, where the crucial case is that of names of extensions,[12] it seems obvious that explicit definition is impossible.

Frege regarded identity contexts of an object-name as those whose sense it is most important to specify. His procedure in the *Foundations* can be regarded as an effort to do this. These identities are of three forms:

(1) the number of *F*'s = the number of *G*'s,

(2) the number of *F*'s = 7,

where '7' could be replaced by any other such expression which we take to refer to a number.

(3) the number of *F*'s = . . .

where '. . .' represents a name of a quite different type, such as 'the moon', 'Socrates', or 'the extension of the concept *prime number*'.

The truth-conditions of identities of form (1) are determined by the above-mentioned principle (A): (1) is to be true if and only if there is a one-to-one correspondence of the *F*'s and the *G*'s. This in turn can be defined without appealing to the concept of number, as we shall see.

Frege analyzed identities of form (2) by defining individual numbers as the numbers belonging to certain particular concepts, so that they are in effect assimilated to identities of form (1). Thus 0 is the number of the concept *not identical with itself* and turns out to be the number of any concept under which nothing falls. $n + 1$ is the number of the concept *member of the series of natural numbers ending with n.* I shall not raise now any problems concerning such identities or Frege's way of handling them.

Thus we can regard the explicit definition mentioned above at the beginning of section III as necessary only to handle identities of form (3), as Frege intimates.[13] By supposing that the notion of the extension of a concept is already understood, Frege provides that the sense of (3) is to be the sense of:

---

of this argument to my discussion. A detailed analysis of the argument occurs in his Ph.D. thesis "Frege's Methodology."

[12]More generally, *Werthverläufe* of functions; see note 5 above.

[13]*Foundations*, p. 116.

the extension of the concept *numerically equivalent to the concept*
$F = \ldots$,

and the sense of the latter is already determined. Then, as I said,
(A) follows from the explicit definition, so that this definition also
serves for other contexts.

V

That Frege did not regard the introduction of extensions as essential to his argument is insisted upon by Peter Geach.[14] He seems
to think that the analysis of identities of form (1) is by itself sufficient
to establish the thesis that numbers are objects:

> Having analysed 'there are just as many *As* as *Bs*' in a way that involved
> no mention of numbers or of the concept number, Frege can now
> offer this analysis as a criterion for numerical identity—for its being
> the case that the number of *As* is the same number as the number of
> *Bs*. Given this sharp criterion for identifying numbers, Frege thought
> that only prejudices stood in the way of our regarding numbers as
> objects. I am strongly inclined to think he is right.[15]

It would seem that we could deal with identities of type (3) very
simply, by specifying that they are all either nonsense or false. The
first alternative is plausible enough in such examples as 'the number
of planets = the moon' or '(the number of roots of the equation
$x^2 + 3x + 2 = 0$) = the class of prime numbers'. It would, however,
be an abandonment of Frege's position that objects constitute a
single domain, so that functions and concepts must be defined for
all objects.[16] The second is plausible enough when '. . .' is a closed
term; it means rejecting the demand that we identify numbers with
objects given in any other way.

However, both these solutions have a fatal defect: apparently we
must explain the sense of (3) when '. . .' is a *free variable*. We cannot
declare that '$N_x Fx = y$' is to be true of *nothing*, for that will contradict
our stipulation concerning (1) and lead to the consequence that our

---

[14]"Class and Concept," pp. 234-235; Anscombe and Geach, *Three Philosophers*, p.
158.
[15]Anscombe and Geach, *Three Philosophers*, p. 161. It is not clear what Geach
intends to be the final relation between numbers and extensions.
[16]See the references given in note 2 above.

universe contains no numbers. It seems that the only ways to take care of this fact are either to give an explicit definition of '$N_xFx=y$' or to assume we understand what the number of F's is and say that '$N_xFx=y$' is to be true of just that object. But this in effect begs the question of reference. This is a quite general difficulty which does not depend on the fact that numbers are to be in the same universe of discourse with other objects. For it is essential to the use of *quantification* over a universe of discourse which is to include numbers that 'N_xFx' should occur in places where it can be replaced by a variable. And not to quantify over numbers is surely to renounce the thesis that numbers are objects.

However, it seems that explanation of '$N_x\ Fx=\ldots$' where '$\ldots$' is a *closed* term (in Frege's language, a proper name) might be regarded as sufficient if it is presupposed that the quantifiers range only over such objects as have names in one's formalism. Indeed, if this is so then it seems that we *have* explained for what objects '$N_x\ Fx=y$' is to be true, so that the above objection has no force. The condition, however, cannot be met if the formalism is to express (in the standard way) enough classical mathematics to include a certain amount of set theory or the theory of real numbers, since in that case the universe will contain indenumerably many objects, while the formalism contains only denumerably many names.

In the *Grundgesetze* (I, 45-51), Frege tries to show on the basis of what seems to be a generalization of the contextual principle that each well-formed name of his formal system has a reference. This is, in particular, the only attempt he makes at a direct justification of his introduction of *Werthverläufe* and his axiom (V) about them. The principle on which he operates is that an object name has reference if every name which results from putting it into the argument place of a referential first-level function name has reference. Similarly, a function name (of whatever level) has reference if every result of putting in a referential name of the appropriate type has reference. Since any sentence (i.e., name purporting to refer to a truth-value) containing a given object name can be viewed as the result of applying a function name to the given object name, the first principle is a generalization of Dummett's principle that a proper name has reference if every sentence in which it occurs has a truth-value.

In applications it is sufficient to show that simple names have reference. The difficult case, and that which is of interest to this discussion, is that of abstracts. The problem of showing that '$\hat{x}(\ldots x\ldots)$' has a reference reduces to that of showing that

'$\hat{x}(. . .x. . .) =$ ———' has a well-determined truth-value, whatever object the name '———' represents.

However, the argument fails for two reasons. First, the principles do not require him to determine the truth-value in the case where '———' is a free variable, so that the same problem arises in interpreting quantification. The other difficulty arises from impredicative constructions. Frege argues that if '. . .x. . .' is referential, then so is '$\hat{x}(. . .x. . .)$'. But to show that '. . .x. . .' is referential we need to show that '. . .———. . .' is referential if '———' is *any* referential object name. But one of these is '$\hat{x}(. . .x. . .)$', if everything turns out right. So that it is not at all clear that the rules exclude circularities and contradiction, as indeed they do not if we let '. . .x. . .' be '$x \in x$' or '—$(x \in x)$'.

Frege could meet both difficulties, at the sacrifice of some of classical mathematics, by restricting himself to predicative set theory. This would, however, also not be in the spirit of his philosophy. It seems incompatible with Frege's realism about abstract objects to admit only such as have names and not to allow quantification over all of them (as would happen if predicativity were realized in the traditional way, by arranging the variables in a ramified hierarchy). Since the same considerations would dictate a predicative interpretation of quantification over concepts and relations, it seems there will no longer be a single relation of numerical equivalence, so that the notion of cardinal number will diverge from the (Cantorian) one which Frege intended. However, the elementary arithmetic of natural numbers would not be affected by this. So the thesis that numbers are objects might be sustained on this basis, if its application were restricted to natural numbers. But the divergence from Frege's intentions which this possibility involves justifies us in not pursuing it further.

From his realistic point of view, Frege cannot complete the specification of the senses of numerical terms and class abstracts. He cannot avoid making some assumptions as to their reference which could in principle be denied. The situation can, however, be viewed as follows: The information given about the *sense* of such terms by the principle (A) and the axiom (V) is not useless, since it enables one to eliminate the reference to numbers and classes from some contexts at least, and to decide the truth-values of some propositions referring to them. We can call (A) or (V) or similar principles *partial contextual definitions*. They give some justification for the assumption of entities of a certain kind. But they are no guarantee even against contradiction, as is shown by the fact that an instance of (V) gives rise to Russell's paradox.

I shall make some further remarks about explicitly defining 'the number of *F's*' in terms of classes. The objection mentioned above loses much of its force if one drops the idea that it is the possibility of definition of this kind which guarantees that numbers exist. This does not mean that such an explicit definition cannot be used, even in a theory which is to serve as a philosophically motivated "rational reconstruction." The point is that the kind of general explanation and justification of the introduction of classes, to which that of numbers can be reduced by a definition, could be done directly for numbers. The identification of numbers with classes could still serve the philosophical purpose of showing that numbers are not more problematic entities than classes. There is still, however, a general difficulty about abstract entities which is illustrated by the fact that infinitely many different definitions, or perhaps none at all, are possible. This is that the concepts of different categories of abstract entities, even in highly developed mathematical theories, do not determine the truth-value of identities of entities of one category with those of another. There is perhaps some presumption in favor of regarding them as false or nonsense, but this weakened by the fact that if it is disregarded for the sake of simplicity, no harm ensues to the logical coherence of the theories, and in practical application the worst that occurs is misunderstanding which can be dissipated easily. There is a similar difficulty even within categories, which is illustrated by Frege's observation that if X is a one-one mapping of the universe of discourse onto itself, then if (V) holds and we set

$$\bar{x}Fx = X(\hat{x}Fx)$$

then

$$\bar{x}Fx = \bar{x}Gx \,.\, \equiv (x)(Fx \equiv Gx)$$

also holds. So that the only condition he assumes about extensions does not suffice to determine which individual objects they are.[17]

Thus we still have a weaker form of our earlier difficulty: numbers and classes are regarded as definite objects while we seem able to choose freely betwen infinitely many incompatible assumptions about their identity and difference relations. One may regard this as just a failure of analogy between abstract and concrete entities[18] or regard it as resolvable by some kind of appeal to intuition.

[17]*Grundgesetze* I, 17. Frege lessens this indeterminacy by identifying certain *Werthverläufe* with the True and the False.

[18]Similar difficulties may arise in the concrete case, e.g., between physical and psychical events.

VI

I shall now discuss the thesis that arithmetic is a part of logic. We are led into this by reminding ourselves that Frege's identification of numbers and extensions was part of the argument for this thesis. He regarded *Werthverläufe* as the most general kind of "logical objects"; the passage from a concept to its extension was the only way of inferring the existence of an object on logical grounds.[19] I shall give reasons for denying that set theory is "logic" once commitment to the existence of classes is introduced, but it is certainly a significant fact that in formal mathematical theories Frege's program of replacing postulation of objects by explicit definition in terms of classes can actually be carried out.

I do not know quite how to assess it. Dummett suggests that taking an equivalence relation of entities of one kind as a criterion of identity for entities of a new kind is the most general way of introducing reference to abstract entities, and if this is so then all abstract entities can be constructed as classes. This may not be enough to make axioms of class-existence logical principles, but if true it is striking and important.

The first part of Frege's argument for the thesis that arithmetic is a part of logic consists in his analysis of the general notion of cardinal number: he argues that the cardinal numbers belonging to concepts must be objects satisfying the principle (A) and then gives the above-mentioned explicit definition of '$N_x Fx$' so that (A) becomes provable. In order to give an analysis of the notion of natural (finite) number, he must pick the natural numbers out from the class of cardinals.

We might ask whether the analysis so far is sufficient for the general notion of cardinal number. So far, I have talked as if what is essential to the notion of cardinal numbers is that they should be objects with the identity condition given by (A). Beyond that, it does not matter what objects they are, for example, whether a cardinal number is even identical with an object given in some other way. The emphasis of the *Foundations* suggests that Frege thinks this is all that is essential, but I doubt that he really does think so. To be clear about this, I shall consider a possible justification of (A) arising from the question, What is it to *know* the number of F's for a given F? It could hardly be simply to know the name of an

---

[19]See in addition to the quotation above (p. 154) *Grundgesetze* II, 148-149, and Frege's letter to Russell of July 28, 1902, *Wissenschaftlicher Briefwechsel*, p. 223.

object with the given identity conditions, for 'the number of F's' is already such a name.

The basis of indisputable fact on which Frege's analysis rests can be brought out by the following considerations. In the finite case, we know the number of F's if we can name, in some standard fashion, a *natural number n* such that the number of F's is n. The primary way of obtaining such a number is by *counting*. To determine by counting that there are n F's involves correlating these objects, one by one, with the numbers from 1 to n. Thus we can take as a necessary condition for there to be n F's that it should be possible to correlate the F's one-to-one with $1 . . . n$, i.e., that F and the concept *number from 1 to n* are numerically equivalent. The role of one-to-one correspondence in explaining the notion of number can be developed from this condition and the following mathematical considerations:

(1) If the F's can be correlated one-by-one with the numbers from 1 to n, then the objects falling under the concept G can be correlated with the numbers from 1 to n if and only if they can be correlated with the F's.

From (1) and our condition for there to be n F's, it follows:

(2) If there are n F's, then there are the same number of F's as G's if and only if there is a one-to-one correspondence of the F's and the G's; that is, the principle (A) holds in the finite case.

(3) That a relation H establishe a one-to-one correspondence of the F's and the G's can be expressed by a formula of the first-order predicate calculus with identity. The condition is

$$(x)[Fx \supset (\exists!y)(Gy . Hxy)] . (y)[Gy \supset (\exists!x)(Fx.Hxy)]$$

where '$(\exists!z) Jz$', which san be read as 'there is one and only one z such that Jz, is an abbreviation for

$$(\exists z)[Jz . (x)(Jx \supset x = z)].$$

Thus, since F and G are numerically equivalent if and only if there is a relation H establishing a one-to-one correspondence of the F's and the G's, '$Glz_x(Fx,Gx)$' can be explicitly defined in the secord-order predicate calculus with identity, in particular without appealing to the concept of number.

(4) For each natural n, a necessary and sufficient condition for there to be n F's can also be expressed by a formula of the first-

order predicate calculus with identity. If, following Quine,[20] we write 'there are exactly n objects x such that Fx' as '$(\exists x)_n Fx$', then we have

(a) $(\exists x)_o Fx \equiv - (\exists x)Fx$

(b) $(\exists x)_{n+1} Fx \equiv (\exists x)[Fx \cdot (\exists y)_n (Fy \cdot y \neq x)]$,

so that for a particular n, the numerical quantifier can be eliminated step by step. Thus '$(\exists x)_2 Fx$' is equivalent to

$(\exists x)\{Fx \cdot (\exists y)[Fy \cdot y \neq x \cdot - (\exists z)(Fz \cdot z \neq x \cdot z \neq y)]\}$, i.e.

$(\exists x)(\exists y)[Fx \cdot Fy \cdot x \neq y \cdot (z)(Fz \supset \cdot z = x \vee z = y)]$.

Frege followed Cantor in taking (A) as the basic condition cardinal numbers had to satisfy in the infinite case as well as the finite. He went beyond Cantor in using it as a basis for the ordinary arithmetic of natural numbers, and in observing that numerical equivalence could be expressed in terms merely of second-order logic. Without thest two further steps, the procedure has, of course, no claim at all to be a a reduction of arithmetic to logic.

As I have said, the remainder of Frege's argument consists in picking out the natural numbers from the cardinals. Although this was not Frege's actual procedure, we can put it in the form of defining Peano's three primitives, 'o', 'natural number', and 'successor', and proving Peano's axioms. 'o' and the successor relation are defined in terms of '$N_x Fx$':

$$o = N_x(x \neq x)$$
$$S(x, y) \equiv (\exists F)[N_w Fw = y \cdot (\exists z)(Fz \cdot N_w(Fw \cdot w \neq z) = x)]$$

Then the natural numbers can be defined as those objects to which o bears the ancestral of the successor-relation, i.e.

$$NN(x) \equiv (F)\{Fo \cdot (x)(y)[Fx \cdot S(x,y) \cdot \supset Fy] \cdot \supset Fx\}$$

From these, Peano's axioms can be proved; it is not necessary to use any axioms of set existence except in introducing terms of the form '$N_x Fx$' and in proving (A), so that the argument could be carried out by taking (A) as an axiom. Lest this statement mislead, I should point out that I am provisionally counting the second-order predicate calculus as logic rather than as set theory; for from Frege's point of view the range of the higher-type variables will be *concepts*, while the extensions with which he identifies the numbers must be *objects* in the same domain with the objects numbered.

---

[20]*Methods of Logic*, §39 (2d ed.), §44 (3d ed.).

The definition of Peano's primitives and the proof of Peano's axioms can be carried out in one way or another not only in Frege's own formal system but also in Russell's theory of types and in the other systems of set theory constructed in order to remedy the situation produced by the paradoxes, which of course showed Frege's system inconsistent. We have seen that it is possible to give these definitions in infinitely many ways. It is sometimes said that what the logicists achieved in trying to prove that arithmetic is a part of logic is the proof that arithmetic can be modeled in set theory. But it should be pointed out that the modeling of the Peano arithmetic in set theory does not need to make use of the facts (1) − (4) above cited or the general analysis of cardinal number in terms of numerical equivalence, if Frege's choice of the sets to represent the numbers is abandoned.

VII

I intend to discuss two main lines of criticism of the thesis that arithmetic is a part of logic. The first points to the fact that the formal definitions and proofs by which the thesis is justified make use of the notion of extension, class, or set, and assume the existence of such entities. It is denied that set theory is logic. It is also denied that a reduction merely to set theory will suffice for a philosophical foundation of arithmetic or for a refutation of the epistemological theses about arithmetic (e.g., Kant's and Mill's views) against which the reduction is directed.

In discussing the thesis that numbers are objects, we found a difficulty for Frege in justifying assumptions of the existence of classes. Indeed, set theory will be logic only if propositions which assert the existence of classes are logical laws. Paul Benacerraf points out[21] that this would not be in accord with the usual definition of logical validity, according to which a formula is logically valid if and only if it is true under all interpretations in any nonempty universe, i.e., regardless of what objects, and how many, there are in the universe.[22] This definition applies to higher-order logic such as the one we have used in formulating Frege's views. But if numbers and classes are to be *objects*, a law which provides for the

[21]*Logicism: Some Considerations*, p. 196n.
[22]The qualification "nonempty" may be regarded merely as excluding a special case for the sake of simplicity.

existence of any at all will require the universe over which the quantifiers range to contain *specific* objects, and if it provides for the existence of enough for even elementary number theory, it will require the universe to be infinite.

Thus there is a clear sense in which the predicate calculus is a more general theory than any set theory, and therefore more entitled to be called "logic." This observation is confirmed by the fact that there are infinitely many possible assumptions of the existence of classes which can be ordered by logical strength, if the notion of strength is construed as what one may prove from them by the predicate calculus. Moreover, the stronger assumptions are in many ways more complex, obscure, and doubtful. Each well-defined system of such assumptions is incomplete in a strong sense and extendable in a natural way, in contrast to the completeness of the predicate calculus. But it is hard to see that the principles involved in such extensions are self-evident or logically compelling. This is particularly true when they allow impredicatively defined classes. Thus the existential commitments of set theory are connected with a number of important formal differences between it and the predicate calculus, and it seems that the predicate calculus is much closer to what was traditionally conceived as formal logic.

As a concession to Frege, I have accepted the claim of at least some higher-order predicate calculi to be purely logical systems. Our criticisms of Frege hold under this condition. The justification for not assimilating higher-order logic to set theory would have to be an ontological theory like Frege's theory of concepts as fundamentally different from objects, because "unsaturated." But even then there are distinctions among higher-order logics which are comparable to the differences in the strength of set theories. Higher-order logics have existential commitments. Consider the full second-order predicate calculus, in which we can define concepts by quantification over *all* concepts. If a formula is interpreted so that the first-order variables range over a class D of objects, then in interpreting the second-order variables we must assume a well-defined domain of concepts applying to objects in D which, if it is not literally the domain of *all* concepts over D, is comprehensive enough to be closed under quantification. Both formally and epistemologically, this presupposition is comparable to the assumption which gives rise to both the power and the difficulty of set theory, that the class of all subclasses of a given class exists. Thus it seems that even if Frege's theory of concepts is accepted, higher-order logic is more comparable to set theory than to first-order logic.

It is also sometimes claimed that the concept of class is intrinsically more problematic than that of numbers, so that the reduction of arithmetic to set theory is not a suitable philosophical foundation for arithmetic. There are several reasons which can be cited to support this contention—the multiplicity of possible existence assumptions, questions about impredicative definitions, the paradoxes, the possible indeterminacy of certain statements in set theory such as the continuum hypothesis. However, it should be pointed out that a quite weak set theory suffices for elementary number theory. Most of the difficulties arise only in the presence of impredicatively defined classes, and for the development of elementary number theory we do not have to suppose that any impredicative classes exist. If the modeling in set theory of a part of mathematics requires the existence of such classes, this can only be because the mathematics itself involves impredicativity, so that this is not a difficulty about the reduction to set theory.

However, it still seems that in order to understand even the weak theory, one must either have a general concept of set or assume it to be restricted in some way which involves the notion of number. The theory may be such that only finite sets can be proved to exist in it, but the reduction is not very helpful if the quantifiers of the theory are interpreted as ranging over finite sets. It seems that for a proposition involving quantification over *all* sets to have a definite truth-value, it must be objectively determined what sets exist. I think one might get around this, but it is certain that the assumption that any such proposion *has* a definite truth-value, which seems to be involved already in applying classical logic to such propositions, is stronger and more doubtful than any principle which needs to be assumed in elementary number theory.

VIII

The second criticism which I want to consider denies that arithmetic is reducible to set theory in the most important sense. This objection is as old as those concerning the foundations of set theory, but on the surface at least independent of them. It is to be found in Brouwer and Hilbert, but was probably argued in greatest detail by Poincaré.[23] Recently it has been taken up by Papert.[24] It is closely related to the criticisms of Wittgenstein[25] and Wang.[26]

This objection holds that the reduction is circular because it makes use of the notion of natural number. This obviously does not mean that the notion of natural number, or one defined in terms of it, appears as a primitive term of any set theory by which the reduction could be carried out, for this would be obviously false. Rather, the claim is that we must use the notion of natural number either to set up the set theory, to see the truth of the set-theoretical propositions to which number-theoretical propositions are reduced by explicit definitions, or to see the equivalence of the set-theoretical propositions and their number-theoretical correlates. This can in fact be seen by a quite simple argument. Inductive definitions, especially, play an essential role both in setting up a system of set theory and in establishing the correspondence between it and the system of number theory. For example, typically the definition of *theorem* for each system will be an inductive definition of the following form: Certain axioms and rules of inference are specified. Then to be a theorem is either to be an axiom or to be obtainable from theorems by a single application of one of the rules of inference. Then the model of number theory in set theory is established by defining for each formula A of number theory a set-theoretical translation $T(A)$. To prove that if A is a theorem of number theory then $T(A)$ is a theorem of set theory, we first prove this for the case where A is an axiom. Then suppose A follows from, say, B and C by a rule of inference in number theory, where B and C are theorems. We then show that $T(A)$ is deducible from $T(B)$ and $T(C)$ in the set theory. By hypothesis of induction, $T(B)$ and $T(C)$ are theorems of set theory. Therefore so is $T(A)$. By an induction corresponding to the definition of theorem of number theory, $T(A)$ is a theorem of set theory whenever A is a theorem of number theory.

Although the observation on which this objection is based is true, this is not sufficient to refute the reductionist. For the latter may maintain that the notion he defines is capable of *replacing* that of natural number (and equivalent notions involving induction) in all contexts, in particular those uses which are involved in describing the logical systems and in establishing the correspondence. So that we can imagine that in both the number-theoretical and the set-

[23]*Science et méthode*, chap. 4.
[24]"Sur le réductionnisme logique."
[25]*Remarks on the Foundations of Mathematics*, part II of 1st ed.
[26]"Process and Existence in Mathematics."

theoretical formal systems 'A is a theorem' is defined, say, as 'A belongs to every class of formulas which contains all axioms and is closed under the rules of inference', as is done in some writings of Tarski. Whenever natural numbers are used ordinally for indexing purposes, they are to be replaced by their set-theoretical *definientia*.

We might consider in this connection a form of Poincaré's objection due to Papert (see note 24). Papert says in effect that the Frege-Russell procedure defines *two* classes of natural numbers, such that mathematical induction is needed to show them identical. For we give explicit definitions in the set theory of 'o', 'So', 'SSo'. . . and *also* define the predicate '$NN(x)$'. How are we to be sure that '$NN(x)$' is true of what o, So, SSo, . . . are defined to be and only these? Well, we can prove '$NN(o)$' and '$NN(x) \supset NN(Sx)$'.

If $NN(S^{(n)}o)$'[27] is the last line of a proof, then by substitution and *modus ponens*, we have a proof of '$NN(S^{(n + 1)}o)$'. By induction, we have a proof of '$NN(S^{(n)}o)$' for every n.

We can likewise prove by induction that every x for which '$NN(x)$' is true is denoted by a numeral. For o clearly is. If $n = S^{(m)}o$, then $Sn = S^{(m + 1)}o$. So by induction (the derived rule of the set theory), if $NN(x)$, then $x = S^{(m)}o$ for some $m > o$.

What we have proved is a metalinguistic proposition. The reply to Papert would be to say that we can define the class of symbols $S^{(m)}o$ by the same device; i.e., if $Num(x)$ is to be true of just these symbols, we define:

$$Num(x) \equiv (F) \{F(\text{'o'}) \cdot (x)[Fx \supset F(\text{'S'} \cap x)] \cdot \supset Fx\}.[28]$$

Then the first of the above two inductive proofs will be an application of this definition, just as the second is an application of the definition of $NN(x)$.

Papert can raise the same question again. We can prove formally that if $NN(x)$ holds, then x is denoted by an object y such that $Num(y)$. Moreover, we can prove formally that if $Num(y)$ then the result of substituting y for the variable 'x' in '$NN(x)$' is provable. But how do we know that the extension of '$Num(x)$' consists just of 'o', 'So', 'SSo', etc.?

This is clearly the beginning of a potentially infinite regress. What is happening is this. If we replace an inductive definition by an explicit one, it takes an inductive *proof* to show that they define

---

[27]'$S^{(n)}o$' is an abbreviation for 'o' preceded by n occurrences of 'S'.

[28]If x and y are expressions, $x \cap y$ is their concatenation, that is, the result of writing x followed by y.

coextensive concepts. *This* difficulty can be avoided by denying that we have an independent understanding of inductive definitions. That is, if we ask Papert what he means by ' "o", "*So*", "*SSo*", etc.', he would have to reply by an explicit definition which would turn out to be equivalent to '*Num(x)*'. But this last reply to Papert depends on the claim that the apparatus in terms of which such explicit definitions as that of '*NN(x)*' are given can be understood independently of even the most elementary inductive definitions. This is implausible since the explicit definitions involve quantification over all concepts. It is hard to see what a concept is, or what the totality of concepts might be, without something like the inductive generation of linguistic expressions which (on Frege's view) refer to them.

There is another difficulty which Papert's exposition shows he has in mind.[29] It is independent of the question whether inductive definitions are replaced by explicit ones. Consider some numeral, say '*SSSo*', and a proposition of the form '*F(SSSSo)*'. If we have proved '*Fo*' and '$(n) [Fn \supset F(Sn)]$', we have two independent ways of proving a proposition '*F(SSSSo)*'. We might infer '$(n)Fn$' by induction and '*F(SSSSo)*' by universal instantiation. Or we might prove it without induction, by successive applications of *modus ponens*:

$$\frac{\dfrac{\dfrac{\dfrac{Fo \quad Fo \supset F(So)}{F(So) \quad F(So) \supset F(SSo)}}{F(SSo) \quad F(SSo) \supset F(SSSo)}}{F(SSSo) \quad F(SSSo) \supset F(SSSSo)}}{F(SSSSo)}$$

We say that the proof by way of induction gives us the assurance that we *can* construct such a proof by successive applications of *modus ponens*. And the complexity of such proofs is unbounded. (But how do we see that we *always* can? By induction!) The application of the procedure involves the iteration of certain steps n times for the proof of $F(S^{(n)}o)$. Neither the reduction of induction to an explicit definition nor the Wittgensteinian doctrine[30] that '$Fo . (n)[Fn \supset F(Sn)]$' constitutes the *criterion* for the truth of 'for all natural numbers n, Fn' gives us an assurance that there will be no *conflict* between the two methods for proving individual cases. When Poincaré said that a step of induction contained an infinity of syl-

[29]Especially in the subsequent essay, "Problèmes épistémologiques et génétiques de la récurrence."

[30]Waismann, *Introduction to Mathematical Thinking*, chap. 9.

logisms, he was saying that it guaranteed the possibility of all the proofs of the instances by reiterated *modus ponens*. He was right at least to this extent: if we do have an *a priori* assurance that there will be no conflict between construction in individual cases and inference from general propositions proved by induction, then this assurance is not founded on logic or set theory.

IX

Let us now return to the question which we left in the air in our discussion of the thesis that numbers are objects: What *does* guarantee that the singular terms of arithmetic have reference?

If we consider two different systems of numerals,

o, *So, SSo,*. . .

o, *To, TTo,*. . .

then the order-preserving correspondence between them fixes when two numerals shall *denote the same number.* Indeed, I do not think there is a more fundamental criterion of this. For this reason I should say that if the critics of logicism intend to deny Frege's assertion that the notion of one-to-one correspondence plays a constitutive role in our notion of number, they are wrong. It also follows that the ordinal and cardinal notions of number are interdependent. One might hold that cardinals are "applied ordinals"; i.e., if we are given the numbers with their order, then the relation to one-to-one correspondence and therefore to numerical equivalence is something that arises only in the application of numbers and does not belong to their nature as pure mathematical objects. However, I do not see how we could be said to be *given* the numbers except through a sequence of numerals or some other representatives of the same type of order. And if it is not part of the concept of number that the sequence in question should be a paradigm which could be replaced by some other equivalent sequence, the unnatural conclusion follows that features of the representative sequence which one would expect to be quite accidental, such as the particular design of the numerals, *are* relevant. But Frege's definitions *do* bring some avoidable complexity into the notion of natural number, because it defines them in terms of *explicit assertions* of the existence of one-to-one correspondences, and this need *not* be part of the sense of statements about natural numbers.

Suppose we regard it as essential to a system of numerals that it be formed by starting with some initial numeral and iterating some basic operation (representing the successor function). Let us try the following model of how the numerals come to have reference: At first they function like indexical expressions, in that in counting each refers to the object with which it is correlated, as is actually true of the expressions 'the first', 'the second', etc. Then by a kind of abstraction, the reference of the numeral is taken to be the *same* on all occasions. What guarantees the *existence* of the number n is the existence of an ordered set in which some object is the nth. For any numeral, the numerals up to that one will be a set. Then no ulterior fact beyond the generation of the numerals is needed to guarantee that they have reference.[31]

In the case of small numbers, the guarantee of their existence is somewhat analogous to that which sense-experience gives of the existence of physical objects. But because any instance of a certain type of order will do, and because such instances can be found in sense-experience, and we cannot imagine what it would be like for us to be *unable* to find such instances, we are inclined to regard the evidence of the existence of small numbers as a priori.

We speak of its being *possible* to continue the construction of numerals to infinity. On the basis of this possibility we can say that numbers exist for which we have no numerals. However, this is an extrapolation from the concrete possibilities, and the only reason for favoring *it* as the guarantee for the existence of large numbers over the hypothesis of an actual infinity of sets, from which representatives of the numbers can be chosen, is that it is weaker. It is comparable to Kant's assumption that the "possibility of experience" extends to regions of space too small or too distant for us to perceive.

With respect to the thesis that arithmetic is a part of logic, our conclusion is that although the criticisms which have been made of it over the years *do* suffice to show that it is false, it ought not to be rejected in the unqualified way in which it appears to have been rejected by Poincaré, Brouwer, and Hilbert. It seems to me that Frege *does* show that the logical notion of one-to-one correspondence is an essential constituent of the notion of number, ordinal as well as cardinal. The content of arithmetic that is clearly additional to that of logic, which Frege failed to acknowledge, is in its existence assumptions, which involve an appeal to intuition and extrapolation. (The same kind of intuitive construction as is involved in arithmetic is also involved in *perceiving the truth* of logical truths.)

---

[31]Cf. Benacerraf, *Logicism*, pp. 162-174, and "What Numbers Could Not Be."

The same appeal is involved in set theory, although if we go beyond general set theory and assume the existence of infinite sets, the extrapolation is of course greater. Concerning the relation of the notion of natural number to the notion of finite set, it seems clear to me that they go together, and that neither can be understood without the other. I should like to argue, although I do not have the space to do so now, that there is a reciprocal relation between such arithmetical identities as '2 + 2 = 4' and the logical truths closely associated with them under the Frege-Russell analysis, so that it will not do to regard one as a mere abbreviation of the other.

Postscript

The treatment in section VIII of this essay of "Poincaré-type" objections to logicism was subjected to extended critical scrutiny by Mark Steiner.[32] Although this criticism brings out unclarities in my exposition, Steiner is not very explicit about when I am presenting Papert's views and when my own, so that his reader could easily get the impression (probably unintended) that I endorse Papert's claims unequivocally.[33] Some of his criticisms are elaborations of moves in the dialectic I engaged in beginning with Papert's points.

However, in the remarks about quantifications over all concepts (p. 170 above) which Steiner quotes on pp. 33-34, I am speaking for myself. There is no doubt that what I say is too cryptic. From Essays 8 and 9, it should be clear how complex I subsequently found such issues. But it is Steiner's "more modest interpretation" (p. 35) that I had in mind. However, I was thinking rather specifically of Frege's notion of a concept, which is tied to language in that his basic explanation of what a concept is is that it is the reference of a predicate. Steiner is right that it is another question whether definitions in the spirit of Frege serve to reduce induction to *set theory*. And that "the notion of a set can precede the realization that language unfolds inductively" (p. 35) is certainly not refuted by any considerations I give, and indeed it is suggested by the idea of a set as constituted by its elements, to which I am generally sympathetic. But inductive generation enters into our understanding of set theory in other respects, for example in the motivation of

---

[32]*Mathematical Knowledge,* pp. 28-41.

[33]Thus on p. 14 he represents as mine a view I explicitly present as Papert's; in the citation for the same view (p. 28), he leaves off the opening part of the first sentence, which is "Papert says in effect that."

the axioms by the "iterative" conception of sets as arising in a well-ordered sequence of stages; see essay 10 below. Moreover, the linguistic considerations that arise in connection with Frege's notion of a concept also arise with respect to the second-order principles of set theory, the axiom of separation, and the axiom of replacement.

Much of Steiner's discussion proceeds on the assumption that the logicist *knows* the truth of his axioms. Papert's and my arguments could be taken as questioning how this assumption could be true given what the logicist admits as logic. Moreover, the assumption does not make the issue about consistency that Steiner discusses on p. 40 (with reference to the last paragraph of section VIII) quite so unproblematic as he makes it appear. For to conclude from it that the logicist knows that his theory is consistent is to assume his knowledge closed under some sort of semantic reflection. Suppose for simplicity that our logicist L has a theory with finitely many axioms $A_1 . . . A_n$. Then for each $i \leq n$, L knows that A_i. But by Gödel's second incompleteness theorem, even the assumption that L knows any logical consequence of what he knows is not sufficient to give him the knowledge that the theory with these axioms is consistent; he would need something like the inference to 'A_i is true' and then some reasoning with the concept of truth.

However, this is not quite the nub of the difficulty, since he can do all this with the help of more set theory, though the assumption that he knows the stronger axioms might get more and more tenuous. I think what Papert had in mind was something like a skeptical doubt: if one proves a generalization with infinitely many instances which can be decided independently of the generalization, how does one know that the independent procedure will not turn up a counterinstance? Any *proof* that it will not will give rise to the same questions. Papert's view seems to be that there is something irreducibly empirical in any assurance one has about such matters. The doubt is of the sort that arises in Wittgenstein's discussion of following a rule.[34] On this subject, I do not have much to add now to the cautious comment in the paper, that doubts of this kind are not resolved by appeals to logic and set theory.

Nonetheless, my view of the force of "Poincaré-type" objections

[34]Indeed, Steiner has pointed out to me that just this doubt, in the case of induction, is discussed by Wittgenstein in *Lectures on the Foundations of Mathematics*, lecture 31. Wittgenstein says that we take the result of the general proof (by induction) as a criterion for the correctness of a result obtained by figuring out a particular case, e.g., deducing in 3,000 steps that a proposition holds for 3,000. What is irreducibly empirical is the general agreement of results that such legislation rests on (see esp. ibid., pp. 291-292).

has changed since writing this paper. The analogy is very close between the kind of circle that arises in the logicist's treatment of induction and the circle that arises in justifying logical laws by means of semantical definitions of the logical operators, such as the characterizations of sentential connectives by truth-tables: to show the most elementary inferences valid, one has to use the same inferences, or others posing the same questions, in the metalanguage. This is just what happens with the possible justifications of mathematical induction that arise (not only the Fregean one). So although the considerations adduced by Poincaré do set a limit on the reducibility of induction to other principles, they do not show that it is not importantly like a logical principle. In a natural deduction formalization of arithmetic, induction can be treated as an elimination rule for the predicate "is a natural number," in analogy with the well-known elimination rules for logical connectives.[35] The matter deserves further exploration.

[35]See, for example, Martin-Löf, "Hauptsatz for Iterated Inductive Definitions," p. 190 (in the context of a general treatment of inductive definitions).

Formulations of constructive mathematics by what are called theories of constructions treat induction and recursion together, but the analogy between the natural number predicate and logical connectives can still be made. See for example Tait, "Finitism," pp. 531-532, 537, and Martin-Löf, "An Intuitionistic Theory of Types," pp. 95-96.

7

Quine on the Philosophy of Mathematics

The "philosophy of mathematics" when it has not concentrated on specific methodological issues in mathematics, has chiefly sought to explain a single impressive gross feature of mathematics: its combination of clarity and certainty with enormous generality. Mathematics shares this feature with logic, but for some reason in the past it was in the case of logic found less impressive. Presumably in earlier times this was due to the much more restricted scope of formal logic and its being combined into a body of knowledge with much reflection on language, inference, and knowledge that is neither so clear nor so certain. More recently, logic has simply been treated together with mathematics; of course they have been considered to be one subject.

Quine's discussion of these matters is in the tradition of considering logic and mathematics together, although in his later writings the distinction between elementary logic and mathematics proper assumes some importance.[1] With respect to clarity and certainty, another distinction has arisen in the last century with the development of set theory: the higher or more abstract parts of set theory have not been found to be so clear and certain as the more ele-

This essay was written for *The Philosophy of W. V. Quine*, edited by Paul Arthur Schilpp and Lewis E. Hahn, Library of Living Philosophers (La Salle, Ill.: Open Court, forthcoming), and will also appear in that volume.

Sections I–V of the essay, in almost their present form, were presented to the Chapel Hill Philosophy Colloquium in October 1975, with Professor Quine as commentator. Quine's reply prompted some local revisions. He also convinced me that his attitude toward "physical or natural" necessity was more negative than I had thought, but there I have let my interpretation stand because of the issues to which it gives rise.

Other revisions have been made in response to comments of Saul Kripke, the late Gareth Evans, and a referee for Cornell University Press.

[1]*Philosophy of Logic*, pp. 64-70; also *The Roots of Reference*.

mentary parts of mathematics, particularly the arithmetic of natural numbers. On the other hand, with the advent of set theory mathematics came to be seen as even more general than before. In earlier times, mathematics was often thought to be applicable only to spatiotemporal entities.

Quine's approach to these features is influenced, though largely negatively, by two characterizations of mathematics of a more technical philosophical nature: the theses that mathematical truths are *necessary* and that mathematical knowledge is *a priori*. He has of course carried out an elaborate criticism of the notion of a priori knowledge, particularly where it was to be explained by the analyticity of the propositions known a priori. I do not want to concentrate on this criticism in this essay, since it has been so widely discussed. Most of the essay will be devoted to Quine's view on the *necessity* of mathematics and logic. In this connection I shall discuss the notion of "mathematical possibility," which Quine has hardly discussed directly. I will then discuss Quine's interpretation of *existence* in mathematics. I shall discuss the bearing of set theory on Quine's views on these matters. In a final section, I shall take up some other aspects of Quine's conception of set theory.

I

In common sense and science, many attributions of necessity are made, and it is clear that a variety of types or concepts of necessity are at stake. Philosophers have held, however, that there is a single "highest" type of necessity, which is often called logical necessity, but "logic" here would have to include more than formal logic. What is necessary in this sense would be necessary in any other sense. For this reason, perhaps the best term for it would be *absolute* necessity. According to tradition, whatever can be known a priori would be absolutely necessary. On the other hand, the tradition would hold that even fundamental laws of empirical science are *not* absolutely necessary.

The truths of logic and pure mathematics have been held to be absolutely necessary by those who have used this notion. There is also a weaker claim that logic and mathematics are necessary in a sense that is more stringent than that in which scientific laws are necessary (at least for the most part). It is this claim that I shall discuss.

Historically at least, the alleged apriority of mathematics has been linked with its alleged strict necessity, although it is not easy to state the connection precisely. We can express the traditional (e.g., Kantian) view of the connection of necessity and apriority by the following two principles:

(1)   If *a* knows that $\Box p$, then *a* knows a priori that *p*.
(2)   If *a* knows a priori that *p*, then $\Box p$.

The truth of these principles is not at all evident if one supposes that necessity of the relevant type is not an epistemic notion.

(1) certainly presupposes a standard of knowledge that knowledge resting on testimony does not meet. But given the strong claim of mathematics to certainty, that seems appropriate enough to the case that concerns us. Quine can be interpreted as denying the apriority of mathematics. If (1) holds, it would follow that mathematics does not possess necessity of the relevant type.

A Quinian argument against the necessity of mathematics would assume (1) and then apply his critique of the a priori: if mathematics is not known a priori, then (1) implies that mathematical statements are not known as necessary, and it is then very implausible that they *are* necessary. In fact, this seems to be Quine's principal argument. He seems to adhere to (1), though perhaps only as an explication of notions that he does not accept. He discusses the necessity of mathematics most explicitly in the popular lecture "Necessary Truth," in which he introduces considerations bearing on analyticity and apriority. Moreover, when he discusses the problems of modal logic, he sometimes assumes that absolute necessity means something like analyticity. His unfriendliness to *de re* modalities derives partly from that source.[2]

Both (1) and (2) have been questioned, notably by Saul Kripke,[3]

---

[2] *From a Logical Point of View*, pp. 134, 149-153; "Three Grades of Modal Involvement"; cf. *Word and Object*, p. 199.

[3] In *Naming and Necessity*. Kripke is probably the most influential recent defender of absolute necessity. He prefers the term "metaphysical necessity." A number of Kripkean examples, such as identities between names, would offer direct counterexamples to (1). His principal examples are not mathematical but he does give (p. 159) the example of a mathematical statement that someone has come to know by carrying out computations with a computer. Of such examples, one might reply that although an individual may know them and not know them a priori, they *can* be known a priori; one may even hold that one does not really know them unless a priori, i.e., by a proof or computation that one has carried out oneself. Where computations of sufficient complexity are involved, the sense in which the statements in question can be known a priori becomes more and more attentuated; it involves applying to the possibility of *knowledge* modes of inference going with the conception

and in the absence of a positive defense of (1) this Quinian argument against the necessity of mathematics has a major gap. The discussion below indicates the diffficulty I see in filling it. If the necessity of mathematics can be defended, then Quine's critique of the a priori is a powerful argument against (1); it would leave (2) vacuous, though not false, for mathematical statements. I shall not consider the interesting question whether the direct consideration of the necessity of mathematics yields an argument for its apriority, thus rehabilitating (1) in the mathematical case. Quine apparently regards as usable a notion of necessity according to which the laws of science are necessary, although that would be an extension of the specific explication of necessity he sketches in "Necessary Truth."[4] There it is *instances* of laws, not the laws themselves, that are necessary. At any rate, he describes the necessity involved as "physical or natural" necessity and says there is no "higher or more austere" necessity.[5] The latter would be a necessity which logic and mathematics would possess but empirical natural science would not.

II

It might seem that Quine should find a specifically logical necessity quite clear. A statement should be logically necessary if and only if it is logically true in the sense of first-order quantificational logic with identity. Then for a statement in canonical notation it would be sharply defined whether it was logically necessary or not, and moreover it would be easy to show that at least some of the basic laws of physics are not logically necessary; indeed even so simple a statement as $(\exists x)\ (\exists y)\ (x \neq y)$ would not be logically necessary.

Such a notion would of course not agree with the intuitive conception of absolute necessity, since on that conception statements

---

of mathematical possibility discussed in the text. (On some problems with this, see Essay 4 of this volume.) Cases in which, in practice, our knowledge of some mathematical theorem depends on computers in such a way that it is not a priori by any reckoning have recently assumed more importance with the proof by Appel and Haken of the four-color theorem. See Tymoczko, "The Four-Color Problem."

Kripke's examples of "contingent a priori" statements might be offered against (2). They seem unlikely to bear on mathematics and are perhaps more problematic in themselves.

[4]"Necessary Truth" pp. 70-71.
[5]Ibid., p. 76.

can be absolutely necessary even though they contain nonlogical predicates essentially.[6] But it does seem perfectly clear. Moreover, if one reads the '$\Box p$' of modal logic as 'it is logically true that p', the theorems without nested modalities of the usual modal propositional logic are true.[7]

It seems that from Quine's point of view this notion is not unclear or incoherent, but perhaps he would not see any need for it: the purposes that talk of logical necessity in this sense would serve are already adequately served by talking of logical truth.

The concept turns out in any case to be a somewhat anomalous notion of necessity and not so clear as it seems at first sight. We have: if 'p' is a first-order sentence,

(3)   'It is logically necessary that p' is true if and only if 'p' is logically true.

But our necessity statement is not itself in the Quinian canonical form of a first-order sentence. Here there are two main alternatives.

The first is to hold to Quine's notion of canonical form and take the right-hand side of (3) as the canonical paraphrase of 'it is logically necessary that p'. Since for a given sentence 'p' ' "p" is logically true' is not a logical truth (however the details of the characterization of logical truth go), it follows that 'it is logically necessary that it is logically necessary that p' is always *false*. The rule of necessitation of ordinary modal logic will fail.[8] Moreover 'it is logically possible that p' will also not be a truth of logic. Logical possibility proves to be a concept of a different character from logical necessity.

---

[6]Since it follows that some truths are absolutely necessary but not logically necesary, there is an apparent conflict with the claim of absolute necessity to be the "highest" type of necessity. The conflict could be resolved by assuming a distinction between "real" and formal, logical, or verbal modalities; absolute necessity is the most general *real* necessity. Thus if it is not logically necessary that p, it may not be *really* possible that $\sim p$ (for example if 'p' is 'Cicero = Tully'), but if it is not absolutely (metaphysically) necessary that p, then it is really possible that $\sim p$. (The example is from Jubien, "Ontological Commitment to Particulars," p. 524.) I am not here claiming either that these notions make genuine sense or that they do not.

[7]It is usually maintained that the logic of logical necessity is S_5, but the modal logics as strong as T and no stronger than S_5 agree with respect to the validity of propositional schemata without nesting of modalities. In the Kripkean semantics, it is easy to see that any such schema either is T-valid or has an S_5-countermodel.

Consider the more informal notion of necessity of Quine's "Necessary Truth." According to this notion, a statement should be logically necessary if it is an instance of a (specifically envisaged) law of logic. But what is a law of logic? A logical truth (other than a pure identity statement) will contain nonlogical predicates and is rather an instance of a law than the law iteself. Quine in fact says that the laws of logic cannot be stated without semantic ascent. Then the law of which

(4)   It is raining or it is not raining

is an instance would be something like

(5)   Any sentence of the form '*p* or not-*p*' is true

which is not a logical truth. Thus on this conception, the laws of logic are not logically necessary.

The second alternative is to depart from Quine's notion of first-order logic with identity as the canonical framework. Then we might treat '$\Box$' interpreted as logical necessity as itself a logical operator. So far it has been explicated only when applied to non-modal sentences, and the extension of the interpretation is not unique. Evidently for '*p*' a nonmodal sentence, '$\Diamond p$' will be true if and only if '$\Box p$' is *not* logically true in the original sense. If our extended interpretation is to make all theorems of S5 true, then '$\Diamond p$' will be logically necessary, but then, since such '*p*' are not recursively enumerable, there will be no proof procedure for logical truth. For Quine that would be a strong reason to reject such an alternative. But an interpretation with these properties can be given. Another definition in the spirit of Quine's definition of the logical truths as the truths containing only logical expressions essentially will yield the result that all theorems of S4, but not all theorems of S5, are logically necessary in the extended sense.[9] For if '*p*' is a consistent first-order sentence with an inconsistent substitution-case, '$\Diamond p$' will be true and '$\Box \Diamond p$' false on this interpretation. But this means statements of logical *possibility* are not truths of logic.

[8]This notion of necessity is one of those considered by Quine in "Three Grades of Modal Involvement"; he makes essentially the above criticism of it (pp. 169-171).

[9]I owe this observation to Saul Kripke, who pointed out confusion in an earlier version of this paragraph.

Another possible expansion of the canonical notation would be to allow, as Quine does not, quantification of sentence and predicate places. The law of which (4) is an instance would be not (5) but

(6) $(p) (p \lor \sim p)$,

which is a theorem of second-order logic.[10] Similarly, a truth which is an instance of a valid first-order schema, say:

$(x) \sim Fxx \ \& \ (x) (y) (z) (Fxy \ \& \ Fyz \ . \ \supset \ Fxz) \ . \ \supset \ (x) (y) (Fxy \supset \sim Fyx)$,

call it $A \ (F)$, would be an instance of the law $(F) \ A \ (F)$, which is again a theorem of second-order logic.

This proposal again has the consequence that statements of logical possibility are not truths of logic. For example, let $B(F)$ be a nonvalid first-order schema whose negation is satisfiable only in an infinite universe. $(F) \ B \ (F)$ would on the above conception express a false "law" of logic. But $\sim (F) \ B \ (F)$ is not a logical truth: it can be true only if the universe of individuals is infinite.

The obstacle all these attempts face is the following: We assumed that if 'p' is a first-order statement, 'p' is logically necessary if and only if it is logically true in the usual sense. We sought to extend the notion of logical necessity to a wider class of logical forms. But our initial condition implies that the first-order statements that are logically possible are just the negations of *nonvalid* ones. It follows that either logical possibility is not certifiable by logic alone, or there is no complete proof procedure for logic in the extended sense. The second-order alternative, if validity is taken in the usual way, has *both* these consequences.

The former conclusion derives plausibility from some simple reflections on the relation of logic and mathematics. The completeness theorem tells us that if a sentence is not logically true (in the first-order sense) then a *mathematical* construction of restricted com-

---

[10]As Henry Hiż reminded me, (6) is already a theorem of truth-functional logic with quantification merely of sentence places. In general, logic with quantification only of sentence places is much simpler than second-order logic in the genuine sense (with quantification of predicate places). Extensionally, the former is naturally interpreted with the bound variables ranging over the two truth-values.

plexity, but which must in some cases involve an infinite universe, yields a counterinstance. Logical truths, since they are true regardless of the size of the universe, seem to have no ontological commitment. But the *absence* of logical truth requires countermodels which may be infinite.

In terms of *mathematical* modalities, we could say that the *logical* possibility of the truth of a statement of the form $B(F)$ is the *mathematical* possibility of a model of the schema. In this sense, the logical modalities are not really more general than the mathematical modalities.

III

Let us now focus on the latter. Quine seems not to challenge the assertion that the axioms and theorems of mathematics have the character of laws rather than accidental generalizations. As we saw above, his objection is rather to the distinction between mathematics and natural science and thus between mathematical and natural necessity. In large part this objection rests on the claim, which I do not want to discuss at the moment, that there is no interesting sense in which mathematics is a priori and theoretical science is not. Where he discusses directly the existence of a distinctive notion of logico-mathematical necessity[11] and in another place where he introduces similar considerations about the difference between mathematics and science,[12] it is not clear that Quine is really trying to give a case against a notion of logico-mathematical necessity independent of his case against apriority.

There is, however, a reason for wanting distinctly mathematical modalities that Quine does not consider. This is that modality is naturally appealed to in explicating *existence* in mathematics and related notions. To make this clear, we need to concentrate not on the necessity of mathematical truths, but on *possibility*.

Corresponding to stringency of necessity is, of course, liberality of possibility. The notion of absolute possibility would allow that states of affairs are possible that are not countenanced by the laws of nature and that must therefore be counted "naturally" impossible. If the mathematical modalities are to be interesting, the same

---

[11]"Necessary Truth," pp. 74-76.
[12]*Philosophy of Logic*, pp. 98-100.

should be the case where absolute possibility is replaced by mathematical possibility.

A simple example from constructive mathematics indicates why mathematical possibilities might be thought to go beyond the "naturally" possible. Whatever the process is by which the natural numbers might be constructed, it is one that *can* be continued to infinity. One implication of this is that at any stage of this construction (at which representatives of a certain initial segment have been constructed) it is always possible to continue. Moreover one assumes a transitivity rule (i.e., '$\Diamond \Diamond p$' implies '$\Diamond p$') from which it follows that at any stage, for any n it is possible to continue for n more steps.[13]

Traditionally, such a construction is thought of as an activity of the human mind. However, it seems we cannot represent this in terms of the actual natural capacities of the human organism: for then there will be a definite bound to the number of possible steps, since certainly there is a limit to the speed with which the construction could proceed and a definite outer limit (certainly less than 175 years) to the possible life span of a man.

The traditional recourse in this situation is to conceive the capabilities of the *mind* in abstraction from those of man as an organism. This course is not at all in accord with Quine's naturalism. One can ask whether it does not amount to defining the capacities of the mind in terms of certain theories about the mind's objects—mathematics in particular.[14] If so, then using it in an argument against Quine's view on the necessity of mathematics would require that mathematical possibility be to some degree antecedently clear.

An alternative apparently more congenial to Quine would be to think of the construction of the natural numbers as a physical process. Then what the "potential infinity" of the natural numbers would amount to is the physicial possibility of the generation of representatives of any initial segment of the natural numbers.

This should remind us of a version of "nominalism" related to that of Nelson Goodman, which at one time attracted Quine.[15] Such nominalism is essentially the above suggestion with the modal element removed. Rather than assume the natural numbers or other mathematical objects directly, one assumes that the physical world has enough structure to represent the system of mathematical ob-

---

[13]Cf. my "Ontology and Mathematics" (Essay 1 of this volume), Section III.

[14]Cf. my "Infinity and Kant's Conception of the 'Possibility of Experience' " (Essay 4 of this volume).

[15]Goodman and Quine, "Steps toward a Constructive Nominalism."

jects that one wants. The difference with the physical interpretation of constructivism alluded to above is the rejection of the modal interpretation of mathematical existence and the distinction of "potential" and "actual" infinity.

Such nominalism has the advantage of avoiding the conceptual complication of modality. Still it may seem that the modalist account is on stronger ground because its hypotheses are weaker. The nominalist has to assume that there actually are infinitely many individuals: that is, he must have, say a two-place predicate which, in platonist terms, describes a one-one mapping of a certain collection of individuals properly into itself. This assumption could hardly be justified by direct empirical evidence. As a hypothesis, its plausibility surely rests on the idea of space and time as infinitely divisible or infinite in extent. But then the existence of such an abstract structure (however "existence" in this case might come to be explained) is more certain than any hypothesis of the existence of a realization of it by actual physical entities such as bodies or particles. This additional assumption seems gratuitous, at least from a purely mathematical point of view. It seems to seek to undo an abstraction already carried out by the Greeks—distinguishing between space and time as abstract forms—one might say of *possible* objects—and the actual physical objects and events, and giving a sophisticated, rigorous theory of the former.

On the modalist view, this assumption of the *existence* of infinitely many physical individuals is dropped. The concrete existence of infinitely many individuals, as opposed to that of any arbitrary finite number, does not even have to be *possible*.[16] But one might ask whether the assumption that the existence of arbitrary finite sequences is *physically* possible is not also gratuitous. Is it any more necessary to mathematics that the laws of the physical universe should permit some *potential* infinity than that some sort of infinity should *obtain* in the actual world?

I shall not inquire further into this question, because an obstacle immediately arises. To inquire in a serious way about physical possibility, one would have to appeal to physical theory. Then the difficulty is that physical theory as it stands is founded on mathe-

---

[16]In the mathematical sense in which, on this view, arbitrary finite sequences are possible. The existence of infinitely many individuals can be shown to be in a sense *logically* possible: Certain theories which intuitively (or in terms of set-theoretic semantics) must have an infinite domain for their bound variables can be shown by constructive (finitary) means to be formally consistent. See, for example, Hilbert and Bernays, *Grundlagen der Mathematik*, I, §6. This case illustrates the remark made above about the relation of logical and mathematical modalities.

matics. One could not inquire into anything by physical theory while treating as genuinely open the question of the possibilities that constitute the existence of the natural numbers.

IV

At this point we should return to the views of Quine. For one is inclined to conclude from the above difficulty that mathematics is prior to physics. If we think of mathematical existence in modal terms, the possibility involved must be more general than physical possibility, either absolute or distinctively mathematical possibility. However, it seems clear that Quine would not accept this conclusion. Mathematics and physics constitute a single body of theory. While it may be relevant to mathematics to consider the possibility of physical realizations of certain structures, this is part of the process of rebuilding the ship of science while it is afloat.

Quine goes further in that he does not demand any specific relation between the existence of natural or real numbers and the possibilities or the facts of physical existence. This does not mean that facts about concrete objects have no relevance to mathematics or mathematical existence. But the relevance is that mathematics belongs to a body of theory that includes empirical science and that relates as a whole to experience. But in order to see how the relation of mathematical existence, possibility, and necessity look from Quine's point of view, we have to consider Quine's view of mathematical existence. I maintain that Quine's view here is somewhat awkward for his view that there is no higher necessity than "physical or natural" necessity. Namely, Quine's view is more "platonist" than the constructivist views alluded to above, not only in the generally conceded sense that mathematical existence and truth are treated as independent of the possibilities of construction and verification, but in the additional sense that they are treated as independent of the possibilities of representation in the concrete.

Quine does not separate mathematical existence from existence in general. The classical objectual existential quantifier is for Quine in effect a primitive.[17] It is "defined" by the theories in which it

---

[17]"Existence is what the existential quantifier expresses. There are things of kind F if and only if $(\exists x)Fx$. This is as unhelpful as it is undebatable, since it is how one explains the symbolic notation of quantification to begin with. The fact is that it is unreasonable to ask for an explanation of existence in simpler terms" ("Existence and Quantification," p. 97).

occurs, the general framework for which is classical first-order quantificational logic. Although Quine makes some use of very general divisions among objects, such as between "abstract" and "concrete," these divisons do not amount to any division of *senses* either of the quantifier or the word "object"; the latter division would indeed call for a many-sorted quantificational logic rather than the standard one. Moreover, Quine does not distinguish between objects and any more general or different category of "entities" (such as Frege's *functions*).[18] There is no difference between saying there is *something* satisfying some condition and saying there are *objects* satisfying the condition.

The effect of this way of proceeding is that what can be said about existence and objecthood in general is common to the most abstract and the most concrete, the most theoretical and the most immediately perceived objects. Moreover, no particular kind of existence has a central or paradigmatic role in canonically formulated theory, although it might from a genetic point of view.

From some phrases in which Quine gives what he considers "intuitive" and "picturesque" characterizations of the difference between singular and general terms, P. F. Strawson derives the view that identifying reference to spatiotemporal particulars is the central, basic form of reference to objects, and that spatiotemporal particulars are "the very pattern" of objects.[19] He goes on to say, "Insofar, then, as things other than spatio-temporal particulars qualify as objects, they do so simply because our thought, our talk, confers upon them the limited and purely logical analogy with spatio-temporal particulars which I have just described."[20]

In his reply to Strawson (in "Replies"), Quine does not discuss this point. Rather, he rejects the demand for an explanation of the distinction between singular and general terms that goes beyond giving such formal criteria as accessibility to quantification and relating the logical apparatus of quantification theory to our own language. His reason is the indeterminancy of translation and the "parochial" character of the apparatus of objective reference. He does not say why that *is* a reason. Moreover, some of the elements of Strawson's account occur in Quine's own genetic speculation.[21]

---

[18]Quine nowhere comments explicitly on Frege's theory of functions, but that he disapproves of it is made clear by his comments on second-order logic (*Philosophy of Logic*, pp. 66-68).

[19]"Singular Terms and Predication," p. 115.

[20]Ibid.

[21]"Replies," pp. 292-293 (to Smart); also *The Roots of Reference*, passim, e.g., p. 88.

Existence thus takes on not only a theoretical but also a formal character. It is emancipated not only from perception but also from spatiotemporarlity and other marks of concreteness. In the case of abstract entities, certain protests against platonism become irrelevant. There is no mysterious "realm" of, say, sets in the sense that they need to have anything akin to location, and our knowledge of them is not based on any mysterious kind of "seeing" into such a realm. This "demythologizing" of the existence of abstract entities is one of Quine's important contributions to philosophy, although it has its antecedent in Frege and Carnap.[22]

It is noteworthy that for Quine this "emancipation" is achieved for existence as such, so that abstract entities are accommodated without the sharp separation of abstract and concrete existence that would be implied by the view that "exists" has a difference *sense* or *meaning* in the two cases. This view Quine firmly repudiates,[23] but given his general account of meaning, the question cannot be fundamental. He seems to characterize the notions of "object" and "existence" in minimal terms so as to apply to all cases of "being," "existence," or "subsistence." Any alleged distinction of senses of "exist" is to be replaced by a suitable distinction of kinds of objects. Since Quine assesses ontology in terms of the *classical* logic of quantification, his conception of existence is not only incompatible with constructivism but can make sense of constructivist theories only by classical models. Although I find that attitude unduly restrictive, I do not wish to pursue the point here. More could be said than I have said elsewhere,[24] but now I am more interested in the consequences of accepting set theory.

---

[22]Frege insisted on distinguishing between existence and actuality (*Wirklichkeit*), and his principle that "it is only in the context of a proposition that words have any meaning" (*Foundations*, p. 73) was applied especially to the case of numbers. This might, however, have suggested the *eliminability* of locutions talking of such objects. The principle is not emphasized in Frege's later writings. Cf. Dummett, *Frege*, pp. 7, 195-196, 495.

Carnap's views about abstract entities are too closely tied to conventionalism about the *truth* of claims about such entities for Quine's and my own taste.

[23]*Word and Object*, pp. 131, 242; "Existence and Quantification," p. 100.

[24]"Ontology and Mathematics," (Essay 1 of this volume); cf. "On Translating Logic." I should emphasize that on my view constructivist and platonist theories of mathematical existence are neither totally reducible to the other and that it is not necessary to make a final choice between them. Set theory is certainly more powerful and comprehensive than any constructive theory developed up to now, and if one *had* to choose between set theory and constructivism, it would be folly to give up set theory.

Quine's arguments for preferring classical logic are partly pragmatic. Pragmatism would allow accepting different theories for different purposes. Why not different logics and "theories of being"?

Some writings on existence in mathematics have given promi-
nence to what, following Leibniz, can be called the "incompleteness"
of mathematical objects. The only properties and relations of such
objects that play a role in mathematical reasoning are those determined by the basic relations of some system or structure to which
all the objects involved belong. This may be the natural numbers,
some other number system, Euclidean or some other space, a given
group, ring, field, or other such structure, or the universe of some
model of set theory. A further natural step would be to say that
the objects just do not *have* properties and relations other than
those derived from the basic relations of some comprehensive structure. However, it seems that to accommodate applied mathematics
we need to consider also certain "external" relations (see below).

If we think of "structures" in the usual algebraic way, then this
view implies that all the "real" properties and relations of mathematical objects will be invariant under any isomorphism of the
comprehensive structure with another. Suppose we are concerned
only with natural numbers. Then we can see them as a structure
consisting of a set N and a binary relation S (the successor relation),
such that the Dedekind-Peano postulates hold.[25] A structure so
conceived can of course have different *realizations*: sets N_0 and relations S_0 satisfying the Peano postulates. If such realizations consist
of mathematical objects, the above thesis implies that the properties
and relations of these objects—including the elements of N_0 and
S_0—derive from the basic relations of some other structure. No-
torious examples are the different construals of natural numbers
as sets, such as the Zermelo and von Neumann numbers, where
the underlying structure is the universe of set theory with the $\in$-
relation.

By "external relation" I mean the sort of relations that arise in
application between the elements of a structure and other objects,
such as the relation between a city x and a natural number n when
x has n inhabitants. Such relations are in general definable in terms
of the basic relations of the structure and others which do not
depend on any choice of realization of the structure. Thus 'x has
n inhabitants' is true if there is a one-one mapping of the inhabitants
of x onto the natural numbers less than n, where "less than" is

---

On Quine's attitude toward "deviant logic," see *Philsophy of Logic*, pp. 80-94; also
Morton, "Denying the Doctrine and Changing the Subject," and my "On Translating
Logic."

[25]Of course I here assume the "background theory" is set theory. First-order
number theory requires additional relations, at least addition and multiplication.

definable in terms of N and S. Particular numbers are also given by notations definable in these terms, e.g.,

$$0 = (\iota x)\ (x \in N\ \&\ \sim (\exists y)Sxy),$$
$$1 = (\iota x)Sx0,\ \text{etc.}$$

Thus these relations can be "construed" in terms of any realization of the structure and they are invariant under isomorphism.

Different ontological morals have been drawn from these facts. The traditional view, most explicit perhaps in Leibniz with the contrast of incomplete objects and individual substances, is that these features are characteristic of mathematical objects or at least abstract objects.[26] The above quotation from Strawson goes a step further in suggesting that such objects are objects only in a derivative or extended sense. Benacerraf goes another step in denying for this sort of reason that numbers are genuine objects at all.[27]

The wide net that Quine's notion of object casts is shown by the fact that he never suggests that the "incompleteness" of mathematical objects is an ontological deficiency or even peculiarity. There are hints that it is characteristic of objects generally. Harman tentatively suggested that the possibility of construing numbers as sets in different ways was an instance of the indeterminacy of translation.[28] Quine has used natural numbers and other mathematical examples to illustrate his thesis of the relativity of reference.[29] In recent writings he has some tantalizing comments about identity which point to the conclusion that identity is just indiscernibility with respect to the predicates of a language and is therefore relative to the richness of its vocabulary. This would imply that the "incompleteness" of mathematical objects does not contrast with anything, or rather that in the case of mathematical objects (and perhaps also some theoretical objects in science) a definite system of relations is given, whereas with ordinary objects the "structure" is much vaguer and more open-ended.[30]

I am not sure how far Quine is ultimately willing to go in this direction. I also shall not criticize these hints. I do not myself have a theory of "concrete objects" or "individual substances" to oppose

[26]Thus, Bernays, "Mathematische Existenz und Widerspruchsfreiheit," and my "Frege's Theory of Number" (Essay 6 of this volume).
[27]"What Numbers Could Not Be."
[28]"Quine on Meaning and Existence," "An Introduction to Translation and Meaning." Cf. "Ontology and Mathematics" (Essay 1 of this volume), Section II.
[29]*Ontological Relativity*, pp. 44-45.
[30]*Philosophy of Logic*, p. 64; "Identity," unpublished.

to them. I do want to say that the general ideas Quine expresses about objecthood are somewhat awkward for his view of necessity and possibility. To see this, let us consider set theory and Quine's views on it.

V

It is only when the higher infinities of Cantorian set theory are introduced that mathematical objects must violate the conditions of representability in the concrete discussed above. Space-time is always construed as a set of $2^{\aleph_0}$ points or at least not of essentially higher cardinality. This puts a definite bound on the number of possible "spatio-temporial" objects, say $2^{2^{\aleph_0}}$. If the "physically possible" is what can in some sense be realized in space and time, then structures of sufficiently high cardinality whose acceptance is uncontroversial among set theorists (e.g., whose existence is provable in ZF) are not physically possible. Although we may think of the existence of sets in modal terms, as the *possible* existence of objects satisfying certain conditions, it hardly makes sense to take such possibilities as possibilities of *physical* objects, where high cardinalities are involved, or even as possibilities of *concrete* objects, whatever the decisive marks of concreteness are taken to be. Whatever convinces us, for example, that "there is" a cardinal number $\beth_\omega$,[31] surely does not convince us that a structure of that cardinality having any but the most bloodless reality is possible: if concreteness demands causal efficacy, or perceivability in some strong sense, or temporality, or even some genuine individuality, we have no reason whatsoever to believe that $\beth_\omega$ can be concretely represented.

This is a sense in which set theory is "platonist" in which even the impredicative theory of real numbers is not. The preceding remarks are meant to indicate that in this case a modal conception of mathematical existence has little force in overcoming nominalist scruples about the objects postulated in higher set theory, although perhaps it does for the objects of pre-Cantorian mathematics. For even modally construed set theory requires a conception of "object" and "existence" so general that objects can exist which have none

---

[31] $\beth_\omega$ is the smallest cardinal of a standard model of the simple theory of types with an infinite number of individuals; it is the least cardinal greater than $\aleph_0$, $2^{\aleph_0} 2^{2^{\aleph_0}}, \ldots$

of the marks of concreteness.[32] What the modal construal does accomplish is to bring out the fact that for pure set theory as for arithmetic the usual distinction of actual and merely possible existence does not apply.

Philosophers have sought to avoid admitting such inclusive conceptions of object and possibility either by rejecting set theory or by not admitting that set theory means what it says—usually by adopting some variety of formalism. Quine, we have seen, grasps the nettle by admitting a conception of object with the requisite generality. But his view has some affinity with formalism. Neither the meaningfulness nor the truth of set theory as it stands has a trustworthy intuitive foundation. Both the axioms and the logic have a certain conventional character. However, in this respect set theory does not differ in principle from other theories, particularly scientific theories. For the latter, a nonconventional factor in determining belief is of course the relation to sense-experience. Because according to Quine mathematics and science constitute a single body of theory, the same connection exists even for set theory. Quine differs from other theorists of mathematics such as Kant, Brouwer, and Gödel in denying that there is any fundamental intrinsically mathematical or logical form of evidence. He also does not place much emphasis on the derivative types of evidence employed in plausible reasoning in mathematics, such as the analogies appealed to in arguing for large cardinal axioms or the consideration of consequences by which Gödel argues for the falsity of the

---

[32]Cf. Putnam, "Mathematics without Foundations," "What Is Mathematical Truth?" In the former paper, Putnam speaks of the possibility of "concrete models" of set theory. He explicitly assumes (p. 57) that there is nothing inconceivable in the idea of a *physical space* of arbitrarily high cardinality. I would certainly agree that there is nothing inconceivable in the idea of a *space* of arbitrarily high cardinality, if what is meant is a structure of the general sort considered in geometry. But what makes such a space "physical"? Putnam seems to require here a distinction between a "necessary" and a "factual" aspect of fundamental physics, such that the cardinality of space-time falls on the factual side. This raises a host of questions. But a prior question is what is gained for the plausibility of the axioms of set theory by the assumption that models of it are possible which are in some way "concrete" or "physical," particularly if one considers that the ideas of set and function are central to the specific conceptions we have of cardinalities higher than the denumerable, even that of the continuum.

In the discussion at a symposium (in December 1965) at which Putnam presented this paper, he qualified his position by expressing doubt whether the structures required by large cardinal axioms really were possible.

In the latter paper, the conception of a concrete model does not occur; perhaps Putnam no longer maintains that the structures whose possibility is in question in the modal interpretation of set theory are concrete.

continuum hypothesis.[33] However, there is no systematic reason why Quine should not regard such considerations as relevant; he would no doubt insist that they are not compelling, but that is generally agreed upon. However, his logical and "class-theoretic" orientation toward set theory (discussed below) is probably at least a psychological barrier.

Quine seems to be in the following awkward position: Both epistemologically, in terms of the evidential basis it rests on, and ontologically, in terms of the nature of its necessity, set theory is on a par with physics. However, the notion of object in set theory, and the structures whose possibility it postulates, are much more general than the notion of physical object or spatiotemporally or physically representable structure. How can he still maintain that these possibilities are "natural" possibilities and that the necessity of logic and mathematics is not "higher"?

Quine would no doubt reply that by "natural" possibility and necessity he means possibility and necessity as it would be assessed in terms of the laws of a theory that includes *both* set theory and physics; the idea that there is no "higher" necessity than natural necessity in effect rests on the claim that there is no definite line between mathematics and empirical science. All possibilities are "natural"; some of these are possibilities of spatiotemporal structures; others are only of sets of perhaps very high rank.

The claim that there is not a clear line between mathematics and empirical science seems to me implausible, at least if it is made independently of Quine's epistemological arguments. In a canonical formulation of scientific theory, the distinction between mathematics and the rest is quite as clear as that between first-order logic and the rest. Quine prefers here formulations in terms of first-order set theory: then the truths of mathematics would be those that contain only '$\in$', '$=$', and the operators of first-order quantificational logic essentially.[34] Moreover, the step from delin-

---

[33]"What Is Cantor's Continuum Problem?" p. 267.

[34]A complication arises from the fact that the set-theoretic framework of a scientific theory would normally allow individuals (*Urelemente*). Then if we add a name 'Λ' for the null set, which clearly should belong to the mathematical vocabulary, we can define 'x is an individual' ('Ix') as '$x \neq \Lambda$ & (y) $(y \notin x)$' and then state in the mathematical vocabulary that there is a certain number of individuals. Such a statement we would not want to count as a mathematical truth.

Clearly, in this setting a mathematical truth must satisfy the additional condition of being indifferent to what individuals there are. If we assume ZF (with Foundation), we can meet the difficulty as follows: Let 'F' be a new one-place predicate. Given a sentence 'p', let 'p^F' be the result of restricting its quantifiers to 'F'. Then

eating a class of mathematical *truths* to using purely mathematical *modalities* is more useful than the corresponding step in terms of

---

we say that 'p' is a mathematical truth if and only if

$$(x)[\sim Ix \supset (y)(y \in x \supset Fy) \equiv Fx] \supset p^F$$

is true and contains only '$\in$', '$=$', 'Λ', and logic essentially.

Semantically, we could obtain the same result by saying that a sentence is a mathematical truth if it is *valid* in the following sense: it is true under any interpretation standard with respect to '$\in$', '$=$', and 'Λ' with the universe a transitive *class U* such that $(x) (x \subseteq U \supset x \in U)$. Of course, this formulation presupposes that the intended range of the variables does not contain proper classes.

The difficulty here is analogous to a simpler one that arises for logical truth: such statements as '$(\exists x) (\exists y) (x \neq y)$' are true and, if identity is counted as logical, they contain only logical vocabulary essentially. This is resolved by the observation that logical truth must be indifferent to the range of the quantifiers.

Note that on the above proposal, '$(\exists x) (\exists y) (x \neq y)$' is a mathematical truth, but '$(\exists x) (\exists y) (Ix \& Iy \& x \neq y)$' is not. This is the right result if, as the formulation envisages, all mathematical objects are construed as sets. One might still object that I am counting as mathematical truths some statements whose truth depends on the structure of the class of individuals, where this might have been otherwise. For a given U, the individuals in U are a subclass of the actual individuals. Hence if P is a property of a class that can be expressed in the language of set theory, such that for any class V, $P(V)$ implies $P(V')$ for any subclass V' of V, if the class of *all* individuals in fact possesses P, it will be a mathematical truth that $P(\{x: Ix\})$.

It is not clear that this *is* an objectionable consequence, but perhaps it is not evident that it is not. A particular case of some interest arises from the axiom of choice. Let $P(V)$ be 'Every subset of V can be well-ordered'. Then P has the above subclass property. If the axiom of choice is stated in the simplest way, that *every* set of nonempty sets has a choice function, then it follows that every set of individuals can be well-ordered; i.e., $P(\{x: Ix\})$. Thus on our criterion, this is a mathematical truth.

Given that set theorists usually leave individuals out of consideration, it is not as clear as it should be that the often expressed intuition that the axiom of choice is "evident" is meant to exclude the possibility of a non-well-orderable set of *individuals*, as opposed to a non-well-orderable pure set. (A set is pure if no individual is a member of it, or a member of a member, or...). To my intuition the axiom of choice does cover the situation with individuals, but it is perhaps not inconceivable that the world should have been like a Fraenkel-Mostowski model in which the axiom of choice obtains for all pure sets but not for sets involving individuals. The above view implies that this is logically possible but not mathematically possible.

It seems to me doubtful that a convincing characterization of mathematical truth could be given which does not have consequences like this, unless mathematical modality or some functional equivalent is allowed in the metalanguage. Quine might object that then the criterion would be circular. It seems to me that this is small comfort for him, because I do not see that the objection to my nonmodal formulation has any force unless we have an intuitive grasp of mathematical possibility to begin with.

I have carried out the above discussion in a metalanguage with class variables, although I assumed that the metalanguge will surely contain predicates of satisfaction and truth for the object language, the required class variables can be *defined*, since the classes needed are all definable in the object language. See for example, my "Sets and Classes" (Essay 8 of this volume).

first-order logic that I discussed above, and it does not have the formally awkward features I mentioned.

In unregimented science, the distinction between pure mathematics and individual sciences seems clear, although in applied mathematics it may not always be sharp. Quine compares the relation of mathematics to physics to that of theoretical to experimental physics.[35] I find this analogy unpersuasive. Theoretical and experimental physics are both about the same subject matter— space-time, energy, physical particles, physical bodies. Experiments are carried out to verify or falsify the theories of theoretical physics; theories are concocted to explain the data yielded by experiments. There is no similar unity of subject matter or interrelation of purposes between mathematics and physics. Quine mentions the applicability of mathematics to other sciences; this is to me an indication of the greater generality of mathematics.

It is clear, however, that Quine is ultimately thinking of the epistemological situation: it is perhaps less important for him to deny the existence of a clear and in some way important *distinction* between mathematics and physics, or even to deny that mathematics has a *content* which is more general, than to challenge the idea that mathematics is disconnected from *observation* in a fundamental sense in which physics is not. This raises a whole range of issues which I have not wished to take up here. My discussion of the necessity of logic and mathematics has tried to proceed independently of their alleged apriority. But it does seem to me that the considerations developed here reflect back on Quine's epistemological position at a few points.

To begin with, from the fact that the statements of set theory contain quantifiers such that only a very restricted part of their range is relevant to observation it seems clear that the role of purely theoretical reasoning in our judgment of the truth or falsity of set theories must be much more central than that of observation, although it is conceivable that the latter could play a role. More specifically, if ability to explain observational data in the ordinary sense were the reason for accepting set theories, then weaker theories (perhaps only some form of second-order number theory) would be preferred to full set theory on grounds of simplicity and perhaps lesser risk of inconsistency. Quine's actual accounts of set theory proceed in a much more abstract fashion (see below).

The following point seems relevant to some of the examples that have been used to argue that logic and mathematics are empirical. A

---

[35]"Necesssary Truth," p. 75.

mathematical theory of an aspect of the empirical world takes the form of supposing that there is a system of actual objects and relations that is an instance of a structure that can be characterized mathematically. The implications of this supposition will vary greatly and may not be altogether clear: an outstanding case is the attribution of a certain geometrical structure to space-time, and this is not necessarily taken to imply the physical reality of the *points* which on the most usual mathematical formulation are the elements of the structure. The point I wish to make is that what confronts the tribunal of experience is not the pure theory of this type of structure but its being supposed represented by a definite aspect of the actual world. If the resulting theory is abandoned or modified, the form is likely to be that of replacing this structure by a different structure from the mathematician's inventory, even if it was developed only for the purpose.

In one sense mathematics may change as a result: the theory of one type of structure may become more salient, that of another type less so. But no proposition of pure mathematics has been *falsified*. If we view mathematics in this light, no proposition of Euclidean geometry is falsified by the discovery that physical space is not Euclidean. One could object that this claim rests on a conception of pure geometry that did not prevail when it was almost universally believed that actual space *is* Euclidean; then the axioms of Euclidean geometry were thought of as at the same time truths of mathematics and statements about real space. However, from the fact that the differentiation of mathematics from physics that I am defending has developed historically, it surely does not follow that it is unsound. I do not even have to assume that it is complete at the present time: there may be aspects of physics or other sciences today of which the mathematical content is not so neatly separable. I do not know of any example, still less an example where such a separation seems impossible in principle.[36] The prevailing methodology of mathematicians, with its emphasis on rigor and abstraction, works for carrying through such a

---

[36]Quantum mechanics may seem to be a case where the development of physics leads to a revision of *logic*. See Putnam, "Is Logic Empirical?"

This is a matter about which I am not competent technically. But I am not convinced that the case is basically different from that of geometry. Quantum mechanics is still a theory formulated with classical mathematics in which the underlying logic is classical. The "propositions" and "connectives" that obey a nonstandard logic are constructions within this classical framework, as in the case of a classical model of intuitionism.

I have to admit that this is an extremely superficial comment about a complex and highly disputed question.

separation in any given case. The point is, however, that on the *general* conception of mathematics I am presenting, the known or easily conceivable examples of empirical falsification of mathematical theories in *science* does not yield any instance or model of empirical falsification of pure mathematics.

VI

In practice, Quine assimilates set theory to physics rather less than some of his general epistemological remarks might suggest. He always related set theory closely to *logic*. Logic has, in turn, been connected particularly closely to *grammar* and to certain grammatical constructions, most explicitly in *Philosophy of Logic*. He is tempted by the characterization of logical truths as "true by grammatical structure" but stops short of adopting it because of its suggestion that logical truth is linguistic in a sense that can be meaningfully opposed to factual truth.[37] Nonetheless, the connection with specific grammatical constructions also extends to set theory, as *The Roots of Reference* shows. Set theory for Quine has always been a theory of extensions of predicates, in that the paradigm of a set is the extension of a predicate, and the axioms of set theory are taken as attributing extensions to certain predicates (to be sure with parameters). I want to close this essay by some comments on this.

By the extension of a predicate 'Fx', I mean an object $\hat{x}Fx$ associated with 'Fx', such that *coextensive* predicates have the same extension; that is, '$(x)\ (Fx \equiv Gx)$' implies '$\hat{x}Fx = \hat{x}Gx$'. Moreover, anything is a *member* of $\hat{x}Fx$ just in case 'Fx' is true of it. Hence we have the principle

$$(7) \qquad (z)\ (z \in \hat{x}Fx .\ \equiv Fz).^{38}$$

[37]*Philosophy of Logic*, p. 95.
[38]Of course I have here simply adapted Frege's characterization of the notion of extension to the situation where we do not assume Frege's theory of *concepts*, entities of a predicative character that are not objects. Frege allows himself second-order logic, so that for him membership is definable: '$x \in y$' can abbreviate '$(\exists F)\ (y = \hat{z}Fz \cdot Fx)$'. (7) follows from his axiom V, which I here state in two parts. I disregard the fact that Frege talks more generally of functions and value-ranges rather than merely of concepts and extensions.

(Va) $(x)\ (Fx \equiv Gx) \supset .\ \hat{x}Fx = \hat{x}Gx \cdot$
(Vb) $\hat{x}Fx = \hat{x}Gx .\ \supset (x)\ Fx \equiv Gx).$

Given (7), (Vb) follows by substitutivity of identity, at least if the extensions exist. (Va) follows from extensionality.

For Frege, (Va) is an instance of the extensionality of concept places in his formal languge. He makes it clear in his discussion of Russell's paradox that it is (Vb) that gives rise to the paradox. See *Grundgesetze*, II, 257.

The category of objects that are extensions of predicates I shall call *classes*, as Quine himself does when he is talking about the elements of the subject.

One can think of the conception of classes as arrived at by steps of three types: nominalization of predicates to yield the noun phrases of which class abstracts are a regimentation, taking such noun phrases as standing for objects, and adopting (for classes as opposed to, say, attributes) coextensiveness of the predicates as the criterion of identity.

In the genetic account of *The Roots of Reference*, these steps are regarded as natural steps to take, and they form the central elements in the genesis of set theory. However, the second step divides into two: first, nominalized predicates are quantified *substitutionally*, and then this quantification "goes objectual."[39]

The manner in which I introduced the notion of extension recalls Frege. I want to claim that Quine's conception of a class or set is Fregean, in that for him the origin of the concept of class lies in predication, and the only requirements on a set theory which are not to a certain degree artificial are that with certain predicates 'F' be associated objects $\hat{x}Fx$ satisfying (7) and that classes obey extensionality: two classes having the same members are identical. This hypothesis about Quine's underlying conception of the subject matter of set theory serves to explain some of the features of Quine's views on set theory that are "deviant" from the point of view of most contemporary thought on the subject.

Quine seems to express this conception in the Introduction to *Set Theory and Its Logic*: "Set theory is the mathematics of classes. Sets are classes." He then introduces the notion of extension, and although he goes on to mention the ideas of "aggregate" and "collection," he seems to understand these as classes and classes as extensions.[40]

It is characteristic of contemporary thought on the foundations of set theory to explain the idea of set in such a way that the universe of sets is taken to have a *hierarchical* structure. The simplest model for this is the simple theory of types: the objects of the theory are

[39]*The Roots of Reference*, §§27-29, 31.

[40]*Set Theory and Its Logic*, pp. 1-2. I say "seems to" because although I believe this interpretation of the passage to be correct, my colleague Mark Steiner does not find it so clear. Peter Geach, in discussion of an earlier version of this paper at the University of Pennsylvania, mentioned a remark in a letter to him from Quine which would be more unequivocal in identifying classes with Fregean extensions.

I am indebted to George Boolos and especially to Steiner for their criticisms of earlier versions of this section.

divided into ground-level objects (individuals), classes, classes of classes, and so on. Russell's procedure was to have different styles of variables for different levels of the hierarchy, but the formal theories used by set theorists today are first-order (or sometimes second-order, where the second-order variables range over classes rather than sets; see below), derived historically less from Russell than from Zermelo. But according to the "standard" theories, every object can be assigned an ordinal number as its *rank*, in such a way that a set has higher rank than any of its elements.[41]

The simple idea that the elements of a set are *prior* to the set already rules out the applications of the comprehension schema (7) (asuming '$\hat{x}Fx$' to denote) that lead to paradoxes such as Russell's. For example, no set can be an element of itself, and hence there can be no universal set and no set of all *non*-self-members.

The explanation given above of the notion of *extension* does not obviously give rise to the priority of the members of the extension to the extension itself, in a sense that would exclude self-member-ship. Even if we have granted, in the wake of Russell's paradox, that not all predicates have extensions, for a contradictory pair, say 'x is a woodchuck' and 'x is not a woodchuck', there is no obvious reason why one should have an extension and the other not. But the natural assumption that extensions are not woodchucks implies that the class of non-woodchucks, if such there be, is a member of itself. Moreover, consider any ordering of some totality of entities acording to some notion of priority. Suppose further that the class of non-woodchucks is among the entities so ordered. Then the only way all its members could be prior to it would be if all the objects that are *not* prior to it are woodchucks. Then if the class of non-woodchucks is posterior to all its members, we can expect other classes (probably including the class of woodchucks) to violate this condition.

This reasoning will of course be unpersuasive if one thinks of extensions from the beginning as *sets* and of sets in terms of stand-ard axiomatic set theory. Then 'x is not a woodchuck' will have no extension, at least not in the strict sense of a set containing abso-lutely everything except the woodchucks. For if there were such a set it would have a certain ordinal α as its rank, but then there are objects, such as the ordinal $\alpha + 1$, which could not belong to it and yet which are not woodchucks.[42]

[41]The rank of a set is the least ordinal greater than the ranks of all its elements. Of course to show that such a rank always exists, one needs the axiom of foundation.
[42]This reasoning requires the axiom of foundation. If we assume that there is a

Quine has long been identified with the position that the hierarchical aspect of ordinary set theory is not essential to it. His own theories NF and ML, although derived in a way from the theory of types, are incompatible with the hierarchical conception of the universe of sets.[43] The reservations that Quine expresses in "New Foundations" about the theory of types apply in part to Zermelo-type theories as well, though they are mitigated when proper classes are added.[44]

In his more recent writings, Quine emphasizes set theories closer to the standard. Thus most of *Set Theory and Its Logic* is devoted to developing what is *de facto* a subsystem of ZF, and the genetic process of *The Roots of Reference* eventuates in the simple theory of types.

Characteristic of all these writings is the conception of extensions as the objects of set theory and reserve toward what I have called the "standard" conception. Quine seems (at least before *The Roots of Reference*) to be motivated by the idea that the most intuitively natural set theory would be based on the idea that every predicate has an extension; since the obvious formalization of that is inconsistent, one restricts the comprehension schema in some way that will avoid paradoxes but still preserve certain desirable features—mainly the derivation of ordinary mathematics. However, it seems even now to be Quine's view that such restrictions have an unavoidably artificial character.[45]

Quine is also somewhat cool to the axiom of foundation, which expresses in first-order set theories the idea that sets form a well-founded hierarchy.[46]

---

set of all *woodchucks*, then it follows trivially in Zermelo-type set theories, without foundation, that there is no set of all non-woodchucks.

[43]"New Foundations for Mathematical Logic," *Mathematical Logic*. Quine discusses these systems in the light of recent research on them in chapter 13 of *Set Theory and Its Logic* and in "Replies," pp. 349-351 (to Jensen).

[44]"New Foundations," pp. 91-92. Donald A. Martin, in his review of the revised edition of *Set Theory and Its Logic*, seems a little unfair in saying that NF is "the result of a purely formal trick intended to block the paradoxes" (p. 113). Quine's own remarks ("New Foundations," pp. 90-91) indicate that stratification was motivated by Whitehead and Russell's idea of *typical ambiguity*: of leaving the types of variables in formulæ of the simple theory of types indefinite, provided that the arguments of '∈' are of consecutive ascending types. However, Quine may have neglected to consider that in applying this idea to a first-order theory, the variables had to range over the whole domain of a model, which is not the case for variables on the "typical ambiguity" reading of a formula of the simple theory of types.

[45]*The Roots of Reference*, pp, 102-103, 122.

[46]*Set Theory and Its Logic*, pp. 285-286; cf. "Reply to D. A. Martin" (i.e., to the review cited in note 44 above).

It seems to me that what often comes out as a formalistic approach to set theory in Quine's writings derives as much from his Fregean conception of what a set *is* as from his epistemology. The extent to which the set theorist's intuitive concept of set is a development of Frege's, or rather derives from different sources, is a difficult question that would deserve a paper for itself. That *some* new conceptual element is needed seems clear. If one sticks as closely to the Fregean explication of extensions as Quine most of the time seems to, then it seems that intuition will indeed carry us no further than Quine admits.[47] Viewing the matter from the other side, the idea that the elements of a set are prior to it is in recent literature justified by the quite un-Fregean idea that a set is in some way formed from its elements.[48]

A little further comment on the axiom of foundation is in order. Set theorists characteristically motivate the axioms of the subject by conceiving of sets as "generated" by transfinite iteration of formation of sets from their elements, beginning either with nothing (yielding the theory of pure sets) or with individuals. On this interpretation, the axiom of foundation (suitably formulated) is evident. The axiom is integral to an account of the structure of the universe of sets that is generally accepted. On the other hand, it is not necessary for the derivation of "ordinary mathematics" in set theory. But the defenders of what is called the "iterative conception" (e.g., Martin) can reply that the reason for this is that the objects constructed for arithmetic (finite and transfinite) and analysis can be generated in this iterative way without the assumption that such a generation yields *all* sets. More technically, the sets yielded by iterating the power set operation, beginning with the empty set, yield an inner model of, say, ZF, which satisfies the axiom of foundation and contains the usual representatives of ordinals, cardinals, real numbers, and so on.

A highly illuminating and clear discussion of the universe of sets is George Boolos, "The Iterative Conception of Set."

A small technical point is relevant to Quine's comments on the axiom of foundation (*Set Theory and Its Logic*, p. 285). The standard formulation of the axiom rules out self-membership and hence rules out identifying individuals with their own unit classes, as Quine did in *Mathematical Logic*. However, it seems to me to bend the general idea that motivates the axiom at most very slightly to see the question whether an individual is to be distinguished from its unit class as one of convention. Quine's idea can be accommodated by reformulating the axiom to exempt sets that have themselves as sole members. If all such objects are assigned the rank 0, then the hierarchy of ranks is still obtained. One does have some slightly unattractive phenomena, such as that a set consisting of two individuals has rank 1, although every proper subset of it has rank 0.

[47]This is admitted by Martin (review of *Set Theory and Its Logic*, p. 112), though he criticizes Quine severely for his neglect of the intuitive basis of hierarchical conceptions of set.

[48]Boolos, "The Iterative Conception"; Shoenfield, *Mathematical Logic*, pp. 238-240; Shoenfield, "Axioms of Set Theory," esp. §2; Wang, *From Mathematics to Philosophy*, pp. 181-182.

On Frege's relation to these ideas, see my "Some Remarks on Frege's Conception of Extension." Frege differs from the now dominant tradition on the elements of set theory in his unequivocal claim that classes are constituted by predicative entities (concepts) and in his firm rejection of any constructive or genetic conception of objects.

Russell's theory of types would suggest, however, that the conception of extension can be developed in a natural way so that extensions form the sort of hierarchy that the standard conception calls for. I do not here want to enter into how Russell's own development of the theory of types bears on this question. In the genesis described by Quine in *The Roots of Reference*, class quantifiers are first introduced as substitutional quantifiers either of predicates or of nominalized predicates, so that the strictures of the theory of types are obeyed: the syntactical difference between predicates and singular terms induces a syntactical difference between (objectual) individual variables and (substitutional) class variables. But then one encounters obstacles with respect to impredicative definitions, just what led Russell to assume the axiom of reducibility. It is then another "leap" to take class quantification as objectual, and Quine interprets this as licensing impredicative classes. However, the simple type hierarchy is preserved if one remembers the substitutional origin of the class quantifiers, for at that stage one was operating with two different types of quantification.[49]

Still, Quine does not quite "see Russell's theory of types as dormant common sense awakened."[50] The reason is that mixing the levels of the type hierarchy is also a possible leap in the extension of the conceptual apparatus. It seems to Quine unnatural to forbid that and to permit, say, impredicativity. Thus, even though types have a certain genetic centrality, Quine maintains his earlier position that the theory of types is at least in part an "artifice" for blocking the paradoxes.[51]

I shall assume that Quine means here by the theory of types a theory of sets or classes according to which they form a well-founded hierarchy, and not more specifically a formal theory in which different levels of the hierarchy are indicated by different syntactical styles of variables. To discuss adequately whether Quine on this interpretation would be right, one would have to examine more deeply the intuitive basis of hierarchical conception.

Without such an examination, however, one can at least observe that the assumption of the priority to a set of its elements blocks the paradoxes in a simple and natural way, and that no nonhier-

---

In *The Roots of Reference*, Quine holds to the first at the very least in that talk of classes originates by nominalization of predicates. The genetic psychological framework, and in particular the derivation in it of the theory of types, represents a move on Quine's part away from his previous agreement with Frege on the second point.

[49]*The Roots of Reference*, pp. 120-121.
[50]Ibid., p. 121.
[51]Ibid., p. 122.

archical theory has been developed that is remotely comparable to standard set theory. Then one could reply to Quine's contention that "mixing the levels" of the type or rank hierarchy is no more impermissible than impredicativity by observing that accepting the latter and not the former has led to a powerful theory that has shown its workability through long experience.

In the sense that standard set theory should be regarded as an established theory, I would accept this argument. However, quite apart from the general fact that established theories can, by later turns in the history of science, be modified or abandoned,[52] it is not the last word. For there is still a temptation toward a non-hierarchical theory of extensions, not as a substitute for standard set theory but as a supplement to it. Even in the development of set theory, one is not content to rest with the view that all extensions are sets in the sense of the usual conception. For this reason, set theorists talk of "proper classes," classes that are not sets. Among such proper classes are the complements of sets, such as our class of all non-woodchucks.

It is not clear that the set theorist's talk of classes really commits him to *extensions* as a kind of object different from sets, not obeying the general restrictions the concept of set imposes. The actual use of classes by set theorists is predicative or at least in accord with the simple theory of types (with sets as individuals), so that it does not violate the basic idea of hierarchy. Moreover, there are ways of viewing the classes of set theory so that they are either not new objects at all or "really" or "ultimately" sets.[53]

Nonetheless, the interpretation of the concept of class in set theory is disputed, and there are those to whom a hierarchical theory of classes has seemed unsatisfying and incomplete.[54] Why this should be so can be brought out by recalling the connection between the notion of extension and the notion of *truth*. Briefly, the notions of class and attribute serve to generalize predicate places in a language, while the notions of truth and proposition serve to generalize sentence places. Since a predicate is just a sentence with an argument place, the notion of truth *of* (satisfaction) does the jobs of generalizing both for sentences and for predicates. Thus

[52]In order to avoid the imputation of inconsistency with what I have said above, I should say that in the case of pure mathematics, theories might be modified or abandoned on *mathematical* grounds—contradictions, unclarity in the conceptual foundations, or perhaps counterintuitive consequences of abstract axioms.

[53]See section IV of "Informal Axiomatization, Formalization, and the Concept of Truth" (Essay 3 of this volume); also "Sets and Classes" (Essay 8).

[54]For example, Donald A. Martin, "Sets versus Classes."

there is an elementary parallelism between predicative theories of classes and Tarski-style satisfaction theories.[55]

In the use of the notions of truth and truth of, what corresponds to the principle of hierarchy is, roughly, Tarski's hierarchy of language levels; truth at a given level is predicable only of statements at lower levels. Indeed, the theories of truth that arise directly from Tarski's ideas are predicative. But many intuitions concerning the possibilities of thought and expression, particularly in relation to natural language, suggest to some that the notion of truth is not essentially predicative or even hierarchical. Thus what one says using the word "true" seems to be true or false. More generally, talk in a natural language involves much that is of a "metalinguistic" character; in particular, it seems just an evident fact that the word "true" is applicable to statements made in English, including those that contain the word "true" itself. Even the sentences that express the Liar and other semantical paradoxes cannot be ruled out of the language on any obviously convincing grounds, even if in some stricter sense they fail to express propositions.

For this reason, the program of a "non-Tarskian" account of the concepts of truth in natural languages has seemed attractive to some.[56] A similarly non-Tarskian account of *satisfaction* would give rise to a theory of classes of a nonhierarchical sort. Paradoxes would have to be avoided by some other device. The one most often suggested would not be congenial to Quine, since it involves a deviation from standard logic: allowing truth-value gaps.[57] I do not myself have anything to contribute to such a program; indeed, elsewhere I have criticized it.[58] It does seem to be in the spirit of the point of view about *classes* that I have attributed to Quine. It may be that some of Quine's earlier ideas have relevance to it.

The motivations from natural language for a nonhierarchical semantics carry over to set theory, all the more since satisfaction and truth for formalized mathematical theories are themselves mathematical notions. Thus the carrying out of the program indicated above would have a tangible advantage for set theory: one would then have a strong, though still incomplete, theory of truth for set theory within set theory. However, there may be more spe-

[55]"Sets and Classes," section I.

[56]For discussions of several such views, see Robert L. Martin, ed., *The Paradox of the Liar*; cf. also Herzberger, "Dimensions of Truth."

[57]Thus van Fraassen, "Presupposition, Implication, and Self-Reference" and "Truth and Paradoxical Consequences," and Skyrms, "Return of the Liar" and "Notes on Quantification and Self-Reference;" also Donald A. Martin, "Sets versus Classes."

[58]"The Liar Paradox" (Essay 9 of this volume).

cific advantages; for example, the concept of proper class plays a role in motivating some strong axioms of infinity. I would say that no non-Tarskian account of satisfaction and truth, or nonhierarchical theory of classes, has up to now been put forth which is intuitively satisfying enough for it to compete with hierarchical accounts.[59] But in this area, where we try to deal with the outer limits of thought and expression, the last word has surely not been said.

[59]I learned of Saul Kripke's very elegant work on the concept of truth when this paper was substantially complete. (See "A Theory of Truth I, II" and "Outline of a Theory of Truth.") Kripke combines a hierarchical approach with truth-value gaps. It seems clear that theories of proper classes could be constructed on the basis of his ideas. The most obvious way of doing so would lead to a theory with a certain predicative character, which would apparently not realize the ideas of Martin (note 54 above). See now also Feferman, "Toward Useful Type-Free Theories, I."

PART THREE

Sets, Classes, and Truth

8

Sets and Classes

You are undoubtedly familiar with the fact that some systems of axiomatic set theory distinguish between two different kinds of objects, sets, and classes, such that among those classes which are not sets are classes whose admission as sets would give rise to familiar set-theoretic paradoxes. What may be less well known is that talk of classes and some sort of distinction between sets and classes is a quite standard part of the conceptual apparatus of set theory. It is used by workers in the field even when the *formal* theory they work with (for example, the familiar Zermelo-Fraenkel system ZF) does not distinguish sets and classes and avoids the paradoxes by an ostensibly different device: affirming the nonexistence of the 'paradoxical' sets.

I want to present some simple, even simpleminded, reasons why talk of classes which postulates more than can consistently be construed as *sets* is so readily adopted and to lay out some technical considerations, in essentials well-known to specialists, which imply that the difference between set theories with classes and set theories without classes is not so very great. Finally, I want to address the question whether the distinction between sets and classes is a fundamental ontological one. Two different answers to this question will be seen to follow from two different ways of looking at the language of set theory. The interest of the investigation lies in part in looking at the connection between set-theoretic and semantical concepts.

This paper was presented to a meeting of the Western Division of the American Philosophical Association in St. Louis, April 25, 1974, and published in that connection in *Noûs*, 8 (1974), 1-12. It is reprinted by permission of the editor of *Noûs*.

Some account is taken elsewhere in this volume, in essays 7 and 10, of the very interesting remarks of my commentators at the meeting, George Boolos and Donald A. Martin.

I

I shall not offer extensive documentation of the claim that classes are part of the standard apparatus of set theory. I shall concentrate on one recent textbook, Takeuti and Zaring's *Introduction to Axiomatic Set Theory*. A large part of this book is devoted to developing set theory in ZF. Yet in the informal discourse of the book the language of classes is used throughout, and quite a number of theorems that uncritical intuition might think of as theorems about sets are stated as theorems about classes. Greater generality is certainly supposed to result from this procedure, since all sets are to be classes and not all classes are sets.

The statement can of course not be made in the language of ZF, which has ϵ and = as sole predicates, and on the intended interpretation of which everything has the characteristic properties of a set, such as belonging to other sets and having a power set, some of which are contrary to the intuitive notion of a proper class.

By an artful use of metalanguage, Takeuti and Zaring combine rather precise presentation of formal proofs with making statements about classes. They combine the use of the notation of *virtual* classes (made familiar by Quine's *Set Theory and Its Logic*) with a rather sly procedure of expressing in the material mode general statements about provability in ZF (but see below). For a variable y and formula ϕ, $y \in \{x: \phi\}$ can be treated as just an abbreviation of $\phi(x/y)$ (the result of substituing y uniformly for x, if it is defined). We can also define:

(1) $\quad y = \{x : \phi\}$ for $\forall z(z \in y \equiv \phi(x/z))$
(2) $\quad \{x: \phi\} \in y$ for $\exists z(z = \{x: \phi\} \wedge z \in y)$.

Thus, one can use the abstract $\{x : \phi\}$ for any formula ϕ, even though for some formulae of ZF (such as $x \notin x$) $\exists z(z = \{x: \phi\})$ will be refutable.

Thus, one formulates in ZF statements which "mention" particular "classes," which may depend on set parameters. Logically, such abstracts behave typically for singular terms that may not denote: $\exists y \psi$ is implied by $\psi(y/\{x : \phi\})$ and $\exists y(y = \{x: \phi\})$ but not in general by $\psi(y/\{x: \phi\})$ alone.

Takeuti and Zaring present theorems as flat general statements about classes, but the "class variables" have been explained as syntactic variables ranging over terms (variables and abstracts; p. 12).

The strict intended interpretation is evidently about provability in ZF. For example, theorem 7.19 reads:

$$A \subseteq On \rightarrow Ord(\cup (A)),$$

which one would naively read as saying that if A is a class of ordinals, then its sum class is an ordinal. The proof lapses into the material mode, in effect taking the claim this way. But of course the strict reading is:

If A is a term, then $\ulcorner A \subseteq On \rightarrow Ord(\cup(A)) \urcorner$ is provable in ZF.

It would be easy enough to set up proofs in such a way that only such "strict" statements would occur (as is done for another system in Quine's *Mathematical Logic*).

I am not criticizing Takeuti and Zaring's method of exposition, which in fact succeeds in giving in an intuitive way quite precise instructions for the construction of formal derivations. But is seems clear that it is meant to convey to their readers something beyond what is expressed by the strict statements. In general terms, it is pretty obvious what this is. The authors of the book want their readers to learn not just facts about provability in a certain formal system, but also the facts about *sets* expressed by such formulae when understood in the natural way.

But then the actual statements made by the material-mode versions of such theorems as the above-cited theorem 7.19 are ambiguous. One reading would be to suppose that the authors actually believe in the existence of classes and intend these statements as literal generalizations about classes. Then a formalization of their informal language would involve quantification over classes.

The other reading would take such generalizations as assertions that every statement in the language of ZF of which it is an instance is *true* (or true of all values of its free variables). Then the formalization of the generalizations would require a truth or satisfaction predicate but nothing else not already renderable in ZF.

Now, I want to say that the introduction of the notion of class answers to a general need to generalize on predicate places in the language. In nonmathematical contexts, this need is met by talk of classes and relations and in addition by the use of terms of a more or less intensional character—property, concept, attribute, quality. Within set theory, of course, such a need is met to a considerable extent by sets: a general truth about all sets will have as instances truths about $\{x : \phi\}$ for every open sentence ϕ which *has* a set as its extension. But the inconsistency of the unrestricted comprehension axiom implies that such generalizations will fail to cover some predicates of any language with a fixed interpretation. Postulating classes allows one to state generalizations that cover *all* predicates in the-language of set theory (without classes), so long as their arguments are sets.

Now we can see the concept of *truth* as primarily a means for generalizing *sentence* places.[1] Such generalization is in natural languages mediated by nominalizing transformations, such that intuitively the resulting noun phrases stand for *propositions*, perhaps for facts.[2] The paradoxes of set theory exhibit limitations on our ability to generalize predicate places. Similarly, the liar paradox exhibits limitations on our ability to generalize sentence places. Just as the one limitation can be interpreted as the nonexistence in some cases of a set or class to be the extension of a given predicate, so the other can be viewed as the nonexistence in some cases of a proposition expressed by a given sentence.[3] If truth is instead taken as a predicate of sentences (and such contextual parameters as may be necessary to accommodate indexical expressions), there is still a close connection between the problems about truth and the possible nonexistence of extensions.[4]

Clearly, the notion of *truth of* (satisfaction) accomplishes for generalizing predicate places what truth *simpliciter* does for sentences. Just as for truth simply, the manner in which truth *of* can cover all the predicates of an interpreted language is by being outside the language; this does not necessarily require additional vocabulary but requires at least an interpretation that is in some way more comprehensive than the interpretation the satisfaction predicate itself formulates.

Since our hypothesis is that classes and satisfaction serve much the same purpose, we might expect some sort of equivalence between extensions of a theory by a satisfaction predicate and extensions of it by a theory of classes, where classes are understood as classes of objects in the domain of the given theory. In fact, it is easy to see that such an equivalence obtains.

Consider a first-order theory T with finitely many primitive pred-

---

[1] Cf. Quine, *Philosophy of Logic*, chaps. 1 and 3.
[2] We leave out of account the fact that some nominalized sentences are thought to denote events or actions.
[3] See "The Liar Paradox" (Essay 9 of this volume), section II.
[4] Ibid., section III.

icates (no names or functors), which can express its own syntax and the theory of finite sequences of its objects. T' and T'' will be two extensions of T: T' by (relatively) first-order definable classes and T'' by a satisfaction predicate for T. Then each can be translated into the other.

To obtain T', we add a new sort of variables $X, Y, \ldots$, and admit atomic formulae $x \in Y$. The quantificational logic becomes two-sorted. For any formula ϕ of T' without bound class variables or free Y, we add the axoim

(3) $\exists Y \forall x (x \in Y \equiv \phi)$.

Evidently, we can treat class abstracts as terms of the upper-case sort along with the variables. We can define $R = S$, for such terms R, S, as $\forall x (x \in R \equiv x \in S)$. If T contains set theory, we can provide for the identification of classes with sets by definitions parallel to (1) and (2) above.

The schema (3) can be replaced by finitely many axioms on the model of Group B of Gödel's axiomatization in *The Consistency of the Continuum Hypothesis*. If T has axiom schemata such as induction or replacment, they are implied in T' by single axioms. If T is ZF, the resulting theory is essentially the set theory *NB* developed by Gödel. These single axioms are generally not provable in T'.

T'' is obtained by adding a new primitive predicate $Sat\ (n, s)$, where n is to be (the Gödel number of) a formula and s a finite sequence of objects assigned to the free variables of n. We add as axioms inductive conditions for the satisfaction relation, for example:

(4) $n = \ulcorner F x_{i_1} \ldots x_{i_k} \urcorner \supset . Sat(n) \equiv F s_{i_1} \ldots s_{i_k}$,

for a primitive predicate F of T' and

(5) $n = \ulcorner \exists x_i \phi \urcorner \supset . Sat(n, s) \equiv \exists x\ Sat\ (\ulcorner \phi \urcorner, s_j^{i,x})$,

where $s_j^{i,x} = s_j$, for all $j \neq i$, and $s^{i,x} = x$.[5]

Then in T' we can define $Sat(n, s)$ by a formula saying that there

---

[5] $\ulcorner \phi \urcorner$ is the Gödel number of ϕ, or whatever else codes it in the universe of T; x_i is the ith variable in a standard enumeration; s_i, the ith term of s, is the object assigned to x_i. We can choose once and for all some object from the domain to assign to x_i if $i >$ length of s.

is a class Y satisfying the above conditions (with $Sat(m, s')$ replaced by $\langle m, s' \rangle \in Y$) for all formulae of logical complexity less than or equal to that of n, such that $\langle n, s \rangle \in Y$. This formula can be proved in T' to satisfy the above conditions and also to satisfy, for any formula ϕ of T the Tarski biconditional

$$(6) \quad Sat(\ulcorner \phi \urcorner, s) \equiv \phi(x_0 \ldots x_k / s_0 \ldots s_k),$$

where k is large enough so $x_{k+1} \ldots$ do not occur free in ϕ.

In T'' we can construe classes as pairs $\langle n, s \rangle$, where n is a formula of T (with, say, x_0 as distinguished free variable) and s is a sequence of objects assigned to the other free variables. $\exists Y(\ldots Y \ldots)$ would be translated as $\exists n \exists s(\ldots \{x : Sat(n, s^{0, x})\} \ldots)$, where the abstract is purely virtual and can be eliminated.

Parallel to this translation is the possiblility of interpreting the class quantifiers of T' by the generalization of substituional quantification I discussed in "A Plea for Substitutional Quantification,"[6] which I would now call relative substitutional quantification. An objectual interpretation of T' would normally involve a notion of satisfaction by sequences of objects assigned to both the lower-case and the upper-case variables; in a relative substitutional interpretation, the notion is satisfaction by sequences for the lower-case variables; terms are substituted for the upper-case variables, which however may contain lower-case parameters. Thus, we have a predicate $Sat_1(n, s)$ with the same first-order clauses as in the axioms of T'', and in addition a clause for class quantification which would read as follows:

$$(7) \quad n = \ulcorner \exists X \phi \urcorner \supset \; . \; Sat_1(n, s) \equiv \exists \ulcorner \psi \urcorner \exists s'$$
$$(\psi \text{ is a formula of } T \wedge \phi \; (X / \{x : \psi\})$$
$$\text{is defined} \wedge \forall j(x_j \text{ occurs free in } \phi \supset s_j = s'_j) \wedge$$
$$Sat_1 \; [\phi(X / \{x : \psi\}), \; s']),$$

where ψ may contain free lower-case variables not free in ϕ, and s' will have to assign an object to such a variable although s need not.[7]

---

[6]Essay 2 of this volume.

[7]In *The Roots of Reference*, pp. 106-110, Quine discusses a different relative substitutional interpretation of class quantification, in which the term $\{x : \psi\}$ of (7) is not allowed to contain free variables additional to those of ϕ. The anomalies Quine points out for his interpretation do not hold for ours. I am indebted to Quine for correspondence clarifying the difference of our conceptions. [See also note 10 to Essay 2 above.]

There are two remarks to be made about these relations. First, most of you know that NB is a conservative extension of ZF; that is, theorems of NB in the language of ZF are theorems of ZF. A detour through classes cannot serve to prove formulae only about *sets* that cannot be proved without classes.

This can be proved finitistically by standard cut-elimination methods, which apply also in our more general setting. For this reason, set theorists do not view quantifying over classes as a substantial addition to their commitments.

Second, a theory of truth for a given system is thought to be a stronger theory than the system itself. But the translatability of T'' into T' shows that T'' is a conservative extension of T. This is so because T'' is a weak theory of satisfaction. If T has axiom *schemata* (for example, replacement in ZF), instances of the schemata for formulæ containing *Sat* do not in general become provable in T''. Thus, if T is ZF or elementary number theory Z, the inductions needed to prove that every closed theorem is true (and thus that T is consistent) are not available in T''. Even quite elementary laws of truth and satisfaction are not provable in T''.[8]

A more natural extension of ZF by a satisfaction predicate would be one in which formulæ containing *Sat* would be allowed in axioms of replacement. Then induction on formulæ containing *Sat* would be derivable. The same mutual translatability would exist between this theory and the theory NB^+ which is like NB except that replacement is allowed for formulæ having bound class variables.[9]

In spite of this equivalence, there is a sense in which the theory of classes is more general than the theory of satisfaction. The axioms of T' prescribe that the classes that there are are closed under first-order definition of the language of T, but they place no definite limit on what classes of elements of the domain of T there might be. Even if we think of classes as always the extension of predicates, the axioms of T would allow them to be extensions of predicates in any language, whether we can now specify it or not, provided they are closed under first-order logical operations. Moreover, if the domain of T is a set, the class variables can be taken to range over *all* subsets of the domain. (On this interpretation, we could allow arbitrary formulæ, even with bound class variables, as instances of (3). Such a theory is called an impredicative class theory.)

The usual satisfaction theory (T'' or the extension mentioned two

[8]"Informal Axiomatization, Formalization, and the Concept of Truth" (Essay 3 of this volume), section I.

[9]Ibid., section IV.

paragraphs back) lacks this generality because it is based on the syntax of the language of *T*. Since it deals with predicates with parameters, it does allow as many "predicates" as there are objects in the domain of *T*. We could imagine throwing into the language covered by the syntax a whole lot of predicates as yet uninterpreted, so that to interpretations of these predicates would correspond models of the satisfaction theory which allowed for construing a more comprehensive domain of classes. However, I see no way of reaching an equivalent of the impredicative class theory without rather obvious cheating, such as postulating a one-place predicate for "every" class. The satisfaction theory seems inherently predicative in a way in which the class theory is not.

II

The elementary considerations presented above can be summed up as the thesis that classes are extensions of predicates. Then one form of the question whether there is a basic difference between sets and classes is whether *sets* are essentially extensions of predicates. This is not a very precise question, and it turns as much on the conception of extension as on that of set. The above discussion, particularly the appeal to truth and satisfaction, should have made clear that it is at least not *obvious* that *extension* and *set* are just one concept.

The question should be divided into two. First, can we reduce sets to extensions? Second, can we reduce extensions to sets? I would answer the first question in the negative. About the second I am not so sure. But I shall at least argue that in one way of taking it, the question is equivalent to a difficult question about quantifying over *all* sets.

The argument against reducing sets to extensions is that there is not, so far as I can see, any convincing reason for not restricting the notion of extension so that it remains predicative, unless one begs the question by appealing to set theory. Put another way, no considerations about generalizing predicate positions in a language could provide an independent justification of impredicative second-order reasoning.

One can generalize on predicates by first effecting a nominalizing transformation such as class abstraction, or directly as in second-order logic, or by semantic ascent as in the above satisfaction theory. But it seems clear that except for intuitions derived from *other* types of explanation of the concept of set (see below), we do not have the independent understanding of what predicates or abstracts denote, or what class or second-order variables range over. It follows that "all extensions. . ." will, unless set-theoretic notions are imported, only mean "the extensions of *all possible predicates.*" And it seems evident that the "totality" of possible predicates is irremediably potential, and more radically so than the natural numbers are on the intuitionistic or other constructive conceptions, since in the former case there is no rule which "generates" *all* predicates on an infinite domain, even modulo extensional equivalence.

When one thinks of "all predicates," one naturally thinks of a language for which the vocabulary and its interpretation are specified in advance, as with such formalized languages as number theory. Then "all classes" will be all *definable* classes (relative to parameters). But the class quantifier immediately yields new predicates, and a diagonal argument immediately shows that it yields new extensions as well. This would tempt us to introduce new class variables for the extensions of the predicates of the expanded language. This procedure can be iterated, yielding the familiar ramified hierarchy.

Talk in a nonramified way of all classes will on this view either presuppose an arbitrary stopping place in some hierarchy such as the usual ramified hierarchy, or, more interestingly, mean all classes that *might* be defined, independently of any specification of the means. To what extent can we regard such possibilities as determinate at any given point? It seems that we have no conception of a totality of all possible languages and interpretations on the basis of which we might claim such determinacy. The ramified hierarchy can in certain special cases be shown to "close out" at some transfinite point to yield no new classes. But to show this—and even to show that the ordinals involved exist—requires set-theoretic reasoning.

I am not here rejecting or even criticizing set theory. My thesis is rather that the two assumptions that get real set theory off the ground—the extensional definiteness of quantification over all subsets of a given set, and the existence of the power set—are not derived from considerations about predication. From Cantor on, set theorists have thought of sets as in some way constituted by their elements.[10] Although Cantor spoke of the elements of a set

---

[10] Thus Cantor writes, "Every set of well-distinguished things can be viewed as a *unitary thing for itself*, in which those things are parts (*Bestandteile*) or constitutive elements" (*GA*, p. 379).

as connected by a law (*GA*, p. 204), this aspect of the idea of a set has tended to disappear as the force of indenumerability has come to be appreciated.[11]

Bernays in "On Platonism," seems to me to have been right in holding that the conception of an arbitrary subset or arbitrary function as it appears in set-theoretic mathematics is of a *combinatorial* character. We have an intuitive grasp of the subsets of a *finite* collection because, by a mode of conception essentially the same as the arithmetical, we can consider in succession all the possible ways of combining them into sets. It is this which is generalized to the infinite case, although the enumerability of all possible sets or functions is lost:

> Passing to the infinite case, we imagine functions engendered by an infinity of independent determinations which assign to each integer an integer, and we reason about the totality of these functions. In the same way, one views a set of integers as the result of infinitely many independent acts of deciding for each number whether it should be included. (p.276.)

The metaphor of choice is not what is essential here, rather the independence of the "determinations" and the absence of any specific role for language.

What now of the possibility of reducing extensions to sets? It might seem that the acceptability of the class theory, together with Russell's paradox, proves the impossibility of such a reduction, for does it not follow that there are classes that cannot coincide with sets in the straightforward sense? However, we are of course assuming an interpretation of the language of set theory according to which the quantifiers range over *all* sets, or at least sets of arbitrarily high rank. What exactly is it to talk of *all* sets?

It is a general maxim in set theory that any set theory which we can formulate can plausibly be extended by assuming that there is a *set* that is a (standard) model of it. If by "set theory" we mean an axiomatizable theory, stronger principles are considered, such as

---

[11]Frege's logic, for example as presented in the *Grundgesetze*, can of course be viewed as an attempt to construe sets as extensions. As it stands it fails because of Russell's paradox. But, of course, any finite part of the theory of types can be obtained without axiom V by introducing sufficiently high-level concepts and, where necessary, an axiom of infinity. But I do not see how to reconcile the gulf between different types of concepts postulated by his theory with the possibility of the cumulation needed to obtain transfinite types. Moreover, I do not see that Frege justified impredicative reasoning about concepts.

that there is a set of which the universe is an elementary extension. The first maxim already implies that we could not produce a discourse in the language of set theory such that it could be interpreted as true if and only if the quantifiers range over absolutely all sets. Is it evident that an interpretation of such a discourse that makes the quantifiers range over a set would be incorrect? In an interesting case, it would have to be a set undreamed of in the philosophy of the propounder of the discourse, since on the proposed interpretation the sentence saying that there is such a set is false. But if you know no set theory beyond ZF, on what grounds can I say that in your language "there exists an inaccessible cardinal" is true? Perhaps I can explain to you what an inaccessible cardinal is, and even make it plausible to you that there might be such a thing. Have I persuaded you of the possible truth of something left open by your theory as it was, or have I rather changed your conception of set? We all know of arguments for there being "no fact of the matter" about such a question. However that may be, it is hard to see how your understanding of the quantifiers of set theory could not at least be taken to be *vague*, so that reading them as ranging over sets of rank less than the first strong inaccessible would be an otherwise correct way of making them more precise.

But then we could view the addition of classes to a set theory in the same spirit—particularly since there seems to be no intrinsic objection to the postulation of stronger and stronger properties of classes, so that they are conceived as indistinguishable from another layer of sets. This process would be that of gradually imposing on our discourse an interpretation which makes the original universe a set. Even before the introduction of classes, the applicaton of classical logic to statements about all sets could be taken as a first step in this direction,.

In connection with the liar paradox (see Essay 9), I have insisted on the priority of the use of language over semantic interpretation, and on the "external" or "reflective" character of such semantic steps as assigning a universe to the variables of a discourse. From this point of view, if I take your quantifiers to range over "all" sets, this may only show (from a "higher" perspective) my lack of a more comprehensive conception of set than yours. But then it seems that a perspective is always possible according to which your classes are really sets.

This way of looking at set theory would make the language of set theory systematically ambiguous; no set-like totality of possible interpretations could capture the range of this ambiguity. Uncomfortable as it is, such a conception seems to be required in any case to understand the semantical paradoxes. How or whether we can by being aware of this ambiguity transcent it, is a question I do not know how to answer.[12]

[12]Cantor, in 1899, distinguished sets and what he called "inconsistent multiplicities" (*inkonsistente Vielheiten*) by appealing to the irreducibly potential character of the "totality" of sets; the latter are such that a contradiction results from supposing a *Zusammensein* of all their elements (*GA*, p. 443). I would agree with Cantor's implicit view that it is the unbounded, "absolutely infinite" character of the possibilities of set formation that makes the existence of the sets involved in the paradoxes impossible; it would be compatible with what he says to treat "multiplicities" by a predicative account of classes. But the exact relation between such potentiality and "systematic ambiguity" remains to be explored.

It should be remarked that Cantor applied his distinction to Russell's paradox in a letter to Jourdain in 1904. See Grattan-Guinness, "The correspondence," p. 119.

[These matters are discussed further in "What Is the Iterative Conception of Set?" (Essay 10 of this volume).]

9

The Liar Paradox

Why is it that today, more than sixty years after *Principia Mathe-matica* and nearly forty years after the first publication of Tarski's *Wahrheitsbegriff*, the liar paradox is still discussed as if it were an open problem? What is the difference between semantical para-doxes and the paradoxes of set theory that accounts for the fact that the paradoxes of set theory are nowadays treated as solved? The general outline of an answer to this question seems to me obvious. The problem of the paradoxes of set theory has been seen as a problem of constructing a way of talking about sets that is adequate for certain *theories*, in particular those parts of mathe-matics which have come to be formulated in set-theoretical terms and rely on set-theoretical methods. The liar paradox tends to be seen as much more a problem of analyzing a *given* conceptual scheme, embodied in one way or another in natural languages. But the use of the word 'true' and its translations into other languages seems to embody just the features that are responsible for the

From *Journal of Philosophical Logic*, 3 (1974), 381-412. Copyright © 1974 by D. Reidel Publishing Company, Dordrecht, Holland. Reprinted by permission of the editor and D. Reidel Publishing Company. The Postscript was written for this vol-ume. A condensed version of the Postscript appears with the reprint of the paper in Robert L. Martin, ed., *Recent Essays on Truth and the Liar Paradox* (Oxford Uni-versity Press, 1983).

I am indebted to Hao Wang and the Rockefeller University for a visiting ap-pointment which provided the freedom to do much of the work on this paper. I presented an early version in January 1972 to John Wallace's seminar at Rockefeller and a version closer to the present one in lectures in March 1973 at the University of California, Berkeley and Los Angeles, and Stanford University. Among those from whose comments or discussion the paper has benefited are Rogers Albritton, Tyler Burge, Donald Davidson, Gilbert Harman, James Higginbotham, David Kap-lan, George Myro, Thomas Nagel, and John Wallace. I owe much to the writings of Hans Herzberger. I am grateful to Bas van Fraassen and to the referee for pointing out misunderstandings and questionable interpretations of van Fraassen's work.

paradox, in particular the impredicativity that the paradox most immediately turns on: statements involving the word 'true' are among those that are said to be true or false. For this reason the hierarchical approach that underlies accepted set theories and which was applied to the semantical paradoxes by Russell, Ramsey, and Tarski, has been rejected in much recent literature on the semantical paradoxes.

I want to argue, on the contrary, that the semantical paradoxes should be treated in a way that stresses their analogies with the paradoxes of set theory. What is needed for such a treatment is to make plausible an interpretation of natural language in terms of some hierarchy such as Tarski's language levels. Part of the key to this, in my view, is to look much more critically than is usually done at the interpretation of *quantifiers* in natural languages and at their interaction with semantical expressions, and at indirect-discourse expressions such as 'say' and 'mean.' When this is done, the analogy of natural languages and the "semanatically closed languages" shown inconsistent by Tarski's methods is much less persuasive. The morals I shall draw about the treatment of quantifiers in natural languages have a wider application than just to the paradoxes.

I

An assumption that leads naturally to the conclusion that natural languages are semantically closed in Tarski's sense is that there is a quite definite concept of truth for a natural language that is expressed by a word of that language that we can translate as 'true'. Then it seems that natural languages can express their own concepts of truth. At any rate, this latter principle seems to underlie a considerable recent literature on the liar paradox where either some form of nonstandard semantics and logic is applied, or some weakening is made of the relation supposed by Tarski between ' "p" is true' and 'p', or both. Before I present my own views I should like to discuss critically some of the ideas in this literature.

The most developed and intuitively appealing theory of this kind is that of Bas van Fraassen.[1] I shall discuss this in some detail and then remark briefly on another due to Brian Skyrms.[2] However, the

[1]"Presupposition, Implication, and Self-Reference," "Truth and Paradoxical Consequences," "Rejoinder."
[2]"Return of the Liar," "Notes on Quantification and Self-Reference."

most subtle and interesting of these approaches is probably that of Hans Herzberger,[3] but his views are not yet available in a form complete enough for detailed discussion.

An idea that arises very naturally about the semantical paradoxes is that the sentences involved are semantically deficient, or at least become so in the contexts in which they are used. But since this deficiency comes to light by *arguments* in which they occur, it does not seem that we meet the difficulty by declaring them outrightly meaningless and banishing them from the language. It seems natural to regard a statement that says of itself that it is false as neither true nor false. Van Fraassen gives a formal semantics in which this possibility arises, and from certain assumptions about paradoxical sentences it *follows* that they are neither true nor false.

He considers languages based on the usual propositional and predicate logic, where the semantics allows truth-value gaps but in which the notions of logical validity and implication are the usual classical notions. The actual manner in which this is achieved need not concern us,[4] but the essence of it is that a sentence will be true or false if it is such under *all* of a certain class of interpretations in the usual classical sense. Evidently the valid formulae of classical logic will then come out true.

The typical situation in which the liar paradox can arise in formal semantics is as follows: Sentences A, B, . . .of a certain object language have in the object language *standard names*, say a, b, . . .and can also be denoted by other expressions, say α, β, . . .Suppose 'T' is a predicate of the object language that purports to express truth in that language. Then α may denote the sentence $\sim T\alpha$.[5] Let that sentence be A and let a be its standard name. Then the Tarski biconditional

(1) $Ta \equiv\, \sim T\alpha$

---

[3]"The Truth-Conditional Consistency of Natural Languages," "Paradoxes of Grounding in Semantics," "Truth and Modality in Semantically Closed Languages."

[4]See "Presupposition," pp. 140-142.

[5]In the case of formalisms containing number theory, the standard names will be numerals for Gödel numbers of the expressions involved. The other terms α, β, . . .can in principle be any closed primitive recursive terms, but in actual paradoxical cases they will be terms constructed by substitution functions. $a = \alpha$ will then be a provable numerical formula.

Another frequently discussed type of case is that in which the standard names are quotations, and α, β, . . .are definite descriptions such as "the sentence written on the blackboard in. . . ." The relevant proposition $a = \alpha$ can be verified by observation.

Note that in this section I assume that the primary truth-vehicles are *sentences*; otherwise 'T' should be read as 'expresses a true proposition'.

should be true, and since $a = \alpha$ is true we obtain the contradiction $T\alpha \equiv {\sim} T\alpha$ or $Ta \equiv {\sim} Ta$.

However, it seems that if ${\sim} T\alpha$ can be neither true nor false then (1) need not always be true. In the situation we have sketched it will be *false* in van Fraassen's semantics since its negation is derivable from the truth $a = \alpha$. This raises a question what relation we can suppose between a sentence A and Ta, given that 'T' is to express truth in the language. Van Fraassen's answer is that this relation is *co-necessitation*: if A is true, Ta is true, and vice versa.

It would be natural to say that if A is not true then Ta is *false*: Ta says of A that it is true, and it isn't. But in our example this does not work: if Ta is false, ${\sim} Ta$ is true, but then so is ${\sim} T\alpha$. But that is A, so if A is not true, it is true. But if ${\sim} T\alpha$ is true, Ta is true by the above, and ${\sim} Ta$ is true by substitutivity of identity. So we have a contradiction. In this case (which turns out to be an instance of what van Fraassen calls the Strengthened Liar) we have a case of a sentence A which is not true, but the sentence that says it is true is not false, but rather neither true nor false.

One might find in this result the failure of 'T' to express the concept of truth as it is used in the metalanguage. However, it is in the first instance a matter not of expressing truth but of expressing negation. ${\sim} Tx$ is true of x if Tx is false of x, and false of x if Tx is true of x, but if Tx is neither true nor false of x, ${\sim} Tx$ may also be neither true nor false of x. In order to express 'not true' as we mean it in the *metalanguage*, it seems we need a connective (say '$\neg$') such that $\neg A$ would be *true* whenever A is neither true nor false. But putting this connective plus van Fraassen's assumptions about truth into the object language produces inconsistency.

This limitation on the means of expression of van Fraassen's language has been offered as a criticism.[6] From the point of view of the presumed motivation of the approach it seems serious: it reveals a divergence in sense between the predicate '${\sim} T$' of the object language and the phrase 'not true' of the metalanguage. We have a sentence A such that (it is true to say in the metalanguage that) it is not true, but Ta, the sentence that is supposed to say in the object language that A is true, is not false, and ${\sim} Ta$ is not true. We were supposing that it was just to prevent such divergences that the approach was put forward in the first place: the idea was that for a natural language there is no difference between what one can say *in* the language and what one cay say *about* the language

---

[6]Herzberger, "Truth and Modality," pp. 29-30.

"from outside." But the divergence can so far be blamed on negation rather than on '*T*' and 'true'.

A sharper divergence between '*T*' and 'true' does occur if we put into the *object language* a simple principle about truth that van Fraassen accepts. He regards the schema

(2) $Ta \supset A$

as harmless.[7] In our liar situation (A is $\sim T\alpha$ and $a = \alpha$ holds) we have

$$Ta \supset \sim T\alpha$$

which, with $a = \alpha$, implies $\sim T\alpha$ and $\sim Ta$, both of which are not true.

Perhaps this just shows that (2) is not so harmless after all, as is clear from the literature on reflection principles.[8] Van Fraassen's endorsement of (2) is a passing remark, and it is not central to his approach. Other premises concerning the expression of the theory of truth in the object language lead to a similar result.[9] But it seems that van Fraassen can escape the proof of untrue sentences by admitting a further incompleteness of expression closely related to that concerning negation, but also more elementary than he admits in "Truth and Paradoxical Consequences."

[7]"Truth and Paradoxical Consequences," p. 15n.

[8]Formalisms containing number theory have a formula Bx which expresses 'x in the Gödel number of a provable formula'. The simple reflection schema $Ba \supset A$ expresses in the object language the soundness of the formalism. If α denotes $\sim B\alpha$ then $\sim B\alpha$ is Gödel's undecidable formula. The above argument with '*T*' replaced by '*B*' shows that if $Ba \supset \sim B\alpha$ is provable, then $\sim B\alpha$ is provable, contrary to Gödel's theorem if the formalism is consistent. Hence there is an instance of the reflection schema that is not provable.

[9]In an earlier version of this paper I showed that $\sim T\alpha$ could be deduced from a set of premises, all of which were, I believe, unexceptionable for van Fraassen except

(i) $Ta \supset T(Ta)$

which was to express the fact that A necessitates Ta. In a letter of October 16, 1973, van Fraassen repudiated (i) and remarked that it is not justified by the semantics of "Truth and Paradoxical Consequences," pp. 20-21. It seems that if the necessitation relation N_0 is expressed in the object language then

(ii) $aN_0b \cdot Ta \cdot \supset Tb$

should hold, as should $aN_0(Ta)$. But then (i) follows. It seems that either N_0 does not go into the object language, or '$\supset$' fails at some point to express the conditional.

Thus it seems clear that van Fraassen's approach provides no escape from the problem left by Tarski's work on the paradoxes, of making sense for natural languages of the limitations of the means of expression of formalized theories. Van Fraassen's original paper, "Presupposition," attempts to deal with the liar in complete abstraction from this.[10] But he writes, "The language has, at any stage of its evolution, a certain incompleteness of means of expression, any given aspect of which may be remedied in its further evolution."[11] It seems to me that once one makes this admission, one might as well admit at the outset that a natural language does not at a given "stage of its evolution" express its own concept of truth. Then the way is open for a treatment of the paradoxes by semantical methods that are simpler and more standard than van Fraassen's, as we will show below.

Skyrms's proposal is that we restrict the substitutivity of identity so that if $F\alpha$ is neither true nor false, $F\alpha \cdot \alpha = \beta \cdot \supset F\beta$ need not be true. In our paradoxical case, this enables him to reject the inference from $T\alpha$ and $\alpha = a$ to Ta and thus to affirm $\sim Ta$ ($\sim T\alpha$ is not true) and yet reject $\sim T\alpha$ itself. He is able on this basis to construct a formal system in which a truth-predicate can be applied to sentences containing it, and the restriction on the substitutivity of identity insures consistency.

The claim motivating the restriction is that Ta may be semantically all right while $T\alpha$ is defective, in that $\sim T\alpha$ is self-referential, and indeed viciously so, while neither Ta nor $\sim Ta$ is. This seems to me to attribute too much importance to self-reference: $\sim Ta$ and $\sim T\alpha$ deny the same predicate of the same object, and in neither is there anything problematic about the manner in which the object (namely the sentence $\sim T\alpha$) is referred to, nor is there any difficulty in identifying the denotation of α (see note 5).

Evidently the proposed restriction has the effect that 'T' is no longer functioning precisely as a predicate. (There is no question of anything turning on the denotation of a or α either changing or being different "in some other possible world.") On Skyrms's

---

[10]However, in the letter mentioned in note 9 van Fraassen suggests the contrary. Perhaps he thought of the truth-value gaps as themselves such an incompleteness.

I should remark that none of the incompletenesses that arise in recent literature is compatible with the idea that natural language is 'universal' if this is to be close to semantic closure in Tarski's sense. Cf. Robert L. Martin, "Are Natural Languages Universal?"

[11]"Rejoinder," p. 61.

hypothesis, the truth-value of *Ta* does not depend just on the denotation of *a* and the meaning of '*T*'.[12]

II

In order to explain my own approach to the Liar, I want now to consider two closely related liar sentences which can arise in a language in which we can talk of a sentence *expressing a proposition*, of a proposition's being *true*, and in which we can name sentences. Nothing turns on the particular manner in which a sentence might come to contain a singular term denoting *itself*, except that to speak of a sentence expressing a proposition, without reference to a context, implies that it does not contain demonstratives, so that devices such as 'this sentence' are excluded. But we can use Gödel numbering or other context-invariant forms of reference.

In the general case where indexical expressions are admitted, the predicate 'expresses' clearly needs to have an additional argument place or places for the features of the context that determine what proposition is expressed in that context. In my view we do not have an adequate account for a whole natural language of what these argument places are. The most complete account I know of is Lewis's "General Semantics."

Consider for example:

(1)   The sentence written in the upper left-hand corner of the blackboard in Room 913-D South Laboratory, The Rockefeller University, at 3:15 P.M. on December 16, 1971, expresses a false proposition.

(2)   The sentence written in the upper right-hand corner of the blackboard in Room 913-D South Laboratory, the Rockefeller University, at 3:15 P.M. on December 16, 1971, does not express a true proposition.

Let us suppose that at the time mentioned the blackboard mentioned contained exactly two sentences: (1) in the upper left-hand corner and (2) in the upper right-hand corner. Moreover let '*A*' abbreviate the quotation of (1) and '*B*' that of (2). Then we have

(3)   *A* = the sentence written in the upper left-hand corner

[12]Cf. Fitch, "Comments and a Suggestion."

of the blackboard in Room 913-D South Laboratory, The Rockefeller University, at 3:15 P.M. on December 16, 1971. B = the sentence written in the upper right-hand corner of the blackboard in Room 913-D South Laboratory, The Rockefeller University, at 3:15 P.M. on December 16, 1971.

Now just what should we assume about truth? One possibility would be to assume instances of the schema ' "p" expresses a true proposition $\equiv p$', and perhaps ' "p" expresses a false proposition $\equiv \sim p$'. But then if we have ordinary propositional logic, we can infer

(4)   'p' expresses a true proposition ∨ 'p' expresses a false proposition,

and therefore

'p' expresses a proposition.

But we want to allow for the possibility that some sentences do not express propositions. A more suitable schema might be the following

(5) (x) (x is a proposition · 'p' expresses x.
        $\supset \cdot x$ is true $\equiv p$).

We shall assume that propositions are bivalent; hence we can render 'false' as 'not true', and (5) implies

(6) (x) (x is a proposition · 'p' expresses x.
        $\supset \cdot x$ is false $\equiv \sim p$).

Suppose now x is a proposition and A expresses x. By (5)

x is true $\equiv$ the sentence written in the upper left-hand corner of the blackboard in Room 913-D South Laboratory, The Rockefeller University, at 3:15 P.M. on December 16, 1971, expresses a false proposition.

and then by (3)

(7) x is true $\equiv A$ expresses a false proposition.

Suppose x is not true. Then by existential generalization:

$$(\exists x)\ (x \text{ is a proposition} \cdot \sim (x \text{ is true}) \cdot A \text{ expresses } x),$$

i.e., A expresses a false proposition. By (7), x is true. Since x is arbitrary,

(8) $(x)\ (x \text{ is a proposition} \cdot A \text{ expresses } x \cdot \supset x \text{ is true})$

and hence

(9) $\sim (\exists x)\ (x \text{ is a proposition} \cdot \sim (x \text{ is true}) \cdot A \text{ expresses } x).$

Now suppose y is a proposition and A expresses y. By (8), y is true. But (7) with y for x follows. But by (9), this is a contradiction. Hence there is no proposition y expressed by A. Hence (1) does not express a proposition.

Similarly, (2) does not express a proposition. But a contradiction is avoided because in such cases the condition for the Tarski biconditional in (5) is always false.

We might imagine that in our object language we can form descriptions of the form 'the proposition that p'. It would then be plausible to assume all sentences of the form

(10)   'p' expresses the proposition that p.

This would lead us into contradiction, at least if we use a logic according to which 'the proposition that p' must denote something. But to suppose it does is to suppose that 'p' expresses a proposition, which we do not want always to suppose. It is a matter of indifference whether we assume (10) with 'the proposition that p' a possibly nondesignating singular term, or admit (10) only where 'p' does express a proposition.

In this connection it should be observed that

(11)   the proposition that p is true $\equiv p$

follows from (5), (10), and '$(\exists x)\ ('p'$ expresses $x)$'. But there is no reason to accept (11) if 'p' does not express a proposition.

A difficulty that arises immediately is the following: (1) says of a certain sentence which turns out to be (1) itself, that it expresses a false proposition. We have shown that (1) expresses no proposition. But then (1) seems to say something false. Are we not forced to say that (1) expresses a false proposition after all?

Similarly, (2) says of itself that it does not express a true proposition; since it does not express any proposition, in particular it does not express a true one. Hence it seems to say something true. Must we then say that (2) expresses a true proposition?

In either case we shall be landed in a contradiction. A simple observation that would avoid this is as follows: The quantifiers in our object language could be interpreted as ranging over a certain universe of discourse U. Then a sentence such as

$$(\exists x) \ (x \text{ is a proposition} \cdot A \text{ expresses } x)$$

is true just in case U contains a proposition expressed by A, i.e., by (1). But what reason do we have to conclude from the fact that we have made sense of (1) and even determined its truth-value that it expresses a proposition which lies *in the universe U*?

We can motivate more directly the exclusion from U of propositions expressed by (1) or (2) by observing that they contain quantifiers ranging over propositions, and hence if (1) or (2) is to express a proposition in U, it must be defined impredicatively, i.e., in terms of a totality of which it is itself a member. But although objections to impredicative characterizations of propositions come readily to hand, it is not obvious that the case against them is decisive. We might suppose that propositions are extramental entities whose existence does not depend on the possibility of their being expressed by sentences. In the context of analysis and set theory, such realism has been taken to license some impredicative definitions.

The above version of the liar paradox is stronger than an argument based on impredicativity. It shows that (5), a highly plausible assumption about the concepts of expression and truth, implies the existence of well-formed sentences that do not express propositions in the range of the bound variables. (5) is apparently quite compatible with realism about propositions.

In the language we are discussing, we can define a Tarskian truth predicate for sentences: $T(y)$, say, is defined as

$$(12) \qquad (\exists x) \ (x \text{ is a proposition} \cdot y \text{ expresses } x \cdot x \text{ is true})$$

and then (5) implies

$$(13) \qquad (\exists x) \ (x \text{ is a proposition} \cdot \text{'}p\text{' expresses } x)$$
$$\supset \cdot T(\text{'}p\text{'}) \equiv p$$

for any sentence '*p*'. But on pain of contradiction, we will not always be able to prove the consequent of (13). Then our language will not be 'semantically closed' in Tarski's sense, and the Tarskian truth-predicate covers only a part of the language.

Interpreting (1) and (2) in terms of the usual semantics for predicate logic, in which the quantifiers range over a certain set U, makes it possible to understand them and to determine truth-values for them: (1) is false and (2) is true. We can even talk of the propositions they express. But these are not in U.

Taking (1) and (2) as statements in English, it is tempting to step outside this semantics and to say that the quantifiers do not range over some definite set U but over *absolutely everything*. But then our argument shows that they do not express propositions *at all*. Yet they seem to make perfectly good sense; we still have an argument for the falsity of (1) and the truth of (2). Can we accept the idea of a perfectly reasonable sentence that does not express a proposition? And what can we then be saying when we say that such a sentence is *true*, if not that it expresses a true proposition?

What I want to say about this is that there is a close analogy between this situation and the paradoxes of set theory. That a is the extension of a predicate '*Fx*' is usually taken to mean that the condition

$$(14) \qquad (x)\,(x \in a \equiv Fx)$$

holds. Analogously to (5) we might have

$$(15) \qquad (y)\,(y \text{ is extension of } `Fx' \supset (x)\,(x \in y \equiv Fx)\,).$$

Taking '*Fx*' as '$x \notin x$' we can deduce

$$\sim (\exists y)\,(x)\,(x \in y \equiv x \notin x)$$

and hence

$$(16) \qquad \sim (\exists y)\,(y \text{ is extension of } `x \notin x').$$

The same alternatives for the interpretation of the quantifiers in (14)–(16) present themselves as in the case of the liar sentences. If the variables range over a universe U, (16) must be interpreted to mean that '$x \notin x$,' taken relative to U, lacks an extension *in* U. But in our metalanguage we can probably show the existence of the relevant set (e.g., by Zermelo's Aussonderungsaxiom) and there-fore prove that it is not in U.

This has the consequence that U falls short of containing "all" sets. Although such an interpretation of set theory is often used in foundational studies, when one considers models of set theory, both set theorists and laymen entertain the idea of set theory as a theory about all sets in an absolute sense. This is (for the usual set theories) incompatible with interpreting the quantifiers of set-theoretic state-ments as ranging over a set.

If we take this interpretation seriously then we must say that the set of *all* things that are not members of themselves does not exist at all.[13] However, this does not detract in any way from the mean-ingfulness of the *predicate* '$x \notin x$'.

The analogy I wish to draw should now be clear: A language may contain perfectly meaningful predicates such that, in a given theory formulated in that language, they cannot be said to have extensions. One would readily concede that the same must be said about the correlations of intensions to such predicates: from within, they do not have attributes as their intensions. What the Liar shows in the first instance is that the same situation arises for *sentences*: A theory expressed in a given language cannot always correlate to a sentence a proposition as its intension, even though the sentence is well-formed and may even be provable.[14] The same should be said about their "extensions": their truth-values.

That some liar sentences are not semantically deficient in the sense that assertive utterance of them must fail to come off, or in the sense that they cannot figure in proofs, or in the sense that it is impossible to make sense of them "from outside" is obscured by the fact that the most typical such sentences, such as 'What I am now saying is false' are naturally interpreted so as to contain a

---

[13]Set theorists usually consider the set of all *sets* that are not members of them-selves. But then the answer is available that although the predicate 'x is a set · $\sim (x \in x)$' indeed has some entity as its extension, this extension is not a set but a "proper" or "ultimate" class. The point is that if a is this object, '$x \in a \equiv x \notin x$' needs to hold only for sets, and hence it is of no moment that it fails for a. A "set" for this purpose can be any entity capable of belonging to classes.

This way out is not available when one considers whether there can be a class or set of absolutely all objects that are not members of themselves.

[14]That a sentence such as (1) or (2) might express a proposition "from outside" that is not in the range of its own quantifiers might be suggested by the following remark of Kneale ("Propositions and Truth," p. 243): "The lesson to be learnt from the Liar paradox is nothing specially concerned with truth or falsity, but rather that ability to express a proposition can never depend on ability to *designate* it." However, Kneale seems to think the paradox disposed of by the observation that these sen-tences do not express propositions.

nondenoting singular term, as 'the proposition expressed by this utterance is false'. But this semantic deficiency (whether or not it debars the statement from having a truth-value) is nonessential. (1) and (2) do not suffer from it.

The view about the Liar paradox here presented has to meet two objections: first, that it presupposes the dubious notion of sentences as expressing propositions, and of propositions as the primary bearers of truth or falsity; second and more important, it seems that in a natural language one can utter, say, (2), intending its quantifiers to be absolutely unrestricted. But then on the proposed account we should be debarred from saying, after having deduced (2) from (5), that (2) is true, or that it expresses a true proposition. But there is no indication that the conventions of English prohibit such an inference, or that a semantical theory about English would not say that (2) *entails* something like 'What (2) says is true'. But then we should be landed in contradiction. Is our proposal just another proposal for linguistic reform?

We shall postpone considering the latter objection to Section V. In answer to the first, I would plead that it is a plausible interpretation of talk of truth and falsity in natural languages that they are predicates of intensional entities that can be construed as propositions.[15] But I would not insist on this, since the approach can also be formulated in the situation where truth-values are attributed to sentences as I shall do in the next section.

III

I want now to consider the situation where truth is considered as a predicate of sentences. Where indexical and other context-dependent expressions are present, of course the truth-value of what is said depends not just on the sentence but on features of the context of utterance. Here the truth-predicate, like the expression predicate of the last section, would require additional argu-

---

[15] A persuasive case for this is made in Cartwright, "Propositions." As Davidson, in "On Saying That," makes clear, logical analysis of indirect discourse in terms of propositions does not require that *we* have a substantive criterion of propositional identity. However, as Tyler Burge and James Higginbotham have made me aware, the logical coherence of ordinary discourse so construed requires that the relevant relations of "saying the same thing" be reflexive, symmetric, and transitive. Transitivity in particular cannot be taken for granted in view of the clearly vague character of such relations. Higginbotham ("Some Problems," chap. 7) raises other difficulties, which however may not be relevant to the most ordinary discourse.

ment places. But I want to continue to exclude this complication, so that 'true' will be treated as a one-place predicate of sentences.

We can construct examples like (2.1) and (2.2), using 'is false' instead of 'expresses a false proposition' and 'is true' instead of 'expresses a true proposition'.[16] Then we can have singular terms α and β such that, as a matter of fact, α denotes the sentence

(1) α is false,

and β denotes the sentence

(2) β is not true.

However, what principle is to play the role of (2.5)? Of course we obtain a contradiction if we replace (2.5) by the usual Tarski schema. I propose to replace (2.5) by

(3)   'p' is true ∨ 'p' is false · ⊃ · 'p' is true ≡ p.

If we define 'x is false' as 'the negation of x is true', then (3) implies

(4)   'p' is true ∨ 'p' is false · ⊃ · 'p' is false ≡ ∼ p.

Taking (3) as the analogue of (2.5) amounts to taking *being either true or false* as the analogue of *expressing a proposition*. I shall defend this analogy in the course of this section. (3) follows logically from (2.5) if 'true' is defined by (2.12) and falsity is defined as truth of the negation.

However, whether or not (3) is a proper analogue of (2.5), it seems to be a plausible axiom. I know of no good reason for rejecting it. Another consideration in its favor is the following: Consider a formalized theory that contains (perhaps by Gödel numbering) its own syntax, and which also contains a truth-predicate (satisfying the Tarski schema) for a *part* of the language of the theory. If the truth-predicate is so defined that it is false of any object other than (the number of) a sentence of the relevant part, then (3) holds for *all* sentences.

An argument parallel to that showing that (2.1) and (2.2) do not

---

[16](2.1) is formula (1) of Section II. This is a paradigm for our references to numbered formulæ outside the section where they occur.

express propositions shows that (1) and (2) are neither true nor false. Since α denotes (1) and β denotes (2), we can then infer

(5) $\sim$ (α is true $\vee$ α is false)

(6) $\sim$ (β is true $\vee$ β is false)

and hence by propositional logic, (2) and the negation of (1) follow. Thus we have the same temptation to say that (1) is after all false, while (2) is after all true.

We must acknowledge a difference in sense between 'true' in the object language and 'true' in the metalanguage, because in the latter sense (2) is true and (1) false, while in the former sense, by (5) and (6), both are neither true nor false.[17]

In the last section, when we supposed the quantifiers in (2.1) and (2.2) to range over a definite universe U, we obtained such a divergence in the sense of 'expresses a true proposition': (2.1) and (2.2) expressed no proposition in the object-language sense (i.e., in U), but perhaps with a larger universe, (2.1) interpreted over U expresses a false proposition and (2.2) a true one.

But so long as we used quantifiers with the same range as in (2.1) and (2.2) themselves, we had to say that (2.1) and (2.2) express no propositions. But this was not a matter of their being syntactically or semantically ill-formed but was rather to be viewed in the same light as a predicate's failure to have an extension. Can we give a parallel explanation of the failure of (1) and (2) to be either true or false? A way of doing so is provided by following Frege and thinking of the truth-value of a sentence as an instance of the general notion of reference. We argued that just as a theory cannot correlate intensions (attributes) to all its predicates, so it cannot correlate propositions as intensions to all its sentences. Likewise, a theory cannot correlate extensions (classes) to all its predicates, and equally it cannot correlate extensions or references (truth-values) to all its sentences. The same will be true for singular terms if certain term-forming operators, such as abstraction, are present in the language.

There is a difference between the standard case of classes, on

[17]Harman ("Logical Form," p. 52) sketches a theory according to which liar sentences are neither true nor false in an object-language sense of 'true' and yet there is a metalanguage sense of 'true' that allows no truth-value gaps, so that these sentences obtain truth-values. As with us, (3.1) is false and (3.2) true in the metalanguage sense.

the one hand, and the present case of truth-values on the other that points up an ambiguity in our talk above of a theory failing to *correlate* intensions or extensions to its expressions. The case of classes seems to reflect an unavoidable ontological lack in a theory: the universe of an interpretation of it will just fail to contain sets or classes satisfying certain conditions. But for classical theories there are only two truth-values: any nontrivial theory can hardly be unable to name two distinct objects to represent them. The difference doees not disappear if we find the identity of abstract entities such as sets arbitrary: Any attempt to identify sets of members of a universe with members of that universe must break down at some point, since there can be no one-to-one correspondence of the members of a class and all its subclasses.

The difficulty arises with respect to the *function* that assigns to each sentence its truth-value, or perhaps to each open sentence and sequence of objects its satisfaction-value. This function cannot be built up from within a theory by assigning to each singular term, predicate, and sentence an extension or reference, because the universe of the theory cannot contain an extension for every predicate, and in some cases it will not be able to contain a reference for every singular term.

I want now to develop this point in a more technical direction. The assignment of extensions to the expressions of a language requires a certain amount of set theory or some functional equivalent, and, for the usual set theories, presupposes that the universe of discourse exists as a set or class. It is instructive to consider the case of a theory which contains the usual first-order set theory and where the interpretation proceeds simply by assigning as universe of discourse a certain set W and otherwise sticking as close to the "homophonic translation" as possible, by assigning to each primitive predicate $Fx_1. . .x_n$ the set

$$\{\langle x_1, . . .,x_n\rangle: x_1. . .x_n \in W \cdot Fx_1. . .x_n\}.$$

We assume W is large enough to contain the denotations of all the primitive singular terms of the language. In such a language we can define a two-place predicate $T(w,x)$ which can be read to mean

> x is the Gödel number of a formula which is true when its quantifiers are interpreted to range over the set w.

Let A be a closed formula and let $A^*(w)$ be the result of restricting the quantifiers of A to w. Then if n is the Gödel number of A, the formula

(7) $T(w, \bar{n}) \equiv A^*(w)$

will be provable in a rather weak set theory.

$T(W, x)$ will be the truth-predicate for the interpretation above sketched. Assuming the theory consistent, it will not yield the full Tarski biconditionals for all formulæ of the language. If t denotes the number of $\sim T(W, t)$, then if n is this number we can in the usual way prove

(8) $\sim [T(W\ \bar{n}) \equiv \sim T(W, t)]$

but by (7) we have

$$T(W, \bar{n}) \equiv (\sim T(W, t)\ )^*(W)$$

and (8) then yields

$$\sim [T(W, t) \equiv T(W, t)^*(W)].$$

Thus the truth-predicate $T(W, x)$ is provably not equivalent to its own relativization to W. In other words, the interpretation proposed for taking $T(W, x)$ as a truth-predicate, and for the truth of instances of (7), is not the same as the one $T(W, x)$ itself formulates. This should not surprise us, since the latter interpretation takes W as the universe, while the "intended interpretation" underlying the construction of the truth-predicates requires W to be a set *in* the universe.

$T(W, x)$ does yield a truth-predicate for the full language, not of course satisfying Tarski's schema, but satisfying (3). (7) does yield the Tarski biconditional for the formulæ provably equivalent to their relativization to W (i.e., provably absolute for W), given a concept of provability for our theory. Then (3) holds if we define 'x is true' as

> x is the number of a closed formula A such that $A \equiv A^*(W)$ is provable, and $T(W, x)$.

Falsity is, as before, truth of the negation. This has the effect that formulæ that are not provably absolute for *W* are neither true nor false.

To conclude this section, I must make a correction of the discussion of the last section. There we tried to assume as little as possible about the concept of proposition. In fact, on some accounts the failure of (2.1) to express a proposition cannot be accounted for by the absence of an appropriate proposition from its universe. Rather, the concept of expression behaves like the concept of satisfaction in the above account.

For example, suppose propositions are construed as functions from possible worlds to truth-values. Suppose further (unrealistically in my view) that there is a definite set *P* of possible worlds, which is constant for all the interpretations we consider. But then if the truth-values are identified with 1 and 0, 2^P is the set of *all* propositions, and the universe can be taken as a set containing 2^P and hence containing all propositions. Then just as in the extensional case, the failure of a sentence to express a proposition cannot be due to the fact that the proposition it expresses is not in the universe.

Consider an interpretation of this sort for which (2.5) is necessarily true. Consider a sentence C like (2.2) except that the definite description which is its subject and which denotes it is a rigid designator. Then 'C expresses no proposition' will be true in all possible worlds. But from outside, C expresses the necessary proposition, and if the universe *U* contains 2^P, this proposition (identified with the constant function 1 on *P*) belongs to *U*. But the pair ⟨C, the necessary proposition⟩ is not in the extension of 'expresses' for any possible world. Thus the difficulty arises not because the universe does not contain enough propositions but because the interpretation does not capture its own relation of expression. It should be remarked that the notion of expression here does the lion's share of the work of the Tarskian truth-predicate. The truth-predicate for propositions is essentially just functional application.

We can describe the situation in a way close to the way we described the extensional situation. An interpretation can here be construed as an assignment to certain linguistic expressions of *intensions*. Thus it is a generalized expression-relation. Of course the universe cannot contain an intension for every predicate. Thus there is the same difficulty in building up the relation of expression as there was in building up satisfaction and truth.

It should be remarked that even on the possible-worlds construal of propositions, the universe can contain all propositions only when the number of possible worlds is small compared to the number of individuals. Perhaps this is not very realistic. But an analogous situation can arise for higher-order languages.[18]

We can summarize the position of this section as follows: the use of 'true' as a predicate of sentences presupposes an interpretation. Such an interpretation can be an *object* of discourse only if it involves

---

[18]For example, the system of Montague, "Pragmatics and Intensional Logic." There propositions are of a different type from individuals. The system has a formation rule to the effect that if T is a propositional term, $T[\ ]$ is a formula, which might be read 'T is true'. 'The proposition that p' can be rendered as '$(\iota F)\square(F[\ ] \equiv p)$'. Then its existence is provable for all 'p', and moreover so is (2.11).

Consider now a theory formulated with this logic that can express its own syntax and formulate self-referential sentences. Suppose it contains a predicate for the relation of *expressing* between a sentence and a proposition. Then if α denotes the sentence 'p', we should expect

$$\text{the proposition that } p = (\iota F) (\alpha \text{ expresses } F)$$

that is

(i) α expresses the proposition that p

(ii) $(F) (\alpha \text{ expresses } F \supset F = \text{ the proposition that } p)$.

But then the Liar paradox can be obtained in the usual way, say if α denotes the sentence

(iii) $(\exists F) (\alpha \text{ expresses } F \cdot \sim F[\ ])$.

Thus in any model of our language, for any two-place predicate of the requisite type (in Montague's terminology $\langle -1, 0 \rangle$) either (i) or (ii) must fail. However, in a standard second-order model, where the propositional variables range over 2^p, we cannot attribute the failure to the absence of a proposition expressed by (iii) in the range of these variables; if $(\exists F) (\alpha \text{ expresses } F)$ is necessarily false, then the identically o function will be such a proposition. Rather, we must suppose that the extension of 'expresses' falls short of the expressing relation "from outside."

If we add to the language a truth-predicate 'T_1' such that for any sentence 'p' with standard name a the formula

(iv) $\square(T_1(a) \equiv p)$

is provable, then we can *define* 'x expresses F' as

$$x \text{ is a sentence} \cdot \square(F[\ ] \equiv T_1(x)\ ).$$

Then (i) with a as α follows immediately from (iv). (ii) also follows if '$F = G$' is defined as $\square(F[\ ] \equiv G[\ ])$.

Note that if arithmetic is necessary and the numbers are the same for all possible worlds then arithmetized syntax will be rigid: all syntactical truths will be necessary. From (iv) and these definitions it will follow that the expression relation is also rigid, that is

$$(x) (F) (x \text{ expresses } F \supset \square(x \text{ expresses } F)\ ).$$

something like an assignment of extensions to the parts of the sentences of the language. But then the universe of discourse ofthe interpretation must also be an object. But then the interpretation can no longer cover the discourse in which it is itself formulated. In this setting, a version of the distinction of object language and metalanguage imposes itself.

This point of view cannot be the last word, because it is also possible to define truth-predicates for formalized languages that do not proceed by assignment of extensions and that, in particular, do not treat the universe as a set. The relevance of this possibility will be taken up in the next section.

IV

So far our primary emphasis has been on interpretations in which the quantifiers range over some *set*, with the consequence that in the metalanguage we acknowledge entities not in the universe of the object language. I did consider one reason why that might not be the most relevant interpretation of the language of set theory. Moreover, a user of a language being interpreted in this way might always protest that by 'everything' he means *everything*; his quantifiers are not restricted unless such a restriction is made explicit or is clearly required by the context of an utterance. The interpretations of the type we have chiefly considered require the rejection of such claims to quantify over absolutely everything, or at least their reinterpretation in such a way that their absoluteness is lost. (The italicized 'everything' in the protest is taken to range over a sufficiently large set, so that the protest becomes true.)[19]

The idea of discourse about absolutely everything or even about

[19]George Myro pointed out that one might interpret the quantifiers of this entire paper as ranging over some sufficiently large set and thus produce a discourse to which the analysis of the Liar paradox here given would not apply. The same remark could be made about general set-theoretic discourses, including discourses about all models of set theory. In each case the "proponent" could come back and extend his discourse so that it would cover the newly envisaged case.

Leaving aside the possibility of an "absolute" interpretation, the generality which such a discourse as this paper has which transcends any particular set as range of its quantifiers must lie in some sort of systematic ambiguity, in that indefinitely many such sets will do. But one cannot express wherein the systematic ambiguity lies except in language that is subject to a similar systematic ambiguity. This is an illustration of the priority of the use of language to semantic reflection on it. It is also congenial to Quine's thesis of the indeterminacy of ontology.

absolutely all sets or propositions, poses problems that themselves could be the subject of a paper. But apart from the possibility that the intended meaning of quantifiers might be thus 'absolute', an interpretation or a truth-predicate need not involve treating the universe as a set or require the metalanguage to acknowledge entities not in the universe of the object language.

The relevance to natural languages of the object-language-metalanguage distinction is one of the issues at stake in the discussion of the Liar. But we can certainly distinguish between a discourse of a metalinguistic character, in which for example truth and falsity are predicated of what someone *says*, and the discourse that it is about, even if it is not a priori excluded that they are sometimes identical. But it is hardly evident that in such a case the metadiscourse presupposes a richer ontology than the object discourse, at least not in the most straightforward sense.

In fact, the relation of ontological enrichment and semantic ascent is complex and subtle, as some examples I have discussed elsewhere should indicate.[20] However, the situation where a theory is so interpreted that its universe is a set can be viewed as just one model for the situation where a sentence lacks a truth-value on a less comprehensive scheme of interpretation but obtains one on a more comprehensive scheme. The latter situation is more general. Consider for example a first-order language with finitely many primitive predicates which is then enlarged by adding a satisfaction predicate with the usual inductive definition. For any sentence of the original language the Tarski biconditional becomes derivable. If we construe the satisfaction predicate as false for formulæ of the enlarged language not in the original language, then, as we indicated above, we obtain a "truth-predicate" satisfying (3.3) in the enlarged language, in which the formulæ that get truth-values are those of the original language, i.e., those not containing the satisfaction predicate. The sentence that "says of itself that it is false" has of course no truth-value, but on the intended interpretation of the enlarged language it is false; its negation should be derivable. Thus it behaves just like (3.1).

The general point I wish to make is that the *use* of language does not require that one already have a thematized *interpretation* that assigns a truth-value to everything one says. Of course one might talk of an "intended interpretation" and of certain statements as

---

[20]Cf. "Informal Axiomatization, Formalization, and the Concept of Truth," section IV, and "Sets and Classes" (Essays 3 and 8 of this volume).

true under it, where one has an idea of how the interpretation would go, or simply where some things one has already asserted are constraints on any interpretation. Thus those interpretations that have been formulated so that 'true' has a definite meaning with respect to a certain class of sentences cannot be expected to cover the discourse in which the interpretation itself is formulated.

There is a consideration that even in this more general setting supports the analogy of having a truth-value with having an extension and of the Liar paradox with Russell's paradox. This is that the addition to a first-order language of a satisfaction-predicate, so that every sentence of the original language gets a truth-value, is essentially equivalent to the addition to the language of predicative class variables, so that every predicate of the original language gets an extension.

A first-order theory with finitely many primitive predicates can be extended either by a satisfaction predicate with the usual inductive definition,[21] or by class variables with predicative class existence axioms like those of Gödel's Group B.[22] Then for each formula A of the original language we can prove in the satisfaction theory the Tarski biconditional

$$(1) \qquad Sat(\langle v_0 \ldots v_{k-1} \rangle, \bar{n}) \equiv A,[23]$$

and in the class theory the comprehension principle

$$(2) \qquad (\exists Y) \, (x) \, (x \in Y \equiv A).$$

In the class theory, by the usual reduction of inductive to explicit definitions we can define a satisfaction predicate and prove the inductive conditions.[24] We can translate the class theory into the

---

[21] Ibid. and "Ontology and Mathematics," section V (Essay 1 of this volume).

[22] *The Consistency of the Continuum Hypothesis*, p. 5.

[23] Cf. Quine, *Philosophy of Logic*, pp. 35-43. $\bar{n}$ is the formal numeral for the Gödel number $\bar{n}$ of A. $v_0 \ldots v_{k-1}$ are the first k variables in a fixed enumeration and include all those free in A.

If the given theory does not contain number theory and a theory of finite sequences of its objects, these must be added before the satisfaction and class theories are constructed. It is sufficient to add the theory of finite sets.

[24] But not *laws* about satisfaction and truth whose proofs would need mathematical induction on formulæ containing these predicates. If we begin with ZF, we do not have such induction in NB, which is stronger than the "class theory" as we have defined it but which is still a conservative extension of ZF. We need the theory NB⁺ of Essay 3 above, note 15, which allows *Aussonderung* and replacement for formulæ containing class variables. The corresponding extension of the satisfaction theory is obtained by allowing *Aussonderung* and replacement for formulæ containing *Sat*.

satisfaction theory by construing a class as a pair of a one-place predicate with parameters and a sequence assigning values to the parameters. Then we can prove the translations of the class existence axioms.[25]

Thus if we go beyond mere truth and think of each sentence of a language as having a satisfaction-value with respect to any substitution for its free variables, this is equivalent to taking any predicate of the language as having, for any substitution for its free variables, a class as its extension. Talk of satisfaction (truth *of*) and talk of classes are ways of generalizing with respect to predicate positions in a language; in this sense they are equivalent ways.

V

Up to now our discussion has been almost entirely an exercise in abstract semantics. We have not undertaken to make our ideas plausible in application to natural languages, and we have hardly taken account of any difference that might exist between a natural language and a first-order formalism.

There is one respect in which we already face a difficulty in applying to natural languages the *formulations* of the Liar paradox that we have discussed. Namely we have used either Tarskian predicates of truth of sentences under a given interpretation or the notion of expressing a proposition. Neither of these concepts is part of "ordinary" language; they are part of the technical language of semantics. Suppose the worst case: that an irresolvable antinomy arose from sentences of the sort we have been considering. Then it seems that it might be possible to attribute the contradiction to the technical semantic concepts and to avoid the claim that ordinary speakers of English are saddled with an incoherent conceptual apparatus.

In connection with the intensional formulations I want to emphasize that the concepts of expression and proposition are causing the difficulty, not the concept of truth. In any language in which sentences express propositions, there is no difficulty in having a predicate 'is true' such that, for any singular term 'α' and sentence 'p', if 'α' *denotes* the proposition that 'p' expresses, then

(1) α is true $\equiv p$

is true (in the metalinguistic sense). The intuition of some theorists

[25]For a little more detail see "Sets and Classes" (Essay 8), section I.

of truth that (1) is tautological is supported by our analysis. We see no help to the solution to the paradoxes in restricting the validity of (1), although we might deny that there always *is* a singular term, even 'the proposition that *p*', which denotes a proposition expressed by '*p*'.

On the other hand, we found that the relation of *expression* between a sentence and a proposition could differ in extension between the object language and the metalanguage, even if the propositions involved were already available in the universe of the object language.

As I said earlier, I believe that the intensional point of view is more directly applicable to natural languages than the purely extensional. In English, what is said to be true or false is what is said, meant, believed, supposed, conjectured, claimed, asserted, and the like. It is certainly regimentation to identify each of these with a *proposition*, particularly in view of the implication of a uniform criterion of identity, say for two utterances to express the *same* proposition. But the lack of uniformity claimed by many writers can exist within a single "propositional attitude," so that it is not taken care of by merely splitting the notion of proposition into that of saying, meaning, belief, etc.

Are there English words that express something like the technical notion of expression? The words 'say' and 'mean' are certainly related, in that they can relate persons, or sentences, or utterances, or occasions of utterance to statements in indirect discourse of *what is said* or meant, and to truth or falsity. Saying figures essentially in the most "ordinary" forms of the Liar paradox, such as 'What I am now saying is false'.

It seems that we would obtain a contradiction by assuming that in English the inference from '*p*' uttered by a person *S* on an occasion *O*, to 'what *S* said on occasion *O* is true' is valid in both directions. For *S* could on occasion *O* say, '*S* is on occasion *O* saying nothing that is true'.

Here we meet the question what could be meant by the charge on these grounds that English is "inconsistent," or, on the other hand, what a "solution" of the paradox is supposed to accomplish in application to English. What is meant by saying, as a descriptive remark about English, that a certain inference is *valid* between English sentences uttered on specific occasions? One thing we might be claiming is that facts about the truth-conditions of the sentences *determine* that they are valid. But an assignment of truth-conditions to sentences of English would have to satisfy some conditions of coherence, since it is not just an account of the beliefs of English speakers but an account of the conditions under which what they say is *in fact* true. But then it is hard to see how such an account could make English inconsistent: Herzberger[26] describes plausible assumptions about an assignment of truth-conditions to sentences of a natural language from which it follows that the outcome of inconsistency is impossible, although it could still be that no such account can be given that can be squared with the data.

Alternatively, we may be trying to describe an aspect of the language by describing inferences that the speakers of the language characteristically make or would accept if others made them. But then the charge of inconsistency just comes to the claim that speakers of English have inconsistent beliefs, which, in general, no doubt they do.

In the specific case of the Liar paradox, the claim would be that speakers are disposed to accept all inferences of the form: from '*p*' said by *S* on occasion *O*, to 'what *S* said on occasion *O* is true', and vice versa. That such a general disposition exists is highly likely. However, the empirical evidence (impressionistic to be sure, mostly based on the behavior of philosophers and their students) hardly bears out the idea that this amounts to a commitment to be honored come hell or high water. Confronted with the Liar paradox argument, almost anyone will recognize that something is fishy. Most would doubt that they fully understand what is being said and suspect nonsense. The difficulty there is in agreeing on a "solution" can be interpreted to mean that speakers do accept these inferences as general principles and do not know what to believe when they see the difficulties that arise in this particular case, unless either indoctrination or their own theoretical reflection prompts them in one particular direction.

This view of the situation gives some scope for theory in dealing with the paradox. It may be that if we treat *all* the dispositions to accept statements and inferences involving 'say', 'mean', 'true', and 'false' as on the same level, the only conclusion we can arrive at is that it is impossible to attribute to all these words at once a coherent meaning. But a theory might be quite coherent and honor *most* of these dispositions, particularly those that come to light in ordinary situations. Then the others can be attributed to confusion, difficulty of understanding expressions used in very abstract contexts, and the like.

[26]"The Truth-Conditional Consistency of Natural Languages."

Suppose *A* says 'I am now saying something false'. Then, parallel to the argument for the claim that (2.1) does not express a proposition, I can prove using only "ordinary" language, that *A* said nothing. Shall I go on to argue that since *A* said nothing, he did not say something false? And since he said that he *was* saying something false, he did after all say something false? Then I would be landed in a contradiction. The possibilities of escaping are manifold; there is perhaps even more than one in the spirit of our general discussion above.

'Say' in this discourse seems clearly to be ambiguous: the claim that *A* said nothing squares ill with the later inference that 'he said that he was saying something false' to 'he was saying something false'. To begin with 'say' seems flexible with respect to the amount of sense required for someone to say something. Thus *A* may have said that he was saying something false, according to a lax standard, while saying nothing according to a stricter standard, needed for truth or falsity. Then the final, paradoxical inference is from '*A* said$_L$ that he was saying something false' and '*A* did not say$_S$ anything false' to '*A* said$_S$ something false'. But '*A* said that *p*' and '~*p*' imply '*A* said something false' only if the first premise is true with the stricter standard.

This observation is only a first approximation to meeting a difficulty we met in the formal situation. *We* say of *A* that he did not say something false, after having shown that *A* said nothing. But then we express our conclusion by a sentence that is (modulo some changes of indexicals to accommodate the changed context) the negation of *A*'s original remark. How can we have said something true or false if *A* did not? And how can we say of either ourselves or *A* that he did or did not without using such a phrase as 'says nothing' involving two of the essential elements, saying and quantification, of *A*'s remark? And I do not see any grounds for proposing that '*A* said nothing' is perfectly meaningful while '*A* said nothing *false*' is not.

Must we admit that we say nothing when we say of *A* that *he* says nothing? The only possible rescue here must come from general considerations. These, however, reveal a further ambiguity or indexical variation in words such as 'say' that is relevant to the Liar paradox discourse.

It would seem that to attribute to *A* a definite general sense of 'say' is to attribute to him a general scheme for interpreting his own words and those of others. In real life, this scheme will be rather vaguely defined and shifting. For example, as regards the range of his own quantifiers, a speaker would generally have no more than certain commitments which put a lower limit on his universe. Most quantifiers in ordinary discourse are restricted and thus would give the interpreter great latitude in choosing a universe. No doubt in many situations where no restriction is explicitly stated, some is nonetheless intended, and a man interpreting himself will take this into account. But in many situations one will translate quantifiers homophonically.

Thus it seems clear that different occasions of use of a word such as 'say' can presuppose different schemes of interpretation. Moreover, it seems clear that the universe for the quantifiers does not have to be taken as constant for an entire language and even throughout a single discourse, so that this is one dimension with respect to which the schemes of interpretation presupposed in uses of words such as 'say' can differ. (Moreover, a general scheme for the semantics of a natural language would allow that the universe for quantifiers can vary with context.)[27]

If all this is so, we can avoid imputing a contradiction to the Liar paradox discourse and yet allow a final inference to 'A after all said something false' if we assume that the latter remark presupposes a more comprehensive scheme of interpretation than the discourse up to that point, which assigns sense or truth-values to utterances not covered by less comprehensive schemes. One way this might be is if the last remark presupposes a larger universe. The discourse up to that point vindicates the sense of A's original remark by, in effect, refuting it. The last remark involves a semantical reflection that could be viewed as involving taking into one's ontology a proposition that had not been admitted before, perhaps because admitting it involves taking the universe of my own and A's previous discourse as an object.

This way of interpreting the discourse attributes to the speakers an implicit theory according to which what is referred to when one talks of 'what is said' belongs to some kind of potential totality which is not exhausted by any set. This is, however, not crucial to distinguishing the interpretation presupposed in A's original utterance and that presupposed in saying that he said something false.

A must presuppose a certain self-interpretation, however vaguely defined. It seems that an equivocation arises at the outset unless, in commenting on what A says, we presuppose the same interpre-

---

[27]This seems clearly to be provided for in Herzberger's notion of "type 3 conceptual framework" ("Paradoxes of Grounding," p. 160).

tation as A's. But both for us and for A, there is a strong temptation not to do this. For since A talks about what he is *saying* and we talk about what A said, A's interpretation becomes part of what we are talking about.[28] Thus our interpretation is likely to involve reflection on A's and thus be more comprehensive.

If this temptation is resisted, the above argument for the conclusion that A said nothing goes through. But then without passing to a more comprehensive scheme of interpretation, which changes the truth-conditions of statements involving 'say', we cannot go on to the paradoxical conclusion that A after all said something false.

If 'said$_1$' presupposes A's scheme of interpretation, and 'said$_2$' presupposes a scheme of ours that gives a sense of A's remark, then we can conclude that A said$_1$ nothing, it is not the case that A said$_1$ something false, and therefore, since A said$_2$ that he was saying$_1$ something false, A said$_2$ something false. But that is not what A said (1 or 2).

On the other hand, suppose that at the outset, when we inquire what A said, we presuppose the more comprehensive scheme. Then we ask whether A said$_2$ something true, but if so we can only conclude

(2) A said$_1$ something false.

From the fact that our scheme is more comprehensive, we might infer

(3) A said$_2$ something false

from which the negation of (2), and thus a contradiction, follow. Thus we can refute the hypothesis that A said$_2$ something true.

But if we now suppose A said$_2$ something false, we do not obtain a contradiction. We can infer the negation of (2), but to make the

[28]However, if we apply this remark literally in all cases of iteration of indirect-discourse operators, we shall arrive at a somewhat counterintuitive theory of them, which would assign to all occurrences of such operators subscripts such that in a context such as "A says that B says that. . ." the outer operator would have to have a higher subscript than the inner one: the sense of "say" for A would have to be more comprehensive than that for B.

I do not have a precise theory that avoids this consequence, but it seems to me it must be possible to construct one, just as it is possible to construct predicative set theories that are not ramified (see Feferman, "Systems of Predicative Analysis," "Predicative Provability"). The passage to a more comprehensive scheme is forced on us in the Liar paradox because of the quantification over "what is said" that cannot be cashed in terms of a particular utterance whose sense is given. [On this matter see the Postscript.]

negation of (3) follow we need some such assumption as that A said$_1$ something true or false, which would yield a contradiction by the analogue of (3.3). Rejecting this assumption, we end up with the "straightforward" conclusion that A did not say$_1$ anything true or false and thus said$_2$ something false. But again, this is not what A said.

In the discourse we have considered, there is a natural relation of the scheme of interpretation at the beginning and that at the end: the latter is more comprehensive. A more ambiguous case is the two-utterance Liar paradox, for example where A says

> B is saying something false

while B says

> A is saying something true.

Each might be thought of as purporting to have a scheme of interpretation that takes in the other's use of 'say' and is thus more comprehensive than the other's. But of course they cannot be simultaneously interpreted in this way. If we attribute to them the *same* scheme, then in those terms neither says anything. However, the sort of reasoning that shows that the typical liar sentence says nothing will not show that *both* utterances say nothing: only that at least one does.[29] This only shows that formal principles such as (2.5) and (3.3) do not force on us the hierarchical point of view we have adopted.

If we attribute to B a more comprehensive scheme (saying$_B$) and

---

[29]The relevance to my discussion of this sort of example was brought to my attention by Thomas Nagel.

Suppose α denotes the sentence
> β expresses a false proposition

and β denotes the sentence
> α expresses a true proposition.

Then it is consistent with (2.5) to suppose that neither expresses a proposition (and hence that both are "from outside" false), or that β expresses a false proposition and α expresses none (and hence is "from outside" true).

Similar latitude can arise for some single self-referential sentences. If y denotes the sentence
> y expresses a true proposition

(2.5) allows that y expresses no proposition (and is false from outside) or expresses either a true or a false proposition. Since the truth-condition for (2.1) is contradictory, that condition implies that (2.1) can express no proposition. The truth-condition for y is tautological and hence imposes no constraints. But any account of (2.1) and y is likely to treat them the same way.

to A a less comprehensive one (saying$_A$), then we can show that both A and B say$_A$ nothing while A says$_B$ something false. So what B says is false, at least in terms of a more comprehensive scheme. If we assume A's scheme comprehends B's, it follows that A and B say$_B$ nothing, B says$_A$ something false, so that from outside what A says is true.

However vaguely defined the schemes of interpretation of the ordinary (and also not so ordinary) use of language may be, they arrange themselves naturally into a hierarchy, though clearly not a linearly ordered one. A scheme of interpretation that is "more comprehensive" than another or involves "reflection" on another will involve either a larger universe of discourse, or assignments of extensions or intensions to a broader body of discourse, or commitments as to the translation of more possible utterances. A less comprehensive interpretation can be appealed to in a discourse for which a discourse using the more comprehensive interpretation is a metadiscourse.

To many the hierarchical approach to the semantical paradoxes has seemed implausible in application to natural languages because there seemed to be no division of a natural language into a hierarchy of "languages" such that the higher ones contain the "semantics" of the lower ones. Indeed, there is no such neat division of any language as a whole. What the objection fails to appreciate is just how far the variation in the truth-conditions of sentences of a natural language with the occasion of utterance can go, and in particular how this can arise for expressions that are crucially relevant to the semantic paradoxes: perhaps not 'true', but at all events quantifiers, 'say', 'mean', and other expressions that involve indirect speech.[30]

However, it would be naive to suppose that the type of account of indexical variation that is now standard in formal semantics can be fully adequate to the context-dependence of these expressions,

---

[30]It is interesting to compare the view of this paper with those of Charles Peirce. Emily Michael has discovered (see her "Peirce's Paradoxical Solution") that Peirce early on realized one of the main difficulties that we have concentrated on: that a statement 'asserting' its own falsity or lack of truth may be assigned a truth-value by virtue of its not expressing a proposition. However, his final conclusion (*Collected Papers* 5.340) is that a sentence such as (2.2) is in a way self-contradictory: what it explicitly states is true (since it does *not* express a true proposition), but it tacitly implies something contradictory to what it states, since any statement tacitly implies its own truth (that it expresses a true proposition).

Such a contradiction would have to lie in the concepts of proposition and truth.

On our view (2.2) could only 'tacitly imply' its own truth on an interpretation more comprehensive than the one it talks about.

which I have rather called "systematic ambiguity" (note 19), which might be related to the "typical ambiguity" of Whitehead and Russell. Herzberger in "Paradoxes of Grounding" already shows the unavoidability of some such ambiguity. In a simple case such as that of the word 'I', we can describe a function that gives it a reference depending on some feature of the context of utterance (the speaker). We could treat the "scheme of interpretation" in this way as argument to a function, but that of course is to treat it as an object, for example a set. But a discourse quantifying over *all* schemes of interpretation, if not interpreted so that it did not really capture *all*, like talk of all sets interpreted over a set, would have to have its quantifiers taken more absolutely, in which case it would not be covered by any scheme of interpretation in the sense in question.[31] We could produce a "superliar" paradox: a sentence that says of itself that it is not true under any scheme of interpretation. We would either have to prohibit semantic reflection on this discourse or extend the notion of scheme of interpretation to cover it. The most that can be claimed for the self-applicability of our discussion is that if it is given a precise sense by one scheme of interpretation, then there is *another* scheme of interpretation of our discourse which applies the discourse to itself under the *first* interpretation. But of course this remark applies to the concept "scheme of interpretation" itself. Of it one must say what Herzberger says about truth: in it "there is something schematic. . .which requires filling in."[32]

Postscript

Since the publication of this essay in 1974, a great deal of new work has been done on the semantic paradoxes. From the point

---

[31]It seems to me that the best way to view the sort of truth-definition envisaged by Davidson in "Truth and Meaning" is as involving such a fixing of the concept of scheme of interpretation. Once the definition itself is translated back into the language, it will of course no longer be complete.

Davidson's remarks (ibid., pp. 314-315) suggest that he does not aim at completeness in this sense.

However, he writes, "But it is not really clear how unfair to Urdu or Hindi it would be to view the range of their quantifiers as insufficient to yield an explicit definition of 'true-in-Urdu' or 'true-in-Hindi.' " What is to prevent a speaker of Urdu from learning enough set theory or semantics and contriving to express it in his language so as to give the lie to such limitation? Obviously Davidson must say that this would change the language (semantically at least). The definition of 'true-in-Urdu' could express a given stage of the development of Urdu and perhaps a hypothesis as to its future development.

[32]"Paradoxes of Grounding," p. 150.

of view of technical formal semantics, the picture as it had presented itself to me has been transformed. The first and largest step in this transformation was the work in 1975 of Saul Kripke.[33] Where the *philosophical* picture stands after this work is not obvious. In what follows I shall try to say something about this.

I

The writers I discussed in section I above advanced "truth-value gap" approaches to the paradoxes as an *alternative* to a hierarchical approach in the spirit of Tarski or Russell.[34] My criticisms took them in this way. Kripke used truth-value gaps to construct a hierarchy that was in many ways more natural than the Tarskian hierarchy in its original form (even assuming some solution to the difficulties Kripke alludes to about transfinite levels). By this the original opposition is in considerable measure *aufgehoben*. In a way, truth-value gaps arise even in my own more Tarskian treatment, since if one assumes the truth schema (3.3) one can prove that liar sentences are neither true nor false. This suggests developing a theory with truth-value gaps in the framework of *classical* logic; see below.

Kripke reproaches a number of earlier writers on the subject with not really having a theory; in practice this seems to mean a developed formal model, on the basis of which questions of truth, falsity, or truth-valuelessness can be given objective answers. As directed against me, the reproach is justified: though in some of my discussion I was running up against inherent limitations of such formal models, the general idea that a hierarchical account of truth

---

[33]"Outline of a Theory of Truth"; see also "A Theory of Truth I, II." References to Kripke are to the "Outline" and are given merely by page number.

An idea similar to Kripke's was discovered independently but not developed to the same extent by Martin and Woodruff, "On Representing 'True-in-L' in L." In connection with the problem of Russell's paradox, constructions like Kripke's had been given by others; for discussion and references see Feferman, "Toward Useful Type-Free Theories, I," esp. §14.

[34]At the time of writing my paper, I did not sufficiently appreciate the affinity of my point of view and Russell's. Russell's approach to the semantic paradoxes by the ramified theory of types was of course the first developed hierarchical approach. Its closeness to Tarski's has been made clear in Church's magisterial "Comparison." For a special case of the close connection between the finite levels of a ramified hierarchy and the finite levels of a Tarskian hierarchy, see the Appendix to Essay 1 above. But a more specifically Russellian aspect of my point of view is the idea of systematic ambiguity.

in natural languages could be given by taking due account of the possibilities of contextual variation certainly called for working out in such terms. The analyses of section V offered only a very sketchy beginning. I have not carried the matter further since, but constructions largely in the spirit of my ideas which come closer to satisfying Kripke's demand for a theory have been carried out by Tyler Burge.[35]

My discussion makes two demands on a theory that seem to pull in different directions. On the one hand, I wanted to avoid too much shifting of the interpretations of indexical elements such as the range of quantifiers over propositions, indirect-discourse terms such as 'say', and 'true' applied to sentences or utterances. That sentences should "seek their own levels," in Kripke's phrase, was a desideratum I hinted at vaguely, without in any way showing how to achieve it. But such shifts were *forced* by some discourses involving paradoxical statements, such as those considered in sections II and III of the paper, in which one reasons by way of the paradox to the conclusion that a sentence *a* does not express a proposition, or is neither true nor false, and then deduces *a* itself, or its negation, from the statement of the deficiency inferred from the paradox. Only with a shift of what I called "scheme of interpretation" can one draw the natural conclusion that *a* is true (or false) without reinstating the contradiction. Following Burge,[36] I shall call such discourses Strengthened Liar discourses.

Kripke's theory enables us to avoid such shifts in many other situations, but not in these. So long as we are dealing with sentences that are grounded by Kripke's definition or another in the same spirit, Kripke's theory obviously provides a more flexible instrument than the more strictly Tarskian paradigm I followed most of the time. A Kripkean fixed-point model could be taken as a model for a "scheme of interpretation" of a certain kind of maximal comprehensiveness. It allows the handling by a single scheme of discourses involving a high degree of iteration of the truth-predicate. In this way, the attribution of "implicit subscripts" to occurrences of semantical terms is minimized.

However, Kripke's theory changes nothing essential in the interpretation of Strengthened Liar discourses. Let us consider how a typical one is interpreted with respect to a fixed-point model. For definiteness, let *L* be a classical interpreted first-order language to

---

[35]"Semantical Paradox" and "The Liar Paradox: Tangles and Chains."

[36]"Semantical Paradox," p. 173. As Robert L. Martin has pointed out to me, Burge's use of this term does not quite agree with earlier usage.

which a truth-predicate 'T' is to be added. Let D be the domain of the interpretation. We assume that L can express its own syntax and that sentences in L have standard names in L. An interpretation of the extended language assigns to 'T' two disjoint sets, its extension and anti-extension. A sentence 'Tt' is true if what t denotes is in the extension of 'T', false if it is in the anti-extension, undefined otherwise. Sentences are then evaluated according to Kleene's three-valued truth-tables.[37] Such an interpretation is a fixed point if any sentence A with standard name a is true if Ta is true, false if Ta is false, undefined otherwise.

Strengthened Liar discourses are pieces of reasoning. The application to them of Kripke's theory therefore raises a question he does not address, how to formulate the principles of reasoning for a language with truth-value gaps.[38] But we can see that however this problem is solved, the fixed-point interpretation fares just the same as earlier truth-value gap proposals. Consider the case where the term s denotes $\sim Ts$. Since $\sim Ts$ is paradoxical in Kripke's sense, it is undefined in any fixed-point model. It follows that the intuitive argument to the conclusion $\sim Ts$, that since s is undefined it is certainly not true, will not go through: being undefined, Ts cannot be derived from true premises. The reason again has to do with the negation of the object language: where A is undefined and therefore not true, $\sim A$ is undefined as well. Within Kripke's logical framework, we can obtain the conclusion $\sim Ts$ by the shift in the extension of T that he calls "closing off" (p. 715): starting with a fixed point model, obtain a new (classical) interpretation by taking T to be *false* of all objects not in its extension with respect to the fixed point.[39]

[37]Kripke, p. 700; Kleene, *Introduction to Metamathematics*, pp. 332ff. For our purposes, we can ignore the fact that Kripke's theory has a schematic character, in that other systems of valuation admitting truth-value gaps can be used.

[38]In interpreting undefined sentences as not expressing propositions and disclaiming the interpretation of his admission of truth-value gaps as the adoption of a deviant logic (p. 700, n. 18), Kripke may seem to brush this question aside. If a discourse only counts as a logical argument if all sentences in it express propositions by this criterion, then of course the deductive apparatus can be a usual classical one. But by this criterion, the Strengthened Liar discourses would not count as logical arguments at all. This would in my opinion throw out the baby with the bath, since it rejects the arguments that bring out the paradoxes themselves.

[39]It is still instructive to look at one deductive apparatus for the Kleene three-valued logic, the sequent calculi of Wang and Scott. (See Feferman, "Toward Useful Type-Free Theories, I,"§10). If a is the standard name of a sentence A, the sequents $Ta \vdash A$ and $A \vdash Ta$ will hold in a fixed-point model. Letting A be the self-referential $\sim Ts$, using the truth $a = s$ we obtain $Ts \vdash \sim Ts$, but the apparently obvious "reductio ad absurdum" inference to $\vdash \sim Ts$ is not sound under the interpretations given of '$\sim$' and '$\vdash$'.

I want also to look at a modification of Kripke's construction presented by Feferman,[40] where we hold to classical two-valued logic and instead of the extension and anti-extension of a truth-predicate 'T', we have the extensions of two predicates 'T' and 'F' of truth and falsity. Monotonicity is obtained by applying the direct semantical evaluation to sentences in which 'T' and 'F' occur positively, that is, sentences built up by conjunction, disjunction, and quantification from atomic formulae in 'T' and 'F' and atomic formulae of L and their negations. A fixed-point model is one in which Ta holds if and only if A^+ does, and Fa if and only if A^-, where A^+ and A^- are "positive approximants" to A and $\sim A$ respectively. In our situation where s denotes $\sim Ts$, we can now derive $\sim Ts$ and $\sim Fs$. But we cannot derive from the former the conclusion Ta suggested by the reflection that $\sim Ts$ is true after all, since $\sim Ts$ is not positive in 'T'. (Evidently $Ta \equiv A$ holds if A is positive.) Once again, we have to distinguish an object-language from a metalanguage sense of 'true'.

Of course Kripke does not in the least dispute the claim that some such distinctions are unavoidable. Where there may be disagreement is on whether something like the Strengthened Liar discourses, interpreted along the lines suggested by me and developed by Burge, belong to "ordinary language." The formulations of section V above avoided any technical terms of semantics; the language involved and even the reflection required for their interpretation seems to me more ordinary than what would be needed to formulate grounded sentences of relatively low transfinite level, say ω^ω.[41] Once we see the relation between the use of language and reflection on it in the way I suggested in the essay, this issue ceases to be at all fundamental. Some semantic reflection certainly belongs to "ordinary" language, but here as in other domains we do not need a sharp line between the reflection that remains ordinary and what goes beyond it. To model *most* ordinary and

---

What '⊢' expresses is similar to van Fraassen's necessitation (section I above), but since a sequent does not enter into further combinations, the sort of reasoning discussed in note 9 above cannot be formulated.

[40]Ibid., §13.

[41]Cf. Burge, "Semantical Paradox," esp. p. 174, n. 9. Kripke remarks that the models he presents "are plausible as models of natural language at a stage before we reflect on the generating process associated with the concept of truth, the stage which continues in the daily life of nonphilosophical speakers" (p. 714, n. 34). If by the "generating process" he means a recursion leading to a fixed point, it should be pointed out that reflection on it does not enter into simple Strengthened Liar situations themselves and does not play an essential role in the type of interpretation Burge and I advocate.

even a lot of nonordinary discourse about truth, we do not need anything mathematically as powerful as the minimal fixed point; carrying a Kripkean or similar iteration through ω or at least ω^ω stages will be adequate.

As a model for a comprehensive scheme of interpretation, I would generally prefer a variant of Kripke's using classical logic, such as the one mentioned above. The inferential naturalness of classical logic is worth the sacrifice of some instances of the equivalence of A and Ta. This equivalence holds in any fixed point of the above framework for any sentence that is grounded in Kripke's sense.[42]

II

The most thorough investigation of the problem of interpreting the Strengthened Liar discourses is that of Tyler Burge (in the papers cited in note 35). Agreeing with me both in discerning contextual shifts in the interpretation of key terms at crucial points in such discourses and in the preference for classical logic, he has offered both formal models and a very instructive system of pragmatic principles for determining the manner in which the interpretation of the semantically relevant terms is fixed by the context. What Burge calls Construction C3, the most permissive of his formal models, might be described in the following way (simplified to consider only truth, not satisfaction).[43] As before, we begin with an interpreted first-order language L that can express its own syntax, and add for each natural number n predicates T_n and F_n for truth and falsity of level n. We obtain an infinite sequence of models M_0, $M_1, \ldots$ as follows: M_0 is the given structure for L, expanded by assigning the empty extension to T_i, F_i for every i. Suppose now that M_n has been constructed. To obtain M_{n+1}, we go through the minimal fixed-point construction for Feferman's classical logic version of Kripke, with T_n, F_n as truth and falsity predicates. If $i < n$, then T_i, F_i are treated as belonging to the base language and given their extensions in M_n; if $i > n$, then T_i, F_i are once again given

---

[42]Relative, of course, to the Kleene valuation. The equivalence does not hold only for such sentences. For example, it holds for a sentence asserting its own truth. That is a case where the equivalence imposes no constraint on the truth-value of the sentence.
[43]"Semantical Paradox," pp. 189-190.

fixed empty extensions.[44] The effect of treating T_i, F_i for $i < n$ as belonging to the base language is that formulæ of the form $\sim T_i t$ and $\sim F_i t$ now count as positive and obtain truth-values and the extensions of T_i and F_i are left fixed in the construction of M_{n+1}. It follows that once T_n, F_n have been treated as truth and falsity predicates, in the construction of M_{n+1} their extensions remain fixed in the construction of subsequent models.

We obtain a final model M_ω by giving T_n and F_n for each n its extension in M_{n+1}. It is then immediate that a sentence is true in M_ω if and only if it is true in M_{n+1}, where n is the maximum k such that T_k or F_k occurs in A. We can define $R_n x$ ('x is rooted$_n$') as $T_n x \vee F_n x$; it means roughly "grounded relative to the underlying model of L and the given extensions of T_m, F_m for $m < n$."

It is possible to show that the extensions of T_{n+1} and F_{n+1} in M_{n+2} include those of T_n and F_n in M_{n+1}. It follows that in M_ω, if $m < n$ then $(x)\,(T_m x \supset T_n x)$ and $(x)\,(F_m x \supset F_n x)$ hold.[45] We can now extend the hierarchy into the transfinite. Suppose we add a new pair T_ω, F_ω of predicates, and put $u \in D$ into the extension of T_ω if it is in that of T_n for some n, and similarly for F_ω.[46] Every

[44]However, we do not treat T_i, F_i for $i > n$ as belonging to the base language; in the definition of A^+ above, we treat them like T_n, F_n; only we do not expand their extensions in the recursion leading to a minimal fixed point. If they were treated as belonging to the base language, the extension of T_n and F_n would be altered in the construction of M_k for $k > n + 1$. I owe this point to Anil Gupta.

We obtain the "liberalized version" of C3 (ibid., p. 190) by treating T_i, F_i for $i > n$ in this construction just as we treat T_n, F_n.

[45]It follows that it is a matter of indifference whether, in giving a recursive construction of M_{n+2}, we begin with the extensions of T_{n+1}, F_{n+1} both empty (as the text suggests) or give them as initial extensions those of T_n, F_n, so that the latter are automatically included in the final extensions of T_{n+1}, F_{n+1}.

Suppose that L is the language of first-order arithmetic. Let Z_0 be the usual axiomatic first-order arithmetic. Let Z_{n+1} be the theory resulting from Z_n by adding the predicates T_n, F_n and the natural generalized inductive definitions of them as truth and falsity at the minimal fixed point with base language that of Z_n, with the associated induction principle. (Cf. Kripke, p. 706, n. 24.) Let Z_ω be the union of all the Z_n. (In all these theories, we assume ordinary induction for formulæ of the extended language.) In this situation, of course, satisfaction can be recovered from truth. Burge's axioms for C3 can be proved in Z_ω. Conversely, if some technical emendations are made to the axioms of C3, then the addition of these axioms to those of Z_0 yields Z_ω.

The theory Z_1 is reminiscent of the theories $PA\rlap{\,\supset}{\subset}$ and $PA[P]\rlap{\,\supset}{\subset}$ of Feferman, "Gödel's Incompleteness Theorems," §§2.2–2.3, which add to first-order arithmetic the introduction clauses of the inductive definition of T_0 and F_0 but which have only ordinary induction. Z_1 clearly includes $PA\rlap{\,\supset}{\subset}$, but its relation to $PA[P]\rlap{\,\supset}{\subset}$ is unclear. Note that Z_ω is an example of arbitrary finite iteration of generalized inductive definitions.

[46]This is not the extensions that T_ω, F_ω take on in the transfinite construction below. But in that construction, we can see as before that the extension of T_ω contains that of T_n for every finite n.

sentence containing only T_m, F_m for finite m is true or false in the sense of T_ω, F_ω. But handling a discourse involving a liar sentence for T_ω or F_ω would require extending the construction further. Suppose that our extended language contains predicates T_α, F_α for all *ordinals* less than some limit ordinal γ.[47] The generalization of our procedure defines M_0 as before, for any α, $M_{\alpha+1}$ is obtained by the minimal fixed-point construction with T_α, F_α as truth-predicates, T_β, F_β for $\beta < \alpha$ as in the base language with their extensions as in M_α, and T_β, F_β for $\alpha < \beta < \gamma$ with fixed empty extensions. For a limit ordinal $\lambda < \gamma$, M_λ assigns to T_α, F_α for $\alpha < \lambda$ its extension in $M_{\alpha+1}$, and for $\lambda < \alpha < \gamma$ empty extensions. If $\alpha < \beta < \gamma$, the extensions of T_β and F_β are proper supersets of those of T_α and F_α.

In extending our construction into the transfinite, we seem to depart from the letter and perhaps also the spirit of Burge's construction. Burge's notation and the absence of any provision for limit levels suggest the thesis that the indexicality of 'true' can be adequately modeled by an ω-sequence of levels. Although in fact Burge does not hold this, some remarks of his suggest an argument for it that is of interest on its own account.[48] The following seems to be the appropriate way of understanding formal models like those Burge presents and our formulation of C3: We are given a discourse to interpret, which for simplicity we will suppose contains only elements that can be straightforwardly translated into our initial language L, plus the words 'true' and 'false'. An interpretation of the discourse will determine for each occurrence of 'true'

---

[47]The assumption that at each stage the structure we are considering contains the syntax of the extended language, together with the assumption that D is a *set*, implies that it is only for an initial segment of the ordinals that D can contain the "notations" needed to represent T_α, F_α. One might claim that we have a more realistic model of natural language if the extended language contains only a single pair T, F of new predicates, interpreted in different contexts. But this will avoid the same conclusion only if the contexts, considered as themselves objects, are outside of D.

We can of course consider the case where D is a proper class. Then we can have "notations" for all the ordinals and continue the construction through all the ordinals.

It is not easy to say what theory of the transfinite iteration of the above-described process of reflection leading to more and more comprehensive schemes of interpretation will yield a realistic model of natural language. It would not be realistic to regard the range D of quantifiers as fixed and determined through an unlimited range of contexts involving such reflection. In fact, for transfinite levels the model is unrealistic because of the absence in the formal language of a notion of truth in a context, or truth under an interpretation (see below).

[48]"Semantical Paradox," p. 192. Burge has informed me (in correspondence) that he did once hold the finite-levels thesis but abandoned it before completing the paper.

or 'false' a level.[49] Because of the indexical character of these words, there is no way of making the levels variables of quantification and thus implicating in the truth-conditions infinitely many levels at once. If we now assume that the discourse is itself finite, the process of interpretation can introduce only finitely many distinct levels. If we assigned transfinite levels initially, the levels could be "pressed down" so that the work of interpretation can all be done by finite levels, in the sense that these can represent all the relations of the levels that count for the discourse in question. Then the translation of any such discourse will require only an ω-level language.

The hypothesis that we are concerned only with finite discourses is essential to this argument. If we allow infinite discourses,[50] then it is a relatively straightforward matter to construct one containing at least one occurrence of 'true' that needs to be read as T_ω. Let S_0 be the sentence: Each of the sentences S_n, for $n > 0$, is true. S_1 is the sentence: S_1 is not true. If $n \geq 1$, then S_{n+1} is the sentence: Each of $S_1 \ldots S_n$ is true, but S_{n+1} is not.

Now our discourse consists of an infinite sequence of stages, as follows:

Stage 0: The speaker utters S_0 and then says, "This is shown by the following argument."

Stage 1: The speaker goes through the usual Liar paradox argument concerning S_1, asserting S_1 not to be true or false, and ends with the conclusion, "S_1 is not true."

Now we assume as inductive hypothesis that for each k, $1 < k \leq n$, Stage k is an argument ending with the conclusion, "$S_1 \ldots S_{k-1}$ are all true, but S_k is not." Thus for each k, $1 \leq k \leq n$, stage k is an argument ending with the conclusion S_k.

Stage $n + 1$: The speaker (taking note of this fact about stages $1 \ldots n$) asserts, "Each of $S_1 \ldots S_n$ is true." Then, by a Liar

[49]"Level" here does not have quite the same meaning that it has in most of the literature on iterative construction of truth-predicates beginning with Kripke. There "level" means the stage of a recursion at which a sentence obtains a truth-value (or, in the Herzberger-Gupta theory, a stable truth-value). If in a Burgean interpretation a sentence is rooted$_n$ but not rooted$_{n-1}$ and thus has "level" n, there is still another ordinal α such that in the recursion generating T_n and F_n as a minimal fixed point, the sentence becomes true or false at stage α. Properly, the level of a sentence might be the pair $\langle n, \alpha \rangle$.

This point seems to be neglected by Gupta in his remarks on Burge's construction, "Truth and Paradox," p. 28.

[50]Burge undertakes to interpret some infinite discourses in "The Liar Paradox: Tangles and Chains."

argument, he refutes the hypothesis that S_{n+1} is true and then concludes, "Each of $S_1 \ldots S_n$ is true, but S_{n+1} is not.

Surely the natural reading of this discourse in terms of Burge's scheme would be to interpret 'true' in the argument of stage $n + 1$, $n \geq 0$, as T_n, i.e., in the model M_{n+1}. The assertion at stage 0 would be an (anticipatory) inference by "infinite induction" from the assertion 'S_n is true' at the outset of each stage S_n. Since there is no upper bound to the subscripts on 'true' in these assertions, we have to assign subscript ω to 'true' at stage 0. If we were to interpret the discourse so as to put an upper bound m on the subscripts of 'true' in the stages from Stage 1 on (so that then we could use subscript $m + 1$ in stage 0), then the stage after the one in which subscript m was reached would have to be interpreted as pathological. This seems to me an unmotivated rejection of the possibility of moving to a higher level; it would certainly not be in the spirit of Burge's pragmatic principles for assigning subscripts.[51] However, we have assumed that *within* each stage 'true' has a constant subscript; otherwise, we could read the conclusion of Stage 2 as, "S_1 is true$_1$ and S_2 is not true$_0$," and there would, so far as I can see, be no reason why we should not be able, in general, to read S_{n+1} within stage $n + 1$ as "$S_1 \ldots S_n$ are all true$_1$, but S_{n+1} is not true$_0$" and then still using subscript 1 for stage 0. Our assumption is justified, it seems to me, by the maxim of not postulating shifts of context beyond necessity, and more specifically by the wording of the sentences S_n: the zeroing of 'true' in the second clause would seem to mandate its being interpreted in the same way as in the first.

It might seem that an example such as this also tells against the above argument to the effect that finite levels will suffice for the interpretation of any *finite* discourse. For the infinitary character of the discourse can be avoided by adding more to the object language, so that we can in effect replace the infinite discourse itself by something like our finite description of it, in which we would be offering a *proof* that each of the sentences S_n is true. However, it is not clear to me how to interpret the modified version without using a notion such as truth at a level, where the subscript is converted into an argument. The reason is that there is no single subscript in terms of which $S_1 \ldots S_n$ are true and S_{n+1} not, and yet S_{n+1} is true, for then we would have an outright contradiction.[52]

[51]"Semantical Paradox," pp. 193-194; "The Liar Paradox: Tangles and Chains," pp. 359-360.

[52]We can, however, by similar means force a nonparadoxical infinite ascent, even far into the transfinite, provided that we assume that the object language can express

Though I am still somewhat skeptical about the argument I do not now have definite grounds for rejecting it. But as I have already indicated, I think the scope of the conclusion is rather limited. Even for the notion of truth as expressed in English by the word 'true', the most the argument shows is that an ω-sequence of levels is sufficient to interpret uses of 'true' applied to sentences (and by parity of reasoning, other truth-bearers) so as to express truth *tout court*. It seems to me, however, that the notion of truth *under an interpretation* is also naturally expressed using the word 'true' and was so before the word 'interpretation' was given a technical sense in the construction of mathematical models. But with the language of truth under an interpretation, or of understanding sentences so that what they say is true, a finite version of our infinite discourse can be constructed and interpreted. I am therefore unrepentant in discerning in the "meanings" of the terms crucial to the semantic paradoxes a richer systematic ambiguity. Because some notion of truth under an interpretation arises quickly, it is connected to that of the language of set theory. But what follows concerning the structure of the levels in a model for the indexicality of these notions is not clear to me and has not been worked out in the literature.

III

Common to Kripke's construction and its variants and to a Burgean analysis of the indexicality of the term 'true' is a certain monotonic character: the accounts analyze stages by which the extension of 'true' (and other semantic predicates) is gradually augmented and truth-value gaps are filled in, but once a sentence is declared true or false it remains so at later stages. We have remained within the very general area of truth-value gap approaches to the paradoxes, even where we have shown a preference for classical logic. The truth schema (3.3) above, which holds at any stage of the classical-logic version of Kripke's recursion to a minimal fixed point, and in Burge's construction for any uniform assignment of levels to the truth-predicate, forces a Liar sentence to be neither true nor

---

enough mathematics. In such a case one would be able to find single (transfinite) subscripts. However, this shows that any scheme that can handle Strengthened Liar situations with only finite levels will be able to accommodate a mathematically powerful object language only if the individual levels are essentially Kripkean fixed points. The case of C_3 shows that this is forced by the assumption that truth or satisfaction is inductively defined by rather straightforward closure properties (see note 45 above).

false.[53] For the valuation schemes Kripke considers, we still have that if $Ta \lor Fa$ is true, then $Ta \equiv A$ is also true (a the standard name of A, as before). In all these situations, liar and other paradoxical sentences are neither true nor false.

Truth-value gap approaches to the Liar have been objected to on the ground that, intuitively, the trouble with liar sentences is that they appear to be both true and false, and this is just denied by a semantics that makes them neither true nor false.[54] Schemes based on classical semantics and forswearing monotonicity have been developed recently by Hans Herzberger and Anil Gupta.[55] They develop an iterative construction like Kripke's but using classical logic without the restriction of the truth-predicate to positive sentences. As before, we begin with an interpreted first-order language L and extend it by a truth-predicate T, but this time an interpretation of the extended language is simply a classical interpretation. We can define an iteration like Kripke's: $L_0(T)$ assigns T some arbitrary extension, say the empty set; in $L_{\alpha + 1}(T)$, an element of the domain D is in the extension of T if and only if it is (the code in D of) a sentence A of the extended language that is true in $L_\alpha(T)$.

Without monotonicity, the rule for limit stages cannot be so simple as before. Herzberger's rule is that if λ is a limit ordinal, an element u of D is in the extension of T in $L_\lambda(T)$ if and only if for some $\alpha < \lambda$, u is in the extension of T in $L_\beta(T)$ for *all* β such that $\alpha < \beta < \lambda$.[56] On this rule, truth-value gaps (i.e., sentences such that neither they nor their negations are in the extension of T) do occur at limit stages. On Gupta's treatment of limit stages, this can be avoided. That the limit rule involves an element of arbitrary choice is well brought out by the discussion of Nuel Belnap, who gives a formulation that includes Gupta's and Herzberger's as special cases.[57]

---

[53]A similar situation obtains if one takes truth as a predicate of propositions and thinks that the trouble with the liar sentence is that it does not express a proposition, for then of course it neither expresses a true proposition nor expresses a false proposition. See the analysis of section II above.

[54]For example, Higginbotham, Review of Martin, p. 401. However, interpreting the truth-value gap as reflecting the failure of the sentence to express a proposition suggests the reply that what the paradox argument shows is that if the liar sentence did express a proposition, that proposition would be both true and false; hence it does not.

[55]Herzberger, "Notes on Naive Semantics," "Naive Semantics and the Liar Paradox"; Gupta, "Truth and Paradox." Each did his main work independently of the other and at roughly the same time.

[56]"Notes on Naive Semantics," p. 68.

[57]"Gupta's Rule of Revision of Theory of Truth," pp. 106-107. Belnap gives a very clear picture of the main ideas of both theories, though his emphasis is on Gupta's.

The liar paradox shows immediately that this construction does not have fixed points, that is, stages at which the extension of T becomes fixed from there on. If t denotes either $\sim Tt$ or Ft (i.e., $T(neg\ t)$), then its truth-value at stage $\beta + 1$ will always be the opposite of its truth-value at stage β. Other paradoxical statements show a similar oscillation of truth-value, perhaps with longer "periods."[58] On Herzberger's and Gupta's limit rules, the construction will reach a stage at which all sentences that will ever attain stable truth-values have already attained them; in fact, Herzberger's version reaches stages, called "alignment points," at which the true sentences are *exactly* those that will remain true forever. Such a point is an analogue of a fixed point, particularly since from then on the construction cycles endlessly.[59]

In this framework, we cannot analyze the Strengthened Liar situations in quite the way we have before. For the interpretation at a given stage, the inference from A to Ta, or vice versa, can fail even with the hypothesis $Ta \lor Fa$, which in fact holds at all successor stages. The paradox is of course a refutation of the equivalence of A and Ta (the role it takes in the proof of Tarski's indefinability theorem); for example, if t denotes $\sim Tt$, we can prove $\sim (Ta \equiv \sim Tt)$ from $a = t$. However, by itself this gives us no way of choosing when we reflect between declaring $\sim Tt$ true and Ta false and declaring the opposite, in keeping with the fact that whatever truth-value a liar sentence has at a given stage, it will have the opposite at the next. There is no straightforward rule that holds generally to license the inference from a statement to the statement of its truth, or vice versa, such as Tarski's schema or its weakening (3.3). In fact, Herzberger rejects the demand that there should be such a rule laid down in advance.[60]

Whether this is a strength or a weakness of "naive semantics" is not easy to decide. One might offer alignment points as schemes of interpretation of a kind of maximal comprehensiveness, like that of minimal fixed points in monotonic approaches. In fact (3.3) holds

---

[58]Herzberger, "Notes," §6. Russell offers a "typically ambiguous" reading of the Liar in terms of his ramified theory of types that exhibits the same oscillation of truth-value (Whitehead and Russell, *Principia*, I, 62).

[59]Herzberger, "Notes," §8, especially theorem 8.1 (p. 79). The term "alignment point" is from "Naive Semantics and the Liar Paradox," p. 494.

The eventual cycling of the construction does not always obtain on the more general formulation of Belnap (see note 57); he argues that it is therefore an artifact of Gupta's and Herzberger's formulations ("Gupta's Rule of Revision Theory," p. 107).

[60]"Naive Semantics and the Liar Paradox," p. 481.

at alignment points. This idea could be the basis of a Burge-type indexical analysis of Strengthened Liar situations, where one uses the alignment points of naive semantics instead of the minimal fixed points of the Kripke-Feferman scheme. For an alignment point to be a plausible model of a scheme of interpretation actually used, however, it would have to be characterizable in the object language in some way, say by a generalized inductive definition. Though I see no reason why this should not be possible, the question will then arise whether it will differ importantly from the outcome of a monotonic approach. It also seems to me that singling out alignment points in this way is not really in the spirit of the approach, and their very existence is an artifact of Herzberger's particular limit rule.

One might offer as another objection to Gupta's and Herzberger's approaches the arbitrariness that seems to infect the limit rule. But it should be pointed out that a similar arbitrariness is not absent from monotonic approaches as well, even if one fixes a valuation scheme, as is shown by the variety of fixed points Kripke's construction gives rise to. The underlying principles of the monotonic approach (e.g., (3.3)) force liar and other paradoxical sentences to be neither true nor false (on the scheme of interpretation that the truth-predicates in them presuppose). But the matter is different, for example, with the "truth teller": a sentence stating its own truth. For such a sentence, the Tarski biconditional is true (in classical logic) but tells us nothing about whether the sentence itself is true. One might say that the logic of truth gives no basis for determining its truth-value. Declaring it neither true nor false (as happens at minimal fixed points) could be defended on the grounds that there is no objective basis for its having a particular truth-value, but this position is certainly not forced on us. However, it may still be thought an advantage of monotonic approaches that they confine this arbitrariness to the bottom of the hierarchy. However, it seems to me that the comparison of the Herzberger-Gupta and monotonic approaches has not up to now been carried very far, so that nothing very definitive can be said about their relative advantages and disadvantages.

IV

The line of interpretation of Strengthened Liar situations common to my essay and Burge's work will seem to some readers to have

been adopted only because of an a priori methodological preference for finding consistency in the use of language. It would exhibit
what Charles Chihara aptly calls "the consistency view of truth," as
I understand it, the view that there is a consistent set of rules fitting
the known facts, which will serve as the set of rules implicitly followed by English speakers in using the word 'true'.[61] Herzberger
seems to take naive semantics to embody rejection of the consistency
view, though not the acceptance of Chihara's alternative "inconsistency view."[62]

My view of this issue was set forth above (pp. 243-245). I do not
hold that if one formulates rules describing the regularities of assertion and inference in the behavior of English speakers using the
word 'true' and related words, one will have this kind of consistency.
But taken by itself, this only shows that speakers of English have
inconsistent beliefs or dispositions toward such. It does not rule
out the possibility of a consistent account of the "meaning" of 'true',
say in truth-conditional terms, so that we could describe, by a consistent set of rules, the conditions under which what speakers say
using the word 'true' is *in fact* true. The interpretation embodied
in these rules would necessarily attribute some sort of consistency
to the language, since it would not say that just anything goes, and
it would probably validate some form of the law of contradiction.

However, whether such an account can be given is in the first
instance an empirical question; there is no a priori assurance that
any proposal of this kind can be squared with the facts of usage.
But the problem is like other problems of interpretation: there is
no hard and fast line between what the meaning of the relevant
expressions determine to be true or warrantedly assertible and what
is widely believed by the community interpreted. To the extent that
an interpretation draws such a line, it will be partly an artifact of
the interpretation.[63] Here I do admit to an a priori preference for

---

[61]"The Semantic Paradoxes," p. 607. Chihara characterizes the consistency view
as "the view that an accurate statement of what 'true' means will be *logically consistent
with all known facts,* and in particular with all known facts of reference." I think this
could not be what he really intends. How could a statement of "what 'true' means"
be "accurate" and yet *not* be "logically consistent with all known facts"? The characterization given in the text seems to me to be closer to the actual object of Chihara's
criticism.

[62]"Naive Semantics and the Liar Paradox," p. 480; cf. Chihara, "The Semantic
Paradoxes," pp. 608, 613.

[63]It seems to me that Chihara neglects this point in his "diagnosis" of the Liar,
for example when he says that "we can assume that there are generally accepted
conventions which give the meaning of 'true' and which are expressed by [*Tr*]"
(ibid., p. 611). [*Tr*] (p. 605) is a principle that Chihara interprets to imply instances

consistency, in the same sense as that in which such a preference is built into the theories of translation and intepretation of Quine and Davidson. We cannot give truth-conditions to the sentences of a discourse containing what we read as logical connectives and quantifiers without attributing to it a logic that we can make sense of.[64] Beyond that, my procedure does embody a preference for finding coherent sense (and then perhaps false belief) over the contrary.

It seems likely that Chihara would reject the "consistency view" even when qualified in the way I suggest. His case against it rests on what he takes to be the implausibility of the hypotheses about natural language required by the solutions that have been offered. The approach I advocate does need to defend itself against objections of this kind. Chihara seems to see Strengthened Liar situations as a principal stumbling block for solutions based on the consistency view,[65] but he does not consider any solution based on discerning indexicality in any of the expressions involved.[66] It seems to me that the case for indexicality in the relevant expressions can be made independently and is made to some extent in my paper. But a more systematic theory is certainly a desideratum even after the advances in Burge's work; particularly, it would be desirable to deal more systematically with the relations of truth, indirect discourse, proposition-like concepts, and the ranges of quantifiers.

Whether the contrary "inconsistency view" offered by Chihara can or should lead to a rival theory, I leave to its advocates to determine. In view of the possibility noted above of integrating it with indexical conceptions, it is not so far clear to me how the Herzberger-Gupta theory is incompatible with a suitably qualified consistency view, although the absence of a "safe" truth schema like (3.3) and the nonmonotonic character itself perhaps do not fit the spirit of that view.

To say that we can attribute to English speakers a *consistent* set of rules in their use of 'true' and related locutions is not to say that

---

of the Tarski truth schema that lead to paradox. I do not know what he means by "meaning" in this statement.

[64]I do not mean to commit myself here to a more far-reaching principle of charity in interpretation such as that advocated by Davidson.

[65]"The Semantic Paradoxes," pp. 611-613.

[66]In another context he does refer to "the fact that English speakers do not attach subscripts to their utterances of 'true' and 'false' " (ibid., p. 614). Speakers of English also do not attach subscripts to different occurrences of pronouns, but *linguists* do in order to mark distinctions that speakers recognize.

Chihara evidently did not have access to Burge's "Semantical Paradox," which appeared when his own paper must have been already in press.

we can describe these rules completely.[67] Clearly in this domain as in related ones in logic, the price of consistency is incompleteness. This shows itself in the inescapable "systematic ambiguities" discussed in the essay. One reason for preferring the verdict of incompleteness over that of inconsistency in our description of natural language is that we know that no rational construction or linguistic reform will avoid the same fate.[68]

[67]Still less does it mean that the set is unique.

[68]Cf. Herzberger, "New Paradoxes for Old."

I am indebted to Nuel Belnap, Robert Martin, and especially Tyler Burge for their comments on an earlier version of this Postscript. Others who have instructed me on these matters since 1974 are Anil Gupta, W. D. Hart, Allen Hazen, Hans Herzberger, and Saul Kripke.

10

What Is the Iterative Conception of Set?

I intend to raise here some questions about what is nowadays called the "iterative conception of set." Examination of the literature will show that it is not so clear as it should be what this conception *is*.

Some expositions of the iterative conception rest on a "genetic" or "constructive" conception of the existence of sets. An example is the subtle and interesting treatment of Professor Wang.[1] This conception is more metaphysical, and in particular more idealistic, than I would expect most set theorists to be comfortable with. In my discussion I shall raise some difficulties for it.

In the last part of the paper I introduce an alternative based on some hints of Cantor and on the Russellian idea of typical ambiguity. This is not less metaphysical though it is intended to be less idealistic. I see no way to obtain philosophical understanding of set theory while avoiding metaphysics; the only alternative I can see

Revised and expanded version of a lecture given in a symposium on the Concept of Set at the Fifth International Congress of Logic, Methodology, and Philosophy of Science in London, Ontario, August 27, 1975; the present version first appeared in R. E. Butts and Jaakko Hintikka, eds., *Logic, Foundations of Mathematics, and Computability Theory* (Dordrecht: Reidel, 1977), pp. 335-367. Copyright © 1977 by D. Reidel Publishing Company, Dordrecht, Holland. Reprinted by permission of D. Reidel Publishing Company and Professors Butts and Hintikka.

The paper of my co-symposiast, Professor Hao Wang, is "Large Sets," in the same volume.

I am indebted to Robert Bunn, William Craig, William C. Powell, Hilary Putnam, and Wang for valuable discussions related to this paper. I regret that time did not permit me to follow up Bunn's remarks on Jourdain's attempt to develop the theory of inconsistent multiplicities.

[1]Wang, *From Mathematics to Philosophy*, chap. 6. In this essay this book is referred to merely by page number.

A widely cited writer whose viewpoint I would also describe as genetic is Shoenfield; see *Mathematical Logic*, pp. 238-240. [See now also "Axioms of Set Theory."]

is a positivistic conception of set theory. Perhaps the latter would attract some who agree with the critical part of my argument.

However, the positive part of the paper will concentrate on the notion of proper class and the meaning of unrestricted quantifiers in set theory. That these issues are closely related is evident since in Zermelo-type set theories the universe is a proper class.

The concept of set is also intimately related to that of *ordinal*. Although this relation will be remarked on in several places, a more complete account of it, and thus of the more properly *iterative* aspect of the iterative conception, will have to be postponed until another occasion.

I

One can state in approximately neutral fashion what is essential to the "iterative" conception: sets form a well-founded hierarchy in which the elements of a set precede the set itself. In axiomatic set theory, this idea is most directly expressed by the axiom of foundation, which says that any nonempty set has an "$\in$ − minimal" element. But what makes it possible to use such an assumption in *motivating* the axioms of set theory is that other evident or persuasive principles of set existence are compatible with it and even suggest it, as is indicated by the von Neumann relative consistency proof for the axiom of foundation.

On the "genetic" conception that I will discuss shortly, the hierarchy arises because sets are taken to be "formed" or "constituted" from previously given objects, sets or individuals.[2] But one can speak more abstractly and generally of the elements of a set as

[2] I shall consider throughout set theories which allow individuals (*Urelemente*); this requires trivial modifications of the most usual axioms, but the choice among possible ways of doing this is of no importance for us. In extensionality and foundation, the main parameters are restricted to sets (or at least nonindividuals, if classes are admitted).

The literature on the foundations of set theory does not sufficiently emphasize that the exclusion of individuals in the standard axiomatizations of set theory is a rather artificial step, taken for the convenience of pure mathematics. An applied set theory would normally have to have individuals. What is more relevant for the present discussion is that some of the intuitions about sets with which set theory starts concern sets of individuals. First-order set theory with individuals is compatible with the assumption that there are no individuals and therefore with the usual individual-free set theory.

[For further comment on the role of the axiom of foundation in set theory, see "Quine on the Philosophy of Mathematics" (Essay 7 of this volume), note 46.]

being *prior* to the set. In axiomatic set theory with foundation, this receives a mathematically explicit formulation, in which the relation of priority is assumed to be well-founded.

For motivation and justification of set theory, it is important to ask in what this 'priority' consists. However, for the practice of set theory from there on, only the abstract structure of the relation matters. Here we should recall that the hierarchy of sets can be "linearized" in that each set can be assigned an ordinal as its *rank*. Individuals, and for smoothness of theory the empty set,[3] obtain rank o. In general, the rank of a set is the least ordinal greater than the ranks of all its elements.[4]

It should be observed that the notion of well-foundedness is prima facie second-order and thus is not totally captured by the first-order axiom of foundation.[5] ZF with foundation has models in which the relation representing membership is not well-founded. However, it can be seen that in such a model there is a (not first-order definable) binary relation on the universe for which *replacement* fails. The axioms of separation and replacement are also prima facie second-order, and the fault for such failure of well-foundedness lies in the fact that their full content is not captured by the first-order schemata. But then the problem of stating clearly the iterative conception of set is bound up with the problem of the relation of set theory to second-order conceptions. This problem was already present at the historical beginning of axiomatic set theory with Zermelo's use of the notion of 'definite property'.

The idea that the elements of a set are prior to the set is highly persuasive as an approach to the paradoxes. If we suppose that the elements of a set must be "given" before the set, then no set can

---

[3]From the genetic point of view, this is an artifice; individuals are presumably given prior to any sets, even the empty set, so that if rank directly reflects order of construction, the empty set should have rank 1. The same holds for the alternative viewpoint I present below.

[4]In a set theory with individuals, some usual theorems about ranks, for example that for every ordinal α there is a set R_α of all sets of rank $< \alpha$, require the assumption that there is a set of all individuals. It follows that there cannot be too many individuals; for example, ordinals cannot all be construed as individuals. The plausibility of this assumption depends on the intended application. It would be a piece of highly dubious metaphysics to assume there is a set of *absolutely* all individuals, if for no other reason because it is not settled once and for all what *is* an individual. Pure mathematics should be independent of this question; for it the individuals can be an arbitrary set, class, or sometimes structure. However, below I shall assume that the individuals constitute a set.

[5]However, in set theories with classes foundation for classes follows from its assumption for sets.

be an element of itself, and there can be no universal set. The reasoning leading to the Russell and Cantor paradoxes is cut off.

However, one does not deal so directly with the Burali-Forti paradox. Why should it not be that all ordinals are individuals and therefore "prior" to all sets, so that there is no obstacle of this kind to the existence of a set of all ordinals? To be sure, once we look at things in this way it becomes persuasive to view the Burali-Forti argument as just a proof that there is no set of all ordinals. Moreover, the conception of ordinals as order types of well-ordered sets would suggest that for any ordinal there is at least one set of that order type to which it is not prior, so that the existence of a set of all ordinals would imply that later in the priority ordering no new order types could arise. But if there were a set of all ordinals, W, then $W \cup \{W\}$ would have just such an order type.

That ordinals need to fit into a priority ordering with sets and indeed be "cofinal" with them seems to have been neglected in discussion of iterative set theory, perhaps because in the formal theory ordinals are construed as sets, so that this happens automatically.

One would like to maintain that the requirement of priority is the *only* principle limiting the existence of sets, so that at a given position "arbitrary multitudes" of objects which are at earlier positions form sets. Although it is difficult to make sense of this, it at least should imply a comprehension principle: given a predicate 'F' which is definitely true or false of each object prior to the position in question, there is *at* the position a set whose elements are just the prior objects satisfying 'F'. In particular, the axiom of separation follows: since the elements of x are prior to x, $\{z: z \in x \wedge Fz\}$ exists and is not posterior to x. But to apply this idea more generally, some way of marking positions is needed. The genetic approach in effect assumes such to be given. Conversely, if set theory is assumed, the ordinals offer such a marking.

II

We have now gone about as far as we can without explaining what I have called the genetic approach. Put most generally, it supposes that sets are "formed," "constructed," or "collected" from their elements in a succession of stages. The first part of this idea has some plausibility as an interpretation of some of Cantor's preliminary remarks about what a set is. Thus Cantor's famous "definition" of 1895:

> By a "set" we understand any collection M into a whole of definite, well-distinguished objects of our intuition or our thught (which will be called the "elements" of M).[6]

If we were to take "collection into a whole" quite literally as an operation, then the priority of the elements of a set to the set would simply be priority in order of construction. Cantor's language suggests rather that "collection" (*Zusammenfassung*) is an operation of the *mind*; in this case the requirement would be that the objects be represented to the *mind* before the operation of collection is performed. However, it will be clear that as it stands this temporal reading is too crude.

It may seem that these notions belong only to the early history of set theory, and in particular that they would have disappeared with the discrediting of logical psychologism at the end of the last century. But the fact is that they are to be found in the contemporary literature. Thus Shoenfield writes

> A closer examination of the paradox [Russell's] shows that it does not really contradict the intuitive notion of a set. According to this notion, a set A is formed by gathering together certain objects to form a single object, which is the set A. Thus before the set A is formed, we must have available all of the objects which are to be members of A.[7]

Although Shoenfield says that we form sets "in successive stages," he does not offer an interpretation, temporal or otherwise, of the stages, although he does use "earlier" to express their order.

Wang writes, "The set is a single object formed by collecting the members together." (p. 181). He recognizes that the concept of collecting is highly problematic and makes an interesting attempt to explain it; he interprets it as an operation of the mind. We shall discuss his views shortly.

We now have the familiar conception of sets as formed in a well-ordered sequence of stages, where a set can be formed at a given

---

[6]*GA*, p. 282. Wang calls this definition "genetic" (p. 188) and speculates that the difference between this and previous ones (in particular, *GA*, p. 204) may be due to Cantor's awareness of the Burali-Forti paradox. It should be remarked that a genetic conception of *ordinals* is intimated in *GA*, pp. 195-196, a text from 1883.
[7]*Mathematical Logic*, p. 238.

stage only from sets formed at earlier stages and from whatever objects were available at the outset.

The language of Cantor, Shoenfield, and Wang invites regarding the intuitive concept of set as analogous to the concepts of constructive mathematics, where one also uses the idea of mathematical objects as constructed in successive stages, and where there is no stage at which all constructions are complete. An immediately obvious limitation of the analogy is that in the typical constructive case (e.g., orthodox intuitionism) the succession of stages is simply succession in *time*, and incompletability arises from the fact that the theory is a theory of an idealized finite mind which is located at some point in time and has available only what it has constructed in the past and its intentional attitudes toward the future. The same interpretation of iterative set theory would require that the stages be thought of as a kind of "super-time" of a structure richer than can be represented in time on any intelligible account of construction in time. It is hard to see what the conception of an idealized mind is that would fit here; it would differ not only from finite minds but also from the divine mind as conceived in philosophical theology, for either the latter is thought of as in time, and therefore as doing things in an order with the same structure as that in which finite beings operate, or its eternity is interpreted as complete liberation from succession.

It may seem that there is a much more obvious conflict between iterative set theory and a constructive interpretation of it: set theory is the very paradigm of a *platonistic* theory. As is customary in discussing the foundations of mathematics, platonism means here not just accepting abstract entities or universals but epistemological or metaphysical realism with respect to them. Thus a platonistic interpretation of a theory of mathematical objects takes the truth or falsity of statements of the theory, in particular statements of existence, to be objectively determined independently of the possibilities of our knowing this truth or falsity. Contrast, for example, the traditional intuitionist conception of a mathematical statement as an indication of a "mental construction" that constitutes a proof of the statement.

Perhaps it would be rash to rule out an interpretation of set theory that would not be platonistic.[8] But in any case it seems that

---

[8]Formalism apart, one ought not to rule out the possibility of an interpretation of set theory along constructivist lines, particularly in view of the broadening of the intuitionist outlook in recent years. ZF has recently been shown consistent relative to some set theories based on intuitionist logic. See Friedman, "The Consistency of Classical Set Theory," and Powell, "Extending Gödel's Negative Interpretation to

a platonistic interpretation is flatly incompatible with viewing the "formation" of sets as an operation of the mind. However, that there is not a direct contradiction should be evident when we observe that we are concerned in set theory with what formations of sets are *possible*. In contrast to the situation with intuitionism, we do not require that a statement to the effect that it is possible to "construct" a set satisfying a certain condition should be itself an indication of a construction. Even if we construe the formation of sets as a mental operation, what is possible with respect to such formation can be viewed independently of our knowledge. Thus there is a prima facie resolution of the difficulty posed by platonism.

However, we have not reckoned yet with the actual content of the set-theoretic principles that seem to require a Platonistic interpretation, such as the combination of classical logic with the postulation of a set of all sets of integers. In an iterative account, the individual steps of iteration are in Wang's word "maximum" (p. 183). Namely we regard as available at any given stage any set that *could* have been formed earlier. We could represent this assumption as that if a set *can* be formed at a given stage, then it *is* formed (or at least that it *exists* at that stage). This of course has effects on what can be formed later, since every *possibility* of set formation at stage α is such that its result is available at later stages and can therefore enter into further constructions.

We can illustrate this by the manner in which these ideas are used to justify the power set axiom. At a given stage α, any "multitude" of available objects can be formed into a set. Let x be a set formed at stage α, and let y be a subset of x. Since the elements of y are all elements of x, they must have been available at stage α. Hence y *could have been formed* at stage α. x is available from stage $\alpha + 1$, on; from our assumption it follows that y is available as well. Thus at stage $\alpha + 1$, *every* subset x is available and $\mathfrak{P}(x)$ can be formed.

We should distinguish two principles that are playing a role here, and which can be confused with one another. One is the "arbitrary" nature of sets, which, following Wang, we have expressed (provisionally) by saying that any "multitude" of available objects can be formed into a set. The other is the principle that allows the transition from possibility of formation at stage α to availability at stage $\alpha + 1$, perhaps by way of existence at stage α. Both principles may be taken to arise from the idea we expressed above that the priority requirement, here interpreted to mean that sets are formed from

---

ZF."

available objects, is the *only* constraint on the existence of sets. But the second principle begins to undercut the idea of sets as *formed* from available objects, since the successive stages of formation are required *only* because a set must be formed from available objects, and not because of any successiveness in the process of formation itself. The question arises whether the interpretation of the priority of the elements of a set to the set in terms of order of construction does not reduce to viewing this priority as a matter of *constitution*: the elements are prior because they *constitute* the set (to use a more abstract phrase than they are its *parts*, which would invite inferences inappropriate to set theory). This view is close to what I advocate below, but it is quite different from the conception of set formation as an operation of the mind.

Wang makes an interesting attempt to develop the latter idea. He says that a multitude can be formed into a set only if its "range of variability" is "in some sense intuitive" (p. 182). I shall for the time being accept the notion of "multitude"; the problems concerning it are related to the question of the notion of *class* in set theory. Wang indicates that to form a set is to "look through or run through or collect together" all the objects in the multitude.[9] Thus a condition for a multitude to form a set is that it should be possible thus to "overview" it. This overviewing is a kind of intuition, presumably analogous to perception. Of course infinite multitudes can be "overviewed."

Clearly Wang does not maintain that human beings have the capacity to "run through" infinite collections. He speaks of overviewing "in an idealized sense" (p. 182). In other words, he has a highly abstract conception of the possibilities of intuition. In constructive conceptions of the arbitrary finite, we already disregard the actual bounds of human capacities, in the sense that if a sequence of steps has been performed we always *can* perform a further step, and any operation can be iterated. Wang's idealized overviewing carries such abstraction further in that finitude and even the limitations posed by the continuous structure of space-time (as the setting of the objects of perception and even of the mind itself) are disregarded.

The question arises what force it still has, on Wang's level of abstraction, to treat the possibilities involved in this kind of motivation of the axioms of set theory as possibilities of *intuition*. The

[9]Wang may be developing the remark of Gödel ("What Is Cantor's Continuum Problem?", p. 272, n. 40) that the function of the concept of set is "synthesis" in a sense close to Kant's.

analogy with sense-perception which is central to the constructive conception of intuition in Brouwer or to Hilbert's distinction between intuitive and formal mathematics seems to be almost totally lost. Consider Wang's remarks on the axioms of separation and power set. The former is stated thus: If a multitude A is included in a set x, than A is a set.

> Since x is a given set, we can run through all members of x, and, therefore, we can do so with arbitrary omissions. In particular, we can in an idealized sense check against A and delete only those members of x which are not in A. In this way, we obtain an overview of all the objects in A and recognize A as a set (p. 184).

The idealization seems to include something like omniscience: A may be given in some way that does not independently of the axiom assure us that it is a set, and yet we can use it in order to "choose" the members of x that are in A. A may of course be given to us by a predicate containing quantifiers that do not range over a set; in deciding whether an element y of x is to be deleted, we cannot "run through" the values of the bound variables as part of the process of checking y against A. It is not clear what more structured account of "idealized checking" would yield the result Wang needs. An alternative would be to view subsets as run through not by verification but by arbitrary selection. But if a predicate is given, how are we to "select" just those elements to x that satisfy it unless we can decide which ones do?[10]

---

[10]Cf. the fact that in intuitionism the classical notion of set splits into those of spread and species.

It is of interest to look at a case, namely the hereditarily finite sets, where something like arithmetical intuition does yield the axiom of separation. Here we can argue by induction on n that there is a w such that

$$(\forall z)\,(z \in w \leftrightarrow z \in \{x_1 \ldots x_n\} \wedge Fz). \ (*)$$

For if $n = 0$, $w = \phi$. Suppose w satisfies (*) and consider $\{x_1 \ldots x_{n+1}\}$. If $\neg Fx_{n+1}$ then w satisfies

$$(\forall z)\,(z \in w \leftrightarrow z \in \{x_1 \ldots x_{n+1}\} \wedge Fz);$$

if Fx_{n+1}, then $w \cup \{x_{n+1}\}$ satisfies the condition. In effect this argument shows us that of the possible subsets of a finite set *one* satisfies the separation condition, without telling us which one.

One might consider interpreting the requirement that a set x can be "overviewed" as meaning that it can be run through in a well-ordered way and then attempt a transfinite analogue of this inductive argument. It seems that to handle limit cases such an argument requires replacement, and of course separation can be deduced from replacement in a more trivial way without assuming x to be well-orderable.

More strain on the concept of intuition appears in Wang's treatment of the power set axiom (p. 184):

> We have. . .an intuitive idea of running through with omissions. This general notion. . .provides us with an overview of all cases of AS [separation] as applied to *x*.

By saying that we have an "intuitive idea" of running through with omissions, he does not only mean that a *case* of such running through is intuitable, for that would not yield the result. Rather, for a given set *x* the concept of such runnings through is intuitive in the sense that "we can" run through *all* cases of it. Something of the content of the idea of intuitive running through seems to be lost here. Clearly in the case of small finite sets of manageable objects, we really do see all the elements "as a unity" in a way that preserves the articulation of the individual elements. Somewhat larger sets can be seen by a completable succession of steps of bringing one (or a few) objects under one's purview. If we consider arbitrary iterations of such steps, there is no longer any limit to how a large a (finite) set can be thus intuited. We also have a simple and clear generative rule for sets such as the natural numbers, though the process of sensibly intuiting is in this case incompletable, so that the *givenness* to us of such a set depends in a more essential way on conception. But regarding the natural numbers as intuitable as a whole amounts just to abstracting from the above incompletability. There is another qualitative leap in dealing with all sets of integers, as has often been remarked on in the literature (as indeed Wang himself emphasizes when he stresses the importance of impredicativity).[11] Here, however we understand the notion of an "arbitrary set" of integers, say by some picture of arbitrary selection, we do not have the conceptual grasp of what the totality contains that would be given by some method of generating them. The divorce from *sensible* intuiton involved in treating this totality as "intuitable" seems complete, unless perception is used only as a source of quite remote analogies. Two mathematical symptoms of this situation are the absence of a definable well-ordering of the continuum and our inability to solve the continuum problem.

I ought to make clear that I understand by "intuition" a quasi-perceptual manner in which an object is presented to the mind. In this I follow Kant. The word "intuition" is also used in the philos-

---

[11]For example the classic formulation of Bernays ("On Platonism," pp. 275-276). Cf. Wang, pp. 183.

ophy of mathematics and otherwise for any manner by which propositions can be known where this knowledge is not largely accounted for by deductive or inductive reasoning. There is a tendency to confuse these two senses. As for the appropriateness of my sense of intuition to Wang, I should point out that the other concept of intuition does not distinguish sets from "multitudes" or other primitive notions that might enter into evident set-theoretic axioms. Moreover, intuition in the latter sense is purely *de dicto*, intuition *that* certain propositons are true, while Wang clearly requires intituition *de re*, intuition of sets.[12]

I no longer understand Wang's talk of "intuitively running through" where it is applied to the set of all sets of integers. In the above I have perhaps connected intuition more closely with the senses (more abstractly Kant's "sensibility") than Wang would find acceptable. But even quite abstract marks of sensibility, such as the structure of time, are lost in this case.

However, there might be an interpretation on which Wang's hypothesis that a "multitude" is a set if and only if it is an object

---

[12]This distinction is explicitly made by Steiner (*Mathematical Knowledge*, pp. 130-131). However, Steiner regards only *de dicto* mathematical intuition as defensible. I have sketched in previous writing an account of arithmetical intuition on Kantian lines (this volume, Essay 6, section IX; Essay 5; Essay 1, section III). Curiously, Steiner cites me and then says, "No one today, however, upholds hard-core intuition—the direct intuition of mathematical *objects*" (ibid.), although he then mentions Gödel as a possible exception. Since I was trying to elucidate the *formal* character of Kantian intuition, perhaps Steiner did not consider arithmetical intuition on my view to be "direct intuition of mathematical *objects*," particularly in Essay 1 above.

Some comment is in order on Gödel's view of mathematical intuition, particularly since he explictly says, "We do have something like a perception also of the objects of set theory" ("What Is Cantor's Continuum Problem" p. 270). This seems to commit Gödel to intuition *de re*. His immediately following remark does not give any argument for this; he says only that it "is seen from the fact that the axioms force themselves on us as being true," which implies only intuition *de dicto*.

However, it seems clear that Gödel holds (ibid., p. 271 bottom), that our ideas of objects of certain kinds contain "constituents" which are *given* (not "created" by thinking) on the basis of which we "form our ideas" of these objects and postulate theories of them. "Evidently the 'given' underlying mathematics is closely related to the abstract elements contained in our empirical ideas" (ibid., p. 272). In the case of set theory, Gödel does not give any indication of wanting to distinguish sets as objects of intuition from other entities (such as "properties of sets" [ibid., p. 264, n. 18]) that the axioms might refer to.

Elsewhere Gödel, in contrast to the above passage, contrasts the intuitive with the abstract ("Über eine noch nicht benützte Erweiterung," p. 281). There he seems to be using "intuition" in a much narrower and more strictly Kantian sense. Of course there he is writing in German; possibly he would not use *Anschauung* in the sense in which he uses "intuition" in "What Is Cantor's Continuum Problem?"

[This is confirmed by the fact that in "On an Extension of Finitary Mathematics," *Anschauung* is translated "concrete intuition." Gödel comments on the matter in note (h).]

of intuition would be defensible. The concept of intuition would be logical and ontological rather than perceptual and epistemic. To be an object of intuition would be simply to be an object rather than a Fregean "concept" or perhaps a property. In other words, Kant's contrast of intuitions as "singular representations" with concepts as "general representations" (*Logik* §1, *Ak.*, IX, 91) would give virtually the only essential mark of intuition.[13] Although this interpretation would bring Wang's hypothesis into accord with the views I express below, I shall not pursue it further in this paper.

I want now to turn to the axiom of replacement, about which Wang has most interesting things to say. Wang writes:

> Once we adopt the viewpoint that we can in an idealized sense run through all members of a given set, the justification of SAR[14] is immediate. That is, if, for each element of the set, we put some other given object there, we are able to run through the resulting multitude as well. In this manner, we are justified in forming new sets by arbitrary replacements. If, however, one does not have this idea of running through all members of a given set, the justification of the replacement axiom is more complex (p. 186).

The picture here is marvelously persuasive; for me, it expresses very well why the axiom of replacement seems obvious. But something like the omniscience assumption of his discussion of separation is present in the remark that "*we put* some other given object there" and "are able to run through the resulting multitude." What is much more revealing is that the objects seem to have no relevant internal structure: Our ability to "run through" a multitude is preserved if we replace its elements by *any* other objects, for example by much larger sets. It is as if the objects were given only as wholes, or at least that any internal structure would not affect the possibility of running through the totality. A model for this (conceptual rather than intuitive) is the case where the objects are given only by names. Wang seems to be making a hypothesis here, although I do not feel the same qualms as in the case of the power set about taking it as a hypothesis about what is intuitable. It is of course the *combination* of the replacement with the power set axiom that yields sets of very high ranks.

[13] According to Hintikka, this is Kant's own view. See "On Kant's Notion of Intuition," also *Logic, Language-Games, and Information*, chaps. 6-9, and *Knowledge and the Known*, Essays 6, 7, 8.

[14] SAR is the statement: if b is an operation and b_x is a set for every member x of a set y, then all these sets b_x form a set (p. 186).

Wang expresses by his picture the idea, present in the earliest intimation of the axiom of replacement (Cantor, *GA*, p. 444; cf. Wang, p. 211) that whether a multitude forms a set depends only on its cardinality and not on the "internal constitution" or relations of its elements. Put in this way, the axiom is not a principle of *iteration* of set-formation, in line with the conclusion of Boolos that it does not follow from the iterative conception of set.[15] The most direct justification of replacement by appeal to ideas about stages seems to me somewhat circular.[16] That of Gödel cited by Wang (p. 186 and n. 5, p. 221) I find less immediate and persuasive.

Although I admit that Wang's picture (apart from the question of omniscience) offers a plausible hypothesis about what is intuitable,[17] it seems to me to be equally plausible as a hypothesis about what can be thought or about what can *be*, and the latter interpretations fit better the case of power set. I want now to pursue the genetic conception of sets in this direction.

III

In the preceding section we saw a number of difficulties with the idea that sets are "formed" from their elements, in particular by

---

[15]"The Iterative Conception of Set," pp. 228-9. Boolos seems, however, to arrive at his conclusion too easily. He seems to assume that the "stages" and their ordering have to be given independently of the concept of set, at least for the expression of *the* iterative conception. His actual axioms about stages (and a further possible one he mentions on p. 227) would permit the stages to be ordered by a very simple recursive well-ordering. It seems to me that sets and "stages" ought to be "formed" together, so that the formation of certain sets should make possible going on to further stages.

However, the most obvious principle of this kind, that if a well-ordering has been constructed then there is a stage such that the earlier stages are ordered isomorphically to the given well-ordering, is weaker than the axiom of replacement.

[16]Thus Shoenfield (*Mathematical Logic*, p. 240) deduces replacement from a "cofinality principle": if "we have a set A, and. . .we have assigned a stage S_a to each element a of A. . . .There is to be a stage which follows all of the stages S_a." (p. 239). However, he justifies this by saying, "Since we can visualize the collection A as a single object (viz., the set A), we can also visualize the collection of stages S_a as a single object; so we can visualize a situation in which all these stages are completed" (ibid.). Here he is assuming that "visualizability as a single object" is preserved by replacement of a by S_a; but that is just the principle of replacement. Wang's picture seems more fundamental than the kind of argument Shoenfield gives.

Wang gives a similar argument (p. 220, n. 4).

The argument does obtain general replacement from the special case where the range of the replacing function consists of stages.

[17]This plausibility is perhaps reinforced by the fact that replacement holds for the hereditarily finite and the hereditarily countable sets.

an activity of running through in intuition. I want now to suggest a more "ontological" view of the hierarchy of sets.

The earliest attempt that we know to explain the paradoxes of set theory and to develop set theory in a way that avoids them is in Cantor's famous letter to Dedekind of July 28, 1899 (*GA*, pp. 443-7). Cantor there presupposes his earlier "many into one" characterizations of the notion of set, such as that of 1895 cited above. He begins (p. 443) with the "concept of a definite multiplicy (*Vielheit*)." What he calls an *inconsistent* multiplicity is one such that "the assumption of a 'being together' (*Zusammensein*) of *all* its elements leads to a contradiction." A consistent multiplicity or set is one whose "being collected together to 'one thing' is possible." It is noteworthy that Cantor here identifies the possibility of all the elements of a multiplicity *being together* with the possibility of their being collected together into one thing. This intimates the more recent conception that a "multiplicity" that does not constitute a set is *merely potential*, according to which one can distinguish potential from actual being in some way so that it is impossible that *all* the elements of an inconsistent multiplicty should be actual.

I am here interpreting Cantor to mean that where there is an essential obstacle to a multiplicity's being collected into a unity, this is due to the fact that in a certain sense the multiplicy does not exist. It does not exist as a totality of its elements; if it did, they would form a unity or could at least be *collected* into a unity. But in the case of inconsistent multiplicities, this is impossible. The sense of this nonexistence needs some further elucidation which Cantor does not supply. The language of potentiality and actuality is not in the text, though Cantor may have been suggesting it in calling an inconsistent multiplicity *absolutely infinite* (p. 443).

What seems to me of interest in the present connection in these hints of Cantor is that he seems to be trying to distinguish sets from "inconsistent multiplicities" without real use of any metaphor of *process* according to which sets are those multiplicities whose "formation" can be "completed."[18] Such a metaphor makes the idea of an inconsistent multiplicity as a merely potential totality rather easy. I suggest interpreting Cantor by means of a modal language with quantifiers, where within a modal operator a quantifier always ranges

[18]However, this is not to say that Cantor now conceives sets as having no intrinsic relation to the mind. Hao Wang has pointed out to me that this would be questionable. For example he characterizes a consistent multiplicity as one the totality of whose elements "can be thought of without contradiction as 'being together', so that their being collected together (*Zusammengefasstwerden*) to 'one thing' is possible."

over a set (not, however, one that is explicitly given or even that exists in the "possible world" it might be taken to range over). Then it is not possible that all elements of, say, Russell's class exist, although for any element, it is possible that *it* exists. As it stands this conception requires it to be meaningful to talk of *any* set (or any object), even though the range of *this* quantifier does not constitute a unity; the elements of its range cannot all "exist together." However, at least some such talk can be replaced by ordinary quantification behind necessity.

What one would like to obtain from this conception is some interpretation of the stages of the iterative conception that also does not depend on the metaphor of process. However, I intend first to look at Cantor's conception of a multiplicity. Wang seems to use "multitude" in the same sense, although he does not use it to translate Cantor's *Vielheit* when he discusses Cantor's 1899 correspondence with Dedekind (p. 211). These notions are among a number which occur in the literature on logic and set theory and which purport to be more comprehensive than the notion of set. The most respectable of these notions is that of (proper) class. We should also mention Frege's "concept," Zermelo's "definite property," and Shoenfeld's "collection." Gödel's "property of sets"[19] presumably also belongs on this list.

Of all these notions, perhaps the most developed from the philosophical side is Frege's notion of a concept; I shall use it for purposes of comparison. What is then striking is that neither Cantor's nor Wang's notion seems to be derived from predication as Frege's is. Since Cantor's notion is one of the prototypes of the notion of proper class, this fact seems to clash with the actual use of the notion of class in set theory (perhaps with exceptions; see below) according to which classes are derived from predication; Zermelo's "definite property" is a more immediate prototype than Cantor's *Vielheit*. I myself have suggested that sets are not derived from predication while classes are.[20]

Cantor in 1899 apprently thought of sets as a species of the genus multiplicity, and then perhaps the nonpredicative (if not impredicative!) character of multiplicities in general was needed in order to preserve the "arbitrariness" of sets against its being restricted by what we might express in language. Frege seems to have obtained the same freedom by his realism about concepts. However, for

---

[19]"What Is Cantor's Continuum Problem?" p. 264, n. 18.
[20]"Sets and Classes" (Essay 8 of this volume).

Frege the nature of a concept could apparently only be explained by appeal to predication (more generally, to "unsaturated" expressions). The sharp distinction between concepts and objects is a shadow of the syntactical difference between expressions with and without argument places. This difference is then "inherited" by concepts that are not denoted by expressions of any language we use or understand.

I want to suggest that predication plays a constitutive role in the explanation of Cantor's notion of multiplicty as well and that at least an "inconsistent multiplicity" must resemble a Fregean concept in not being straightforwardly an object. In the Cantorian context, predication seems to be essential in explaining how a multiplicity can be given to us not as a unity, that is as as set. Much the clearest case of this is understanding a predicate. Understanding '*x* is an ordinal' is a kind of consciousness or knowledge of ordinals that does not so far "take them as one" in such a way that they constitute an object. We might abstract from language and speak with Kant of knowledge through concepts, but whatever we make of this the predicational structure is still present.

The philosophy of Kant might suggest another way in which a multiplicity might be given not as a unity, namely as an "unsynthesized manifold." It seems clear that in the cases Kant actually envisaged, the objects involved would have the definiteness necessary to constitute a Cantorian "multiplicity" only if they are a set. Even if we generalize the notion in some way, I do not see how such a "manifold" can be taken up into explicit consciousness except perceptually (intuitively) or conceptually.

The idea that to be an *object* and to be a *unity* are the same thing is very tempting and has deep roots in the history of philosophy. An object is something whose identity with itself (represented in different ways) and difference from other objects can be meaningfully talked about; it is then subject to at least rudimentary application of *number*. This line of reasoning inclines us to identify Cantor's "multiplicity" with Frege's concept at least in that a multiplicity which is not a set is not an object. Some such assumption seems necessary to cut off the question why there are not multiplicities whose elements are not sets or individuals: multiplicities are multiplicities of *objects*, and under that condition there are no restrictions on the existence of multiplicities (although possibly on the use of quantifiers over them), but if a multiplicity is an object, then it is a set.

However, we have to deal with the fact that in Frege the gulf between concepts and objects comes from the structure of predication itself, so that a concept is irremediably not an object, even if only one object falls under it. Cantor evidently holds that some multiplicities just *are* sets, in particular those that are not too large. This may seem not a very essential difference; if a concept F is such that there is a set y of all x such that Fx, then the distinction between F and y is just the distinction between $(\ ) \in y$ and y.[21] For an inconsistent multiplicity there is not such "reducibility." In view of Russell's paradox, the idea of the predicative nature of the concept will motivate the idea that there should *be* inconsistent multiplicities, but it does not seem to motivate Cantor's particular principles as to what multiplicities are "consistent." For reasons which will become clear later, I do not think we have yet captured the sense in which an inconsistent multiplicity is not an object.

Let us look for a moment at the well-known difficulties of Frege's theory of concepts. The conception has the great attraction that it enables us to generalize predicate places without introducing nominalized predicates that purport to denote objects (classes or attributes)—something that has to be restricted on pain of Russell's paradox. But the temptation to nominalize is irresistible, as Frege himself discovered on two fronts. His construction of mathematics required an "official" nominalization in postulating extensions. "Unofficial" nominalizations cropped up repeatedly in his own informal talk about concepts and gave rise to the paradox that the concept *horse* is an object, not a concept.[22] At the end of his life Frege decided the temptation was to be resisted and that neither the expression "the concept F" nor the expression "the extension of the concept F" really denotes anything.[23] However, perhaps

[21]In a sense $(\ ) \in y$ is F, since coextensiveness for concepts is the analogue of identity for objects, but we cannot say that the concepts are identical.

[22]"Über Begriff und Gegenstand." Only it might be a concept after all, since "is a concept" is syntactically such that it takes object-names as subjects, and is therefore a predicate of objects.

Of course a voluminous literature has grown up on this question.

[23]*Nachgelassene Schriften*, pp. 288-289, a text written in 1924 or 1925. Cf. ibid., pp. 276-277, from 1919. The late evolution of Frege's thought on these matters is discussed in my "Some Remarks on Frege's Conception of Extension."

One can question whether the problem of generalizing predicate places is really solved by Frege's approach. Once we have generalizable variables in predicate places, we have new predicates that are not generalized by the variables in question—predicates which in Frege's semantics denote "second-level concepts." Hence the urge to extend the language by nominalization appears in Frege's context in another form. An ultimate Fregean canonical language would have to be a predicate calculus of order ω of which the semantics can no longer be expressed, unless we admit predicates of infinitely many arguments of different types. Surely *we* understand such a language by a means which from this Fregean point of view is a falsification,

yielding to temptation can in one way or another be legitimized, even where the extensions postulated are not sets (cf.).

Does the Cantorian concept of "multiplicity" have to be understood realistically? To the extent that *sets* are understood realistically, of course "consistent multiplicities" are mind-independent in the corresponding sense. However, obviously it does not follow from the fact that we allow classical logic and impredicative reasoning about sets that we have to allow either about classes or other more general entities. The suggestion made below that such entities are at bottom *intensions* would imply, if we think of an intension in the traditional way as a meaning entertained by, and in some sense constructed by, the mind, that realism about them is inappropriate. However, in view of the interest of impredicative conceptions of classes for large cardinals, both predicative and impredicative conceptions should be pursued.[24]

Let us now return to Cantor's suggestion that the elements of an "inconsistent multiplicity" cannot all exist together. I do not con

---

namely by a recursion in which *in general* variables with argument places of given types range over relations of these arguments, and each type is reached by finite iteration of the ascent from arguments to function. That involves a "unification of universes" that Frege rejected, and which essentially contains nominalization.

Frege's logic contained bound variables only for objects and first-level functions, and free variables for one type of second-level function. He refers informally in at least one place to a third-level function (*Grundgesetze* I, p. 41), which would seem to be required by the *semantics* of his system. Formally, he thought higher levels dispensable because second-level functions could be replaced by first-level functions in which the function arguments were replaced by their *Wertverläufe* (ibid., p. 42). This was untenable because it depended on the inconsistent axiom V. But of course in a less absolute way to replace functions by sets which are objects is just the procedure of set theory, which then does dispense with "higher-level functions" for most purposes. It is only quite recently, with the discussion of measurable and other very large cardinals, that higher than second-order concepts relative to the universe of sets have had any real application. See especially Reinhardt, "Remarks on Reflection Principles," and Wang, "Large Sets."

[24]Analogously to the theory of predicatively definable sets of natural numbers, one can explore mathematically the predicative definability of classes relative to the universe of sets. See Moschovakis, "Predicative Classes."

Wang's discussion of the axioms of separation and power set could lead one to think that impredicative reasoning about "multitudes" is already involved in motivating the axiom of power set. Although this may be psychologically natural, what the power set axiom says is that given a set x there is a set of all sub*sets* of x, not that there is a set of all "multitudes" whose elements are elements of x. Thus being an arbitrary subset of x has to be definite, but the "multitude" of them is defined without quantifying over arbitrary *multitudes*. The axiom of separation tells us that *any* "multitude" of elements of x is a subset of x, so that the "definiteness" of the property of being a subset of x implies that of being a submultitude of x. But we do not need to assume the definiteness of the latter property; indeed if we think of "multitudes" intensionally (see below), it is only their *extensions* that become a definite totality by this reasoning.

clude that an inconsistent multiplicity does not exist in any sense; even the hypothesis that it is not an object will have to be qualified. However, one implication is clear: it is not a totality of its elements; it is not "constituted" in a definite way by its elements. Its existence cannot require the prior existence of all its elements, because there is no such prior existence.

I wish to explicate the difference between sets and classes by means of some intensional principles about them. From the idea that a set is constituted by its elements, it is reasonable to conclude that it is *essential* to a set to have just the elements that it has and that the *existence* of a set requires that of each of its elements. Exactly how one states these principles depends on how one treats existence in modal languages. I shall assume that the truth of $x \in y$ requires that y exists (Ey). Then we have:

(1) $x \in y \rightarrow Ex \wedge Ey$

(2) $x \in y \rightarrow \Box(Ey \rightarrow x \in y)$

(3) $x \notin y \wedge Ey \rightarrow \Box(x \notin y).$[25]

My proposal is that these principles should fail in some way if y is an "inconsistent multiplicity" or proper class. Indeed Reinhardt has suggested that proper classes differ from sets in that under counterfactual conditions they might have different elements.[26] I am endorsing this suggestion as an explication of the intuitions about "inconsistent multiplicities" that I have been discussing.

---

[25]The most natural and elementary application of these principles is in relation to sets of ordinary objects that are the extensions of predicates contingently true of them. I intend to discuss these matters in a paper in preparation; cf. Parsons, "Sets and Possible Worlds" and Tharp, "Three Theorems of Metaphysics."

(1)-(3) exactly parallel familiar principles of identity except that identity is usually treated as independent of existence.

(2) implies that set abstracts are not rigid designators. If 'F' is a predicate that holds of an object x, but not necessarily so, then $\Box(E\{z{:}Fz\} \rightarrow x \in \{z{:}Fz\})$ is true with the scope of the abstract outside the modal operator but false with the scope within. I assume that in any possible world $\{z{:}Fz\}$ is the set of existent z such that Fz in that world.

The free variables in (1)-(3) range over all possible objects, although for the present discussion the appropriate modal logic has *bound* variables ranging only over existing objects. If this treatment of free variables is thought to be too Meinongian, then (3) needs to be replaced by a schema

$$(\forall x)\Box(x \in y \rightarrow Fx) \rightarrow \Box(\forall x)(x \in y \rightarrow Fx)$$

or, in the second-order case, by the corresponding second-order axiom.

[On all these matters see Essay 11 below.]

[26]"Remarks on Reflection Principles," p. 196.

As before, for a given "possible world" we should think of the bound variables as ranging over a set, perhaps an R_α (see note 4); but the sets that exist in that world are elements of the domain, while classes are arbitrary subsets of the domain.

Reinhardt does not use an intensional language; in his formulation the actual world V is part of a counterfactual "projected universe" which is the domain of the quantifiers. He assumes a mapping j on $\beta(V)$ such that if $x \in V$, $jx = x$. We can think of jx as a "counterpart" of x in the projected universe, in the sense of Lewis.[27] Thus sets are their own counterparts and can be strictly reidentified in alternative possible worlds. A set y can have no new elements in the projected universe and its only new nonelements are all x such that $x \notin V$. This accords with (1)–(3).

For a class P, jP can have additional elements in the projected universe, so that it violates (3), although jP agrees with P for "actual" objects (elements of V).[28] Reinhardt's extensional language must distinguish jP from P; hence my thinking of jP as a "counterpart." A complication is that P itself occurs in the projected universe, though now as a set. Reinhardt himself suggests an alternative reading, by which a class x is an *intension*, so that P in the actual world and jP in the projected universe are the "values" in two different possible worlds of the same intension. A formal language in which this reading might be formulated is the second-order modal language of Montague, where the first-order variables range over objects and are interpreted (in the manner usual for modal logic) rigidly across possible worlds, and the second-order variables

[27]"Counterpart Theory and Quantified Modal Logic."

[28]Thus the relation of P and jP does not contradict (2). However, this is due to the special nature of the projected universe: j is an elementary embedding of the (sets and classes of) the actual world into it. (2) is presumably not an appropriate general principle about proper classes.

The explicit application to set theory of a modal conception of mathematical existence and the use of modal quantificational logic to explicate it seem to originate with Putnam, "Mathematics without Foundations." Putnam does not address the questions about "trans-world identification" of sets that our principles (1)-(3) are meant to answer. However, it appears that his suggested translation of statements containing unrestricted quantifiers (p. 58) requires that a "standard concrete model of Zermelo set theory" should have a structure that is rigid, that is, the relations are not changed when considered with respect to an alternative possible world. If this assumption is made, equivalents of (2) and (3) follow from the fact that a standard model is maximal for the ranks it contains (p. 57). On "concreteness," see "Quine on the Philosophy of Mathematics" (Essay 7 of this volume), note 32.

Putnam seems to envisage a first-order formulation, which requires his "models" to be objects. The second-order formulation seems to us more appropriate not only for the set-class distinction but also for explicating the priority of the elements of a set to the set (Section V below).

range over intensions, which in the semantics are functions from possible worlds to extensions of the appropriate type;[29] but note the qualification in the next section.

No doubt what is most interesting about Reinhardt's idea is the impredicative use of proper classes that he combines with it, with the result that the ordinals in V are, in the projected universe, a measurable cardinal.[30] However, in discussing the idea of proper classes are intensions I want to keep the predicative interpretation in mind as well.

IV

Cantor's conception suggest a more radical view than we have drawn from it so far, namely that one can in a sense not meaningfully

[29]See Montague, "Pragmatics and Intensional Logic."

A reformulation of Reinhardt's ideas in an intensional language would be desirable, in particular in order to eliminate the Meinongian ontology of possible non-acutal sets and classes that he uses, especially in "Set Existence Principles." Montague's intensional logic would not be adequate as it stands for this purpose, since his first-order quantifiers range over all possible objects; however, there is no difficulty in reformulating it to fit an interpretation in which bound variables range over existing objects. If the only alternative possible worlds one wants to consider are those with more ranks, then the version of quantified modal logic of Schütte (*Vollständige Systeme*) is applicable. This has the additional advantage that free variables also range over existing objects.

Pure modal logic, however, would not suffice to state Reinhardt's schema (S4) ("Remarks on Reflection Principles," p. 196), since it expresses a condition on a single "possible world" for infinitely many formulae.

The question arises how a class P can recur "in extension" in another possible world such as Reinhardt's projected universe. The answer is that it would be represented by its "rigidification," that is an attribute Q satisfying the condition

$$(\forall x)(Px \leftrightarrow Qx) \wedge \Box \; \{(\forall x)(\Box \; Qx \vee \Box \neg Qx) \wedge (\forall R)[(\forall a) \Box (Qx \rightarrow \Box (\forall x)(Qx \rightarrow Rx))]\}.$$

I assume here that bound variables range over existing objects; otherwise Barcan's axiom would hold and the third conjunct would be unnecessary.

Wang ("Large Sets") formulates Reinhardt's ideas in the opposite direction, by eliminating the intensional motivation and thinking of V not as the "actual world" but as a set which is an "approximation" to the universe. What mathematical interest an intensional formulation would have is not clear; perhaps it would suggest "intuitionistic" approaches to strong reflection axioms.

The statement that every attribute has a rigidification is just the formulation appropriate to this setting (without the Barcan formula) of the axiom E^{σ} (for $\sigma = (e)$) of Gallin, *Intensional and Higher-Order Modal Logic*, p. 78. That the second conjunct above is equivalent to Gallin's $(\forall x)(\Box Qx \vee \Box \neg Qx)$ follows from Barcan's axiom. However, (1)-(3) (p. 286) and the comprehension axiom of p. 295 are inconsistent with the Barcan formula.

[An axiomatization of set theory based on the ideas of this paper was briefly presented in my "Modal Set Theories." See also "Sets and Modality" (Essay 11 below), sections III-IV.]

[30]Reinhardt, "Set Existence Principles," p. 22; Wang, "Large Sets," p. 327.

quantify over absolutely all sets. In essays 3 and 8 above, I sketched all too briefly a "relativistic" conception of quantifiers in set theory. The idea was that an interpretation that assigns to a sentence of set theory a definite sense would take its quantifiers to range over a set (presumably R_α for some large α), but that normally such a sentence would be so used as to be "systematically ambiguous" as to *what* set the quantifiers ranged over.[31] A merit of this view was that it yielded a kind of reduction of classes to sets.[32] Since I have here followed Cantor and Reinhardt in viewing classes as quite different from sets, here I can defend this relativistic position only in a modified form. I shall now present my present understanding of the matter.

What *does* follow from the thesis that the elements of an inconsistent multiplicity cannot all exist together is that quantification over all sets does not obey the classical correspondence theory of truth. The totality of sets is not "there" to constitute any "fact" by virtue of which a sentence involving quantifiers over all sets would be true. The usual model-theoretic conception of logical validity thus leaves out the "absolute" reading of quantifiers in set theory.

If the only constraint on an interpretation of a discourse in the language of set theory is that it should make statements proved in first-order logic from axioms accepted by the interpreter *true*, then the interpreter needs only a minimally stronger theory than that applied in the discourse to interpret it so that the quantifiers range over some R_α.[33] However, it seems clear that this condition is too weak, and it remains so even if the interpreter seeks to capture not just one particular discourse but what the set theorist he is interpreting might be taken to be *disposed* to assent to. This is the case envisaged in Essay 8 above, p. 218, where we suppose that the interpreter takes the quantifiers to range over R_α for an α with an inaccessibility property undreamed of by the speaker.

Let us suppose this inaccessibility property to be P and that the interpreter chooses the least such α. One weakness of the reading

[31]Such a conception is hinted at in Zermelo, "Über Grenzzahlen und Mengenbereiche"; see esp. p. 47.

The conception of quantification over all sets advanced here is close to that of Putnam, "Mathematics without Foundations," except for the addition of the concept of systematic ambiguity.

[32]See Essay 8 above, pp. 218–220.

[33]The existence of an R_α such that V is an elementary extension of R_α is provable in ZF plus impredicative classes (Bernays-Morse set theory; see Drake, *Set Theory*, p. 125). What is essential, however, is not impredicative classes but allowing bound class variables in instances of replacement; one could use the system NB^+ mentioned in Essay 3 above, note 15.

is that it takes $(\exists\alpha)P\alpha$ to be false; although the supposition that the speaker is not disposed to assent to it is reasonable, the result is arbitrary in that we have no reason to suppose the speaker disposed to dissent from it.

A more decisive objection is that if the interpreter gets the speaker to understand P and convinces him that there is a cardinal satisfying it, then he must attribute to the speaker a meaning change brought about by this persuasion: previously his "concept of set" excluded P-cardinals; now it admits them.

The speaker, however, can (going outside the language of set theory and talking of himself and his intentions) question this and say that P-cardinals are cardinals in just the sense in which he previously talked of cardinals; he will presumably reinforce this by assenting to a number of statements that follow directly from the existence of a P-cardinal by axioms or theorems of set theory he accepted previously.

The idea that quantifiers in set theory are systematically ambiguous was meant to meet this kind of objection by saying that the interpretation of the speaker's quantifiers as ranging over a single R_α cannot be an exactly correct interpretation, since it fixes the sense of statements whose sense is not fixed by their use to this degree. However, it still seems to imply that the speaker who is convinced of the existence of a P-cardinal undergoes a meaning change in a weaker sense, in that the ambiguity of his quantifiers is reduced by "raising the ante" as to what degree of inaccessibility an α has to have so that R_α will "do."

We should observe that assertions in pure mathematics are made with a presumption of *necessity*; if we attribute this to our speaker we can see how P-cardinals are immediately captured by his previous set theory, since necessary generalizations are not limited in their force to what "actually exists. We can see the "meaning change" in accepting P-cardinals as analogous to the speaker's considering a different possible world or range of possible worlds.

The force of this analogy is limited, as we can see by a little further reflection on the conception of "inconsistent multiplicities" as intensions. It seems that we cannot consider a proper class as given even by an *intension* that is definite in the sense of, say, possible-world semantics, as a function from possible worlds to extensions. To begin with, it is only by an interpretation external to a discourse that one can speak in full generality of the range of its quantifiers and the extensions of its predicates. The systematic ambiguity of the language of set theory arises from the fact that such an interpretation can itself be mapped back into the language of set theory when stronger assumptions are made. Thus we should think of predicates whose "extensions" are proper classes as really not having *fixed* extensions.[34]

This situation does not change if we enlarge the language of set theory to an intensional language. Here we are able to express the "potentiality of the totality of sets" in that it is necessarily true that the domain of a bound variable *possibly* exists as a set. But however such an intensional language is formulated, it will still be possible to read it in a set-theoretic possible world semantics, and even if on the most straightforward reading the union of the domains for all possible worlds is all sets, the assumption that there is a set that realizes the properties of this union presumably has the same plausibility that other such reflection principles have. In such a model, of course a second-order intension will be represented by a set.[35]

We should not be surprised at this; it is really a consequence of the general nature of true "systematic ambiguity," where there is no general concept of "possible interpretation" which is not either inadequate or infected with the same difficulties as the language it interprets. Otherwise one could resolve the ambiguity as generality (meaning by *A*, "*A* on every possible interpretation") or indexically, by some contextual device or convention indicating which inter

---

[34]Cf. the remarks on the discomfort evinced by use of proper classes in Wang ("Large Sets," n. 8). However, Wang does not make clear whether this discomfort would be removed if we confine ourselves to thinking "of these large classes as extensions or ranges of properties."

The point which I would emphasize is that if the language of set theory with quantifiers read as ranging over "all sets" has a "fixed" or "definite" sense, then it is naturally extended by a satisfaction predicate, and definiteness of sense is preserved. But in the extended language one can of course construe the classes required by the Bernays-Gödel theory. Iteration of the procedure yields more classes.

In Wang's terms, this justification of classes no doubt falls within the conception of them as "extensions or ranges of properties." Still, such an enlargement of the language of set theory seems to be treated with reserve by many set theorists, although the reason could be just that in deductive power it is inferior to stronger axioms of infinity.

Locutions requiring either classes or satisfaction and truth are frequent in writings on set theory (cf. "Sets and Classes," Essay 8 above), but the characteristic informal use is very weak and could be captured by a free variable formalism for classes with very elementary operations on them, as George Boolos's comments on "Sets and Classes" reminded me.

[35]The "straightforward reading" involves replacing "set" by "class" at certain points in the standard model-theoretic account of (modal) logical validity, just as in the case of ordinary logic in set theory. The same should be the case for set theories with intuitionistic logic, which are suggested by the same considerations as suggest the modal language. It should not be thought that changing to intuitionistic logic will remove the fundamental dilemmas about quantification over all sets.

pretation is meant. Russell's "typical ambiguity" was essential in that according to his theory of meaning there was no way of expressing by a single generalization *all* instances of a formula where the variables were understood as typically ambiguous. In "The Liar Paradox" (Essay 9 above) I handled semantical paradoxes by observing that paradoxical sentences could be taken to have no truth-value or not to express propositions on the interpretations presupposed by the semantic concepts occuring in them, while obtaining definite truth-values or coming to express propositions on interpretations "from outside." But at some point there must on this account be systematic ambiguity, or else one could generate "super" paradoxes such as "this sentence is not true on any interpretation".

Thus although it is true to say that a proper class is given to us only "in intension," this statement does not have quite its ordinary meaning. Obviously what is lacking is not just its being given to *the mind* in extension; that is lacking for most sets as well. What is lacking has to do, one might say, with being, and moreover if the underlying intension had the fixed, completed existence a proper class lacks then the class would have it as well. However, some general ideas about intensional concepts do have application to this case. If we think of classes as given only by our understanding of the (perhaps indefinitely extendible) language of set theory, then the assumption that impredicative reasoning about classes is valid is rather arbitrary. This way of looking at classes corresponds to thinking of intensions as meanings and of meanings as constructions of the mind. This is the conception that is appropriate to applying intensional logic to propositional attitudes. Alternatively (and here we have a clearer theory) intensions are thought of as individuated by modal conditions, as "functions from possible worlds to extensions." This is the conception appropriate to modal logic. It seems neutral with respect to the question of impredicativity.

Let me make a final comment on the *predicative* conception of classes. If we understand a second-order language containing set theory in this way then set existence does for us the work of the axiom of reducibility in Russell's theory of types. For predicates which are high in a ramified hierarchy or which more generally are expressible only by logical complex means, the existence of a set $\{x: Fx\}$ provides a simple equivalent $x \in a$ for a a *name* of $\{x: Fx\}$. Clearly it serves as an equivalent only *extensionally*. In the intensional situations envisaged above the equivalence of $x \in a$ and Fx will not be necessary even if the name a has been introduced by stipulation ("a priori" in Kripke's sense),[36] and therefore the two predicates will behave differently in intensional contexts. Thus the license for impredicativity given by assuming the existence of sets does not nullify the predicative conception of intensions even for intensions that have sets as extensions. Of course this "reducibility" does not obtain for predicates that do not have sets as extensions.[37]

In conclusion, I would claim that the above discussion had added something to the explication of the idea that an "inconsistent multiplicity" is not really an object, since even as an intension it is systematically ambiguous. The task remains to explain whether the ideas of the last two sections are helpful in understanding the "stages" of the genetic conception and the underlying priority of the elements of a set to the set.

V

In the last two sections we sought to avoid using either epistemic concepts or the metaphor of process in trying to understand the conditions for the existence of sets. However, we concentrated largely on the distinction between sets and classes or "multiplicities" and on discourse about absolutely all sets.

The idea that any available objects can be formed into a set is, I believe, correct, provided that it is expressed abstractly enough, so that "availability" has neither the force of existence at a particular *time* nor of givenness to the human mind, and formation is not thought of as an action or Husserlian *Akt*. What we need to do is to replace the language of time and activity by the more bloodless language of potentiality and actuality.

Objects that exist together *can* constitute a set. However, we do have to distinguish between "existing together" and "constituting a set." A multiplicity of objects that exist together *can* constitute a

---

[36]Cf. Tharp, "Three Theorems of Metaphysics."

[37]It is commonly claimed that the axiom of reducibility nullifies Russell's ramification of his hierarchy of types. This claim depends on ignoring, presumably on the grounds that nonextensional features of functions are not significant for mathematics, the fact that Russell thought of propositional functions intensionally.

On the other hand it is hard to see what is left of Russell's no-class theory once the axiom of reducibility is admitted. Russell himself says that the axiom of reducibility accomplishes "what common sense effects by the admission of classes" ("Mathematical Logic as Based on the Theory of Types," *Logic and Knowledge*, p. 81), but he considers the axiom a weaker assumption than the existence of classes. The weakness must consist in the restrictions of the simple theory of types.

set, but it is not necessary that they *do*. Given the elements of a set, it is not necessary that the set exists together with them. If it is possible that there should be objects satisfying some condition, then the realization of this possibility is not as such the realization *also* of the possibility that there be a set of such objects. However, the converse does hold and is expressed by the principle that the existence of a set implies that of all its elements.

The same idea would be expressed in semantic terms by the supposition that we can use quantifiers and predicates in such a way that the range of the quantifiers and the objects satisfying any one of the predicates can constitute single objects, but these objects are not already captured by our discourse. However, this way of putting the matter might be taken to rule out too categorically an "absolute" use of quantifiers and predicates. Without returning to an ontological characterization such as the Cantorian language of "existing together," we can say that this is the condition under which quantifiers and predicates obtain definiteness of sense.

Above we suggested that the axiom of power set rests on a sort of principle of plentitude, according to which all the possible subsets of a given set are capable of existing "at once." Against what we have just said one might object that there is no intrinsic reason why the potentiality of a set relative to its elements should not be nullified in our theory by a similar principle of plentitude.

The short answer to this objection is that such treatment would lead to contradictions, Russell's paradox in particular. We could apparently consistently assume (as in New Foundations) that the domain of discourse is a set in the domain, but then of course there will be other "multiplicities" of elements of the domain that are not in it.[38]

A further point is that there seems to be an intrinsic ordering of "relative possibility" in the element-set relation that is lacking for the arbitrary subsets of a given set. A set is an *immediate* possibility given its elements, the sets of which it is an element are at least at another remove. We do of course have conceptions of the "simultaneous" realization even of infinite hierachies in this or-

---

[38]In the case of NF, these additional "multiplicities" would correspond to the proper classes of ML.

If a model of NF is given as a set in the ordinary set-theoretic sense, the domain of the model and the *V* of the model will of course differ. The membership relation of the model will obviously not be the same as the membership relation "from outside."

dering, but such a conception gives the possibility of sets that are still higher.

This observation should remind us that more is involved in the iterative conception of set than the priority of element to set, since in Gödel's words we think of arbitrary sets as obtained by *iteration* of the "operation 'set of'" starting with individuals, and we have not yet dealt with the concept of iteration. To do so adequately would be beyond the scope of this paper. I shall make a few remarks.

First, our strategy has been to use modal concepts in order to save the idea that *any* multiplicity of objects can constitute a set; one makes only the proviso that they "can exist together" and this proviso I take to be already given by the meaning of the quantifiers unless they are used in a "systematically ambiguous" way. One saves thereby the universal comprehension axiom as well, though in a form that hardly seems "naive" any more: In the second-order modal language it would have to be expressed by the statement that for every attribute P there is an attribute Q that is the rigidification of P (note 29 above) and such that

$$(4) \qquad \Diamond\,(\exists y)(\forall x)(x \in y \leftrightarrow Qx).^{39}$$

However, even with the assumptions needed to obtain a version of the power set axiom we do not obtain greater power than that given by a much more traditional way of saving the comprehension axiom: the simple theory of types. It is clear that without some principle allowing for *transfinite* iteration of something like the above comprehension principle we will not obtain even the possible existence of sets of infinite rank, such as the usual axiom of infinity already requires. For the axiom of infinity, the principle needed is one allowing the conversion of a "potential" infinity into an "actual" infinity: we can easily show

$$(5) \qquad \Box(\forall x)\,\Diamond\,(\exists y)(y = x \cup \{x\}),$$

but to use (4) to infer that ω possibly exists, we would need to get from (5) to

$$\Diamond\,(\forall x)(\exists y)(y = x \cup \{x\});$$

[39] Thus if in some possible world $(\forall x)(x \in y \leftrightarrow Qx)$ holds, the elements of y are just the objects that *actually* have P. In many cases they will not be just the objects that have P in the world in question.

in terms of a set-theoretic semantics for the modal language, the possible worlds containing finite segments of ω need to be collected into a single one.

Second, it is clear that there has to be a priority of earlier to later ordinals, whether this is *sui generis* or derivative from the priority of element to set. One could of course assume a well-ordered structure of individuals, within which there would be no ontological priority of earlier to later elements. The axiom of infinity of *Principia* is such an assumption. To make it is natural enough, unless we assume a relation to the mind is essential to the natural numbers. Then it seems that smaller numbers are prior to larger ones by virtue of the order of time, as in Brouwer's (and apparently also Kant's) theory of intuition.

For reasons indicated above, no such structure can represent all ordinals. In fact larger ordinals seem conceivable to us only by characteristically set-theoretic means such as assuming that there is already a *set* closed under some operation on ordinals.

Third, it seems to me that the evidence of the axiom of foundation is more a matter of our not being able to understand how non-well founded sets could be possible rather than in a stricter insight that they are *impossible*. We can understand starting with the immediately actual (individuals) and iterating the "realization" of higher and higher possibilities. It seems that (at least as long as we hold to the priority of element to set) we do not understand how there could be sets that do not arise in this way. Non-well-founded ∈-structures have been described (simple ones already by Mirimanoff in 1917), but we do not recognize them as structures of *sets* with ∈ as the real membership relation, even when they satisfy the axioms of set theory.[40] We are at liberty to say that the *meaning* of "set" is, in effect, "well-founded set," but that does not exclude the possibility that someone might conceive a structure very like a "real" ∈-structure which violated foundation but which might be thought of as a structure of sets in a new sense closely related to the old.

I shall close with a rather speculative comment. The conception of "inconsistent multiplicities" as indefinite or ambiguous raises a doubt about whether it is appropriate to talk of *the* cumulative hierarchy as most set theorists do. The definiteness of the power set is maintained even though the hope of deciding such questions

[40]Perhaps this could be said of trivial variants such as that resulting from identifying individuals with their unit classes. But here of course a slightly modified axiom of foundation holds.

as the continuum hypothesis and Souslin's hypothesis by means of convincing new axioms has not been realized. However, in this case the idea of the "maximality" of the power set gives us some intuitive handle on the plausibility of the hyotheses or of "axioms" such as $V = L$ that *do* decide them.

Maximality conceptions also contribute to the plausibility of large cardinal axioms. Here it seems conceivable in the abstract that we might see the possibility of a cardinal α with a "structural property" P and of a cardinal β with such a property Q, where these properties are not "compossible," that is, we would see (perhaps even in ZF) that such α and β cannot both exist. That would yield two incompatible possibilities of cumulative hierarchies.

This has not happened with any of the types of large cardinals considered in recent years, where it has generally happened that of two such properties one (say P) implies the other, and indeed $P\alpha$ implies the existence of many smaller β such that $Q\beta$. That this is so has seemed rather remarkable; perhaps it is evidence against the views I have advanced.

However, one reason for thinking that "incompatible large cardinals" will not arise is that by the Skolem-Löwenheim theorem both would reflect into the countable sets. If our confidence in the uniqueness of $\mathfrak{P}(\omega)$ is so great as to lead us to reject the possibility of incompatible large cardinals, one would still wish for some more direct reason for doing so.

11

Sets and Modality

I. Rigidity, Sets, Classes, and Attributes

In the previous essay, I defended and applied the idea that set membership is *rigid*, that is (roughly), that if an object x is an element of a set y, it is so necessarily, and likewise, if x is not an element of y, that is also necessary. These were intended as principles for discourse about sets in modal contexts in general, not just of the rather special kind considered there. In particular, these principles apply to the kind of nonmathematical modal context prominently discussed in the literature on the foundations and application of modal logic. With some possible qualifications concerning existence to be discussed presently, this has now become a widely accepted view.[1]

This essay was written for this volume. However, the ideas of section I and versions of the technical work of section III were presented in lectures to various audiences in the period from 1974 to 1977. Since the point of view expressed in section I has become widely current and has in some respects been explored more deeply by others, notably Kit Fine, the main purpose of Sections I and II is expository.

Allen Hazen was kind enough to read a draft of this essay and make valuable comments. I also wish to thank Peter G. Hinman, Harold Hodes, David Kaplan, Ruth Barcan Marcus, Gert H. Müller, William C. Powell, W. N. Reinhardt, and an anonymous referee for comments or assistance at various stages of this work.

[1]Principles like (R $\in$) and (R $\notin$) seem to have been first proposed and defended by Ruth Marcus in "Classes, Collections, and Individuals," which was accepted for publication in 1965. She is there working out a train of thought in her 1963 paper, "Classes and Attributes in Extended Modal Systems," pp. 128-129. But for the special case of time and change, see Sharvy, "Why a Class Can't Change Its Members." They were arrived at independently by me and by W. N. Reinhardt, and no doubt by others. (See my "Sets and Possible Worlds" and, for Reinhardt's use, Essay 10 above, section III.) Other defenses are to be found in Hambourger, "A Difficulty with the Frege-Russell Definition of Number," and Fine, "First-Order Modal Theories I: Sets."

Before I present arguments for these principles, some clarification of their meaning is needed. There is a formal parallel between the rigidity of membership and the rigidity of identity. The usual modal quantificational logic validates the formulæ:

$$(R =) \qquad x = y \rightarrow \Box(x = y)$$
$$(R \neq) \qquad x \neq y \rightarrow \Box(x \neq y).$$

It is proposed that any extension of modal logic to include some set theory should validate the formulæ:

$$(R \in) \qquad x \in y \rightarrow \Box(x \in y)$$
$$(R \notin) \qquad x \notin y \rightarrow \Box(x \notin y).$$

$(R =)$ and $(R \neq)$ are assumed in formulations of modal logic in which the range of the variables includes objects that *might not exist*, as for example in the type of first-order modal logic, going back to Kripke's "Semantical Considerations on Modal Logic" (1963), that admits a possible-worlds semantics in which the domain for the bound variables can vary freely with the possible world, and this domain is understood as consisting of just those objects that *exist* in that world. I will take such a logic as the general first-order modal logic for the theories I discuss. Then the underlying non-modal logic will be a *free logic* in which a logical predicate E of existence is added to the apparatus of first-order logic with identity (see Appendix 1). But in such a setting assuming $(R =)$ and $(R \neq)$ involves treating identity as *independent of existence*. That is, if $x = y$, we regard this as still true in an alternative possible world in which x (and a fortiori y) does not exist. This yields a smooth theory of identity; but it makes an exception of identity if one assumes, as logicians sometimes do, that an atomic predicate cannot hold of an object unless that object exists.

Similarly, in this logical setting the principles $(R \in)$ and $(R \notin)$ treat *membership* as independent of existence. Their intended import is that a set y in another possible world should have exactly the same elements, whether or not these objects exist, and indeed whether or not y itself exists. The idea that in the last essay motivated the treatment of membership as rigid suggests something that may seem incompatible with existence-independence, namely that a set *exists* only if each of its elements exists:

$$(E \in) \qquad (Ey \wedge x \in y) \rightarrow Ex.$$

Motivated in part by this consideration, in the previous essay I went further and proposed an existence-*dependent* treatment of membership, replacing the three principles (R $\in$), (R $\notin$), and (E $\in$) by

$(R_0 \in)$ $x \in y \rightarrow \Box(Ey \rightarrow x \in y)$

$(R_0 \notin)$ $(x \notin y \land Ey) \rightarrow \Box(x \notin y)$

$(E_0 \in)$ $x \in y \rightarrow (Ex \land Ey).$[2]

However, the modal set theories developed either below or elsewhere in the literature either are based on unfree quantificational logic (like my own theories in section III) or treat membership as independent of existence (as does Kit Fine).[3] Moreover, there is in fact no conflict between (E $\in$) and the principles of an existence-independent treatment. I can now see no reason for adopting the existence-dependent treatment other than the desire to maintain without exception the policy that atomic predicates can be true only of existents. For reasons that it would take me too far afield to go into, I find the compromise of that policy in the case of identity and membership (and perhaps in other cases) worthwhile for the smoothness of theory it affords. Therefore I shall largely ignore the existence-dependent treatment in what follows.

The manner in which the above principles have been stated involves free variables including in their range objects that do not exist. This is in fact unavoidable at least in logical axioms in formulations of free modal logic that admit free variable reasoning at all. However, the strength particularly of nonlogical free variable axioms is sensitive to assumptions about the range of the free variables (see Appendix 1). The above principles will capture my intent only if assumptions are made that go with a possible-worlds semantics in which the free variables range over the objects of all possible worlds (in the terminology of Appendix 1, Q3 models). This may seem to be too free in "speaking of what is not." The theories of section III below would lose some of their point in such a logical setting. A stricter approach would, in using (R $\in$), (R $\notin$), and (E $\in$) as axioms, replace them by their closures. If A is a formula of a modal language with free variables $x_1 \ldots x_n$, we write $(\Box)A$ for

---

[2] See above, p. 286. Note that in this treatment, $x \in y$ can be false and yet possibly true, if its actual falsity is due to the nonexistence of y.

[3] "First-Order Modal Theories I: Sets," hereafter cited as "Sets."

$\square \forall x_1 \ldots \square \forall x_n \square A$, and call this the closure of A.[4] Assuming certain free variable axioms may be stronger than assuming their closures (see Appendix 1).

Connected with this complication is the fact that some care is needed in talking of the rigidity of predicates. Two intuitive conceptions are that if a predicate holds of an object (or tuple of objects) then it holds necessarily, and that the predicate has the same extension in all possible worlds. However, these do not always agree. If we think in terms of possible worlds and take $\square$ to mean absolute necessity, that is, truth in all possible worlds, then rigidity in the first sense does imply rigidity in the second; $(R \in)$ expresses the rigidity of membership provided its free variables range over all the objects we consider (i.e., over the outer domain of a Q3 model). However, I do not want to assume that our language contains an operator for absolute necessity. But in a modal set theory with an absolute necessity operator that obeys S5, $(R \in)$ will be the only rigidity axiom needed.

In a more general setting, we will call a predicate *persistent* if it is rigid in the first sense. I will use $Pers(F)$ to abbreviate $(\square)\,(Fx \to \square Fx)$ (similarly for polyadic predicates). I will call a predicate *absolute* if it and its negation are both persistent. $Abs(F)$ will abbreviate $(\square)\,(\square Fx \vee \square \neg Fx)$.[5] The membership relation is absolute if we assume $(R \in)$ and $(R \notin)$.

If the range of the quantifiers were fixed for all possible worlds, then absoluteness would amount to rigidity in the second sense noted above; indeed we could simplify $Abs(F)$ to $\forall x(\square Fx \vee \square \neg Fx)$. Otherwise, absoluteness allows that in another relevant possible world, F might be true of objects not in its actual extension, provided only that these objects do not actually exist. In a Q3 model, the free variable formulation of absoluteness will rule this out, but $Abs(F)$ as it stands will generally not, and moreover we cannot in general do so by a condition expressible in the usual first-order

---

[4]More generally, a closure of A is a closed formula obtained from A by prefixing universal quantifiers and $\square$ symbols. We will take as axiomatization of first-order free modal logic that of Fine, "Model Theory for Modal Logic, Part I," pp. 131-132, dropping axiom (9) ($\lozenge Et$ for any term t) and perhaps adjusting the propositional logic. If the propositional logic is S5 (or if we use unfree logic and it contains S4), then $(\square)A$ implies any other closure in the above sense. Thus this will obtain for the theories explicitly considered. The adjustment of my general remarks needed to cover weaker logics should be obvious.

[5]$Pers(F)$ and $Abs(F)$ will in general fall short of expressing the full generality of the free variable formulæ $Fx \to \square Fx$ and $\square Fx \vee \square \neg Fx$, if taken as generalized over all objects in all possible worlds.

modal language.[6] With an "actuality operator" A, we can express something like what is wanted by adding to absoluteness the condition $\Box\forall x(Fx \rightarrow A(Fx)$); however, what is really needed is that A should put the '*Fx*' it governs outside the scope of the modal operator; we can express the condition we want in the scoped first-order modal logic introduced in section III and explained further in Appendix 1, section 2.

Another way of keeping "new" objects out of the extension of F, if F is a predicate of existing objects, is by a second-order condition based on the Barcan formula: we add the condition $\Box \forall G[\forall x(Fx \rightarrow \Box Gx) \rightarrow \Box \forall x(Fx \rightarrow Gx)]$, i.e., the Barcan axiom *relativized* to F. If we want to assume as an axiom that F has a fixed extension, we can in practice get by with a first-order schema. Let us call a predicate *fully rigid* if it is absolute and satisfies whatever condition we use to rule new objects out of its extension. We do not want to say that the membership *relation* is fully rigid, since there may possibly be sets that are not in the range of our variables as they actually are. The view I do want to defend is that for a given set y, membership in y is fully rigid. Hence, in view of (E $\in$), we should be prepared to assume for existing y

$$(\text{BF} \in) \qquad \forall x(x \in y \rightarrow \Box A) \rightarrow \Box\forall x(x \in y \rightarrow A).[7]$$

It is a consequence of the rigidity of set membership that in general set abstracts are not rigid designators. Since there are not fewer than seven planets, the number 7 belongs to the set of numbers x such that there are at least x planets. If we were to instantiate the abstract {x: there are at least x planets} in (R $\in$) we would obtain

$$7 \in \{x: \text{there are at least } x \text{ planets}\}$$
$$\rightarrow \Box(7 \in \{x: \text{there are at least } x \text{ planets}\}).$$

Taking the equivalence of 'there are at least 7 planets' and '7 $\in$ {x: there are at least x planets}' to be necessarily true, we would obtain the unacceptable conclusion $\Box$ (there are at least 7 planets). But from the rigidity of membership it follows that in a world in which

---

[6]That is, by a free variable axiom in an arbitrary free model (see Appendix 1) and by a sentence even in a $Q3$ model.

[7]In a $Q3$ model in which (R $\in$), (R $\notin$), and (E $\in$) hold for all values of their free variables, $\Box \forall y(\text{BF} \in)$ holds for any formula A. Its *proof* from these axioms in the free modal logic mentioned in note 4 requires Thomason's modal generalization rule R4, unless the propositional logic contains B; see Appendix 1.

there were fewer than seven planets, {x: there are at least x planets} would denote a different set, and since it is therefore not a rigid designator, it cannot be instantiated for a variable inside a modal operator. This is just what we would expect if we treat abstracts as definite descriptions, i.e., {y: A} as an abbreviation for $\iota x[\mathfrak{M}x \wedge \forall y(y \in x \leftrightarrow A)]$.[8]

When I first began to think about these matters, I was tempted by the following picture (cf. Essay 8 above): Since *sets* are constituted by their elements, set membership is rigid, and therefore abstracts are not rigid designators, for two predicates that are actually coextensive but not necessarily so would have to denote the same set in the actual world but different ones in some alternative. But *classes* are essentially extensions of predicates; they are as it were constituted by predicates. But then, however it might be with sets, when we speak of classes in modal context we should regard class abstracts as rigid designators and therefore reject the rigidity of membership for classes.

However, this will conflict with extensionality. Suppose '*F*' and '*G*' are coextensive predicates. Writing '$\hat{x}Fx$' and '$\hat{x}Gx$' for their extensions, we will then have

$$\hat{x}Fx = \hat{x}Gx.$$

But if we take the abstracts to be rigid designators, we will then have by the rigidity of *identity*

$$\Box(\hat{x}Fx = \hat{x}Gx).$$

Assuming extensionality and that the equivalence of '$y \in \hat{x}Fx$' and 'Fy' holds necessarily, we will have to conclude

$$\Box \, \forall x(Fx \leftrightarrow Gx),$$

i.e., that our predicates are necessarily coextensive. Thus in the usual modal logic, if our language contains predicates that are coextensive but not necessarily so (in practice, if it can express

---

[8]When we talk of sets of ordinary objects, we need a set theory with individuals (cf. note 2 to Essay 10 above); hence the qualification '$\mathfrak{M}x$' ('x is a set').

The existence-dependent treatment would not affect the outcome of the above argument, since we are dealing with sets of numbers, which presumably necessarily exist.

contingent truths), either extensionality or the rigid designation of abstracts has to be sacrificed.[9]

This does not mean that in an intensional language we cannot have abstracts that are rigid designators; however, the objects they denote will not obey extensionality but the weaker principle sometimes called "intensionality":

$$\square \ \forall z(z \in x \leftrightarrow z \in y) \rightarrow x = y.$$

I conclude that these objects would in fact be intensions, and in discussing any theory of such objects, I will use the term "attribute" rather than "set" or "class." We shall consider some theories of this kind briefly in section II.

Thus the consideration of intensional contexts does offer an argument for distinguishing sets from classes/attributes.[10] But the conception of the latter objects as essentially associated with predicates forces in the usual intensional languages the replacement of classes (obeying extensionality) by attributes (obeying intensionality). The need to generalize predicate places in set theory is satisfied by extensional classes only because the language itself is extensional, and when intensionality creeps into set theory, as it does in the work of Reinhardt discussed in the last essay,[11] the role of classes

[9]Marcus succumbed to the temptation I am speaking of here, but she then bit the bullet and rejected extensionality for classes ("Classes, Collections, and Individuals," p. 231). My only difference with her is about terminology: there are better terms for the objects she calls "classes": "attribute" and "property."

Other ways of blocking the argument are possible, familiar from the discussion of other modality-collapsing arguments. One might reject the rigidity of identity, as is done in John Myhill's "Intensional Set Theory." This amounts to taking individual variables to range over individual concepts, and "set concepts" are hardly distinguishable from attributes. One might also deny the necessity of $\forall y(y \in \hat{x}Fx \leftrightarrow Fy)$. The only interpretation I can think of that would motivate this would be taking the abstract as always having *wide scope*, i.e., interpreting it even in modal contexts as standing for the class of objects that are *actually F*. In effect, this prevents the question whether class abstracts are rigid designators from arising.

A more interesting possibility is to express intensional logic in a formally extensional language, after the pattern of Church's logic of sense and denotation; see my "Intensional Logic in Extensional Language." Then classes would be represented in "intensional" contexts by class-concepts, again in effect attributes.

[10]Marcus ("Classes, Collections, and Individuals") seems to have been the first to perceive this.

[11]Section III. In the more recent paper "Satisfaction Definitions" Reinhardt develops a modal theory, in which set membership is rigid; on top of the universe of sets there is a hierarchy of intensional "properties," with a theory of them patterned after ZF. One purpose of this is evidently to formulate very strong higher-order reflection principles. Though this theory has points in common with the theories discussed below (including our own), limitations of space and understanding preclude discussing it in any detail.

is then assumed by intensional entities. For the same reason, in second-order theories the analogue of identity for second-order entities is no longer coextensiveness but some form of necessary coextensiveness.[12]

Leslie Tharp pointed out the following apparent anomaly resulting from the rigidity of set membership.[13] Consider again the set {x: there are at least x planets}. We can introduce into our language a *proper name* for this set, say '*P*'. Then

(1)' $P = \{x:$ there are at least x planets$\}$

is true by stipulation, in the usage of Saul Kripke, a priori.[14] Now since '*P*' *is* a rigid designator, and $7 \in P$, by (R ∈) we can conclude that '$7 \in P$' is necessary. But the equivalence of 'there are at least 7 planets' and '$7 \in P$' is obtained by simple mathematics and the stipulation (1) and thus is plausibly a priori. Hence 'there are at least 7 planets' is a priori equivalent to a necessary proposition. This argument can obviously be generalized: if '*a*' is a proper name, then a predication '*Fa*' will be a priori equivalent to '$a \in Q$', where '*Q*' is a proper name of {x: Fx}; then if '*Fa*' is true, (R ∈) implies that '$a \in Q$' is necessary. On the existence-dependent treatment, it will still follow that '$a \in Q$' holds in any world in which Q (and hence all its elements, including *a*) exists.

This seems to me not at all a real difficulty; rather it serves to reinforce the distinction Kripke makes between necessity and apriority. The equivalence of '$7 \in P$' and 'there are at least 7 planets' is not itself necessary, nor is (1), since the abstract is not a rigid designator.

Kripke and many others hold that certain general terms for

---

[12]Both Frege's conception of "concepts" and relations as extensional and his concept of classes as derived from concepts can be preserved in a formally extensional intensional logic. Such a logic can still express what we want to of modality, even modality *de re*, by using a version of Church's logic of sense and denotation augmented on the basis of an idea of Kaplan; see Parsons, "Intensional Logic in Extensional Language," §3. Since these logics are based on the λ-calculus, classes are represented by functions from (say) individuals to truth-values. Inside an intensional operator, an entity has to be represented by a "sense" or "concept" of it (the latter in the Carnap-Church rather than in the Frege sense). As remarked above (note 9) concepts of classes do not differ importantly from attributes. In *de re* modal statements, an entity is represented by its *rigid concept*, that is, a concept that in every possible world is a concept of it. However, the problem of constructing such a theory has so far been solved only for the fixed domain case; see ibid., end of §4.

[13]"Three Theorems of Metaphysics;" cf. the abstract "Necessity, Apriority, and Provability."

[14]*Naming and Necessity*, p. 56.

natural kinds are rigid designators. This view can be interpreted in terms of traditional essentialist ideas, so as to imply that these natural kind terms express essential properties of the objects they are true of.[15] Should we then say that what they designate are sets? That seems counterintuitive. Anyway we can see that they are not fully rigid. Suppose we read '$\in$' as the copula and allow general terms such as 'horse' that we presume to be rigid designators to be instantiated for the variables in the above principles. If understood existence-independently, 'horse' is on our assumption persistent and seems also to be absolute: not being a horse would be an essential property of nonhorses. But (BF $\in$) fails, since there might be more horses than there actually are. Let y be the kind Horse, and let $a_1 . . . a_n$ be names of all the horses that there actually are. Then, by the rigidity of identity, $\forall x[x \in y \rightarrow \Box(x = a_1 \vee . . . \vee x = a_n)]$ holds, but $\Box \forall x[x \in y \rightarrow (x = a_1 \vee . . . \vee x = a_n)]$ surely fails, since it implies that there could be no horses other than the actual ones.[16] We could still construe such a term as 'horse' as rigidly designating a set of *possibilia* (presumably in the context of fixed domain modal logic). Such a reading would be uncongenial to the "actualist" intuitions of many essentialists, such as Kripke and Plantinga, and is not very plausible as a reading of 'horse' in English.

Up to now I have concentrated on analyzing the precise meaning and some implications of the rigidity of membership but have not set forth reasons for holding it. In the last essay, I argued that if sets are constituted by their elements, then it should be essential to a set to have exactly the elements that it has. The same consideration prompts (E $\in$). However, it would be desirable if further arguments could be offered. A consideration that has some force is that in the course of some discussion of these matters, no plausible alternative has been offered that does not compromise extensionality and therefore treat "sets" more like attributes. But one ought to have some explanation of this absence of alternatives.

The priority of the elements of a set to the set might be taken to imply that the identity of sets is derivative from that of their elements. Extensionality is often understood as a "criterion of iden-

---

[15]Kripke himself is not explicit on whether, in a case like 'horse', he would accept this implication.

[16]Similar reasoning shows that in free modal logic $\{x: x \text{ is a horse}\}$ is also not a rigid designator. However, even if 'a' is a proper name of a set, $\{x: x \in a\}$ is in general not a rigid designator, since in a given world it denotes the set of elements of a that exist in that world.

tity" for sets; then treating membership as fully rigid would result from using the same criterion for "trans-world identification" of sets. Suppose, for example, that a set that actually has the elements o. . .9 (say it is $\{x:$ there are at least x planets$\}$) had in some counterfactual situation the elements o. . .5. To say that it is *the very set* that actually has the elements o. . .9 seems to be to attribute to a mathematical object an identity beyond what is determined by the structure in which it has its home. For this reason, I think we cannot follow the example of Kripke's remarks about concrete objects and insist that we can simply stipulate that we are talking about a situation in which that set has elements o. . .5. Moreover, in such a situation no other actual set, such as $\{$o. . .5$\}$, could *also* have just those elements, for that would contradict the rigidity of nonidentity.

We can sharpen the argument in the following way. Consider a finite set of necessarily existing objects whose elements are identified by names, say $\{$o. . .5$\}$. Merely from the rigidity of identity, the assumption that numerals are rigid designators, and the assumption that numbers necessarily exist, we can prove $(R \in)$ and $(R \notin)$ with the abstract $\{x: x = $ o v. . .v $x = 5\}$ instantiated for y. It is hard to see how such an abstract could *not* be a rigid designator. Now consider a predicate 'Fx' contingently coextensive with '$x = $ o v . . .v $x = 5$', say 'x is a natural number no greater than the number of boroughs of New York City'. Then by extensionality,

$$\{x: Fx\} = \{x: x = \text{o v}. . .\text{v } x = 5\}.$$

Then if we do assume that $\{x: x = $ o v. . .v $x = 5\}$ is a rigid designator, it designates a set such that membership in it is rigid. Since $\{x: Fx\}$ is the same set, it follows that membership in it is rigid as well (although $\{x: Fx\}$ is not a rigid designator).[17] One could argue similarly about infinite sets, given an infinitary language, and about sets whose elements exist contingently, given bound variables not restricted to existing objects.[18]

The rigidity of membership seems almost forced on us by possible-worlds semantics together with a clear distinction between sets and attributes. Looking from outside at a certain actual set a, we have for each possible world i a set a_i consisting of the elements of a in i. Given worlds i and j, the relation of a_i to a_j, for all i and j,

[17]In a temporal context, a similar argument is offered by Sharvy, "Why a Class Can't Change Its Members," pp. 310ff.
[18]The existence-dependent rigidity principles $(R_0 \in)$ and $(R_0 \notin)$ can be obtained without this addition, as can $(BF \in)$.

has to be a one-one function, by extensionality and the rigidity of identity and nonidentity. Now considering an *attribute* to be an arbitrary function from possible worlds to sets of objects available in those worlds, then we have singled out as the *sets* of the model a selection of the attributes, with the constraint that no two such attributes ever coincide (that is, are coextensive in some world). Without the rigidity of membership, I see no objective basis for selecting such a system of attributes to identify sets across worlds.[19]

Finally, a comment about (E ∈): as Fine points out, it also has a motivation from extensionality. If one freely admits sets of possible objects, there is no reason to expect a set to be determined by its existing members. Suppose an existing set x has a nonexistent element u. Then it seems that there should be a set $x - \{u\}$; call that y. If the quantifier is read as ranging over existing objects, then $\forall z(z \in x \leftrightarrow z \in y)$ will hold, but $x \neq y$. (E ∈) implies that this cannot happen for existing x, y, provided that extensionality holds with the quantifier ranging over all possible objects.[20]

II. Set Theory in the Context of Absolute Modalities

We now turn to the problem of constructing a formal theory for speaking of sets in modal contexts. Where the modalities in question are absolute, that is, where necessity is thought of as truth in *all* possible situations, theories exist in the literature. However, two approaches should be distinguished at the outset. First, we may have as basic objects of the theory not sets but attributes, and then identify sets with attributes satisfying suitable rigidity conditions. Second, we may try to construct a theory of sets independent of any theory of attributes.

[19]Cf. the discussion of the same issues in Fine, "Sets," pp. 180-181. It might also be remarked that the analogue of the rigidity of membership, namely the rigidity of the application of function to argument, holds in Montague's intensional logic, although apparently Montague was not attending to these issues in constructing his logic. In Gallin's axiomatization IL (*Intensional and Higher-Order Modal Logic*, pp. 19-20), we can easily prove $f_{\alpha\beta}x_\alpha = y_{:\beta} \rightarrow \Box (f_{\alpha\beta}x_\alpha = y_\beta)$. The rigidity of application is built directly into IL by counting a term $A_{\alpha\beta}B_\alpha$ as modally closed if A and B are.

If §4 of Parsons, "Intensional Logic in Extensional Language," the rigidity of application is used as a criterion of trans-world identification for functions, in order to construct such identification from the bottom up, taking it as given only for individuals. See esp. pp. 307-308.

[20]Fine, "Sets," p. 181. Note that the closure of (E ∈) is $\Box \forall x \Box \forall y \Box((Ey \land y \in x) \rightarrow Ex)$; without the inner $\Box$, it would be a trivial consequence of the logical truth $\forall x Ex$. A misprint of Fine's definition of ($\Box$) (p. 185) omitted this inner $\Box$.

The construction of a theory of the first kind seems quite straight-forward if we are prepared to assume a fixed domain of possible individuals. Our language can be a first-order modal (not free) language with identity, with a predicate '$x \, \eta \, y$' read 'the object x has the attribute y' and a predicate I for being an individual. The logic is just ordinary quantificational logic combined with S_5 modal prop-ositional logic.[21] I will use the notation $[x: A]$ for attribute abstraction (possibly virtual in the sense of Quine, *Set Theory and Its Logic*, chap. 1) and continue to use the notation $\{x: A\}$ for set abstraction (again possibly virtual). I will assume that attributes obey Intensionality, that is,

$$\Box \, \forall z(z \, \eta \, x \leftrightarrow z \, \eta \, y) \rightarrow x = y.$$

We can develop an "iterative conception of attribute" parallel to the iterative conception of set. Attributes will arise in a transfinite sequence of stages. Let A_α be the collection of individuals and at-tributes formed before stage α, so that $A_0 = D$, the domain of individuals. At stage α, we form all possible attributes of objects available; let $\Pi(A_\alpha)$ be the collection of all such attributes; then $A_{\alpha + 1} = A_\alpha \cup \Pi(A_\alpha)$. Clearly for a limit stage α, $A_\alpha = \bigcup\limits_{\alpha < \lambda} A_\alpha$. The idea is that the attributes of our theory will be those generated by iterating this process through stages marked by all the ordinals.

It is not difficult to embody this idea in all axiomatic theory whose axioms are for the most part modeled on those of ZFI, i.e., ZF with individuals. Some distinctive features should be mentioned. Let us say that a is *rigid*, $Rn(a)$, if $\forall x[\Box(x \eta a) \vee \Box \neg(x \eta a)]$.[22] Since the Barcan formula is provable in our logic, $Rn(a)$ implies full rigidity. Sets will be identified with rigid attributes; then $(R \in)$, $(R \notin)$, and $(BF \in)$ will be provable (with their variables restricted to sets). We assume that $[x: Ix]$ exists and is rigid. We have axioms of Intensionality, "pair attribute," "sum attribute," "power attribute," and Replace-ment whose formulation is clear, as well as η-induction (corre-sponding to Foundation), and, if desired, Infinity and Choice. We need to assume that every attribute is coextensive with a rigid attribute:

---

[21]E.g., Hughes and Cresswell, *An Introduction to Modal Logic*, pp. 138-141.
[22]Cf. Gallin, *Intensional and Higher-Order Modal Logic*, p. 77.

(E) $\forall a \; \exists b [\forall x(\Box(x \; \eta \; b) \vee \Box \neg(x \; \eta \; b)) \wedge \forall x(x \; \eta \; b \leftrightarrow x \; \eta \; a)].$[23]

(I take the variables $a, b, \ldots$ to be restricted to attributes.) We can prove that for any a, $[x: \Diamond (x \; \eta \; a)]$ exists; we can then derive a strong schema of Separation:

$$\exists a \; \Box \; \forall x[x \; \eta \; \alpha \leftrightarrow (\Diamond (x \; \eta \; b) \wedge A)].$$

Just as in set theory, we can describe the iterative generation of attributes within the theory: $\Pi(a)$ can be defined as $[x: \Box \; \forall z(z \; \eta \; x \rightarrow z \; \eta \; a)]$, and A_α can be defined by transfinite recursion. A_α is a set for every α. By η-induction, we can prove that evrything is in A_α for some α.[24]

This theory of attributes generalizes the higher-order modal logic of Montague and Gallin in the same way that ZFI generalizes the simple theory of types. Taking sets as rigid attributes, one obtains an inner model of ZFI, in which the "individuals" (nonsets) are the individuals and *non*-rigid attributes. The universe of sets built on the individuals of the theory consists rather of the *hereditarily* rigid attributes, that is, whose construction in the iterative process involves only rigid attributes. From a modal point of view, in either case the set theory is trivial, in a way it also is in higher-order modal logic: the structure of sets is the same for each possible world.[25] This can be seen to be a consequence of the fixed domain of individuals: the modes of set formation available in each possible world are the same, and the individuals with which we begin are the same. This might be considered not a very satisfactory result.

We can mitigate this triviality if we recall that the domain of objects of our theory is fixed because we understand it to consist

[23]This corresponds to the axiom E^α of Gallin's higher-order modal logic, ibid., p. 78; cf. Appendix 1, section 3.

[24]From this it follows that every attribute is a subattribute of a rigid one; it is from this that we infer the existence of $[x: \Diamond (x \; \eta \; a)]$. In a theory without η-induction, the latter would apparently have to be assumed.

[25]In Gallin's higher-order modal logic with comprehension and E, possible worlds can be constructed internally as what he calls atomic propositions (*Intensional and Higher-Order Modal Logic*, p. 85). In our theory with (E), the same can be done in essentially the same way. Let ϕ be the null attribute. We can construe propositions as attributes that are necessarily had either by nothing (false) or by ϕ alone (true). Then we can straightforwardly carry out Gallin's proof that there is necessarily a true atomic proposition (ibid., pp. 87-89). Cf. Fine, "Postscript," pp. 120-121; Fine presents the argument informally and uses the perhaps apter term "world proposition." The idea of representing possible worlds within a modal logic by such propositions goes back to Prior; see *Worlds, Times, and Selves*, p. 43, and Fine, "Postscript," p. 119.

of all *possible* objects in some sense. It follows that we should be prepared to admit the predicate E of existence as a *nonlogical* predicate. In fact, if we admit such a predicate just for *individuals*, we can extend it to a predicate E^* of existence for sets built up from individuals (i.e., hereditarily rigid attributes) by using an idea of Kit Fine, that not only does the existence of a set require that of its elements (as (E $\in$) prescribes), but the existence of all the elements of a set is sufficient for that of the set.[26] Given existence E of individuals, clearly we can define E^* satisfying this condition by transfinite recursion. Then, in each world the hereditarily rigid attributes satisfying E^* are just those such that the individuals in their transitive closures exist in the sense of E; these should be the existing sets according to Fine.

However, if we consider the sets in general that are given by this theory, we have to admit that they are *sets of possibles* rather than possible sets in the sense of sets that possibly *exist*. Suppose for example that x and y are two individuals that are separately possibly existent (i.e., $\Diamond$ Ex and $\Diamond Ey$ hold) but are not *compossible*, i.e., $\Diamond (Ex \wedge Ey)$ fails. Then although they have a pair set $\{x, y\}$, this set does not even possibly exist (in the sense of E^*). From the assumption that every individual possibly exists, it does not follow that every set built up from individuals possibly exists. To prevent sets whose existence is impossible from arising, we need to base a modal set theory on a more delicate analysis, such as is given by Fine (see below).[27]

Another reservation about our theory is that it is through and through based on intuitions about sets. In motivating the theory by the "iterative conception of attribute," I assumed that at each stage we could form the totality of *all* attributes of objects available at that stage. If we also assume that we have at the outset a domain I of possible worlds, then we can construe this as just a set-theoretic construction, since we can take the totality of such attributes to be just the set of functions mapping I into $\mathfrak{P}(A_\alpha)$.[28] But it is very

[26]"Sets," pp. 180, 186.

[27]Moreover, we have completely swept under the rug any issues concerning the existence of attributes or propositions. Concerning the former, the existence criterion given by Fine for the simple theory of types in "Properties, Propositions, and Sets," p. 152, might be generalized to cover the present theory (given a structure on the individuals). I have no intuitive sense of whether it is a good criterion. It appears that it might allow some rigid attributes to exist *qua* attributes that, according to (E $\in$), do not exist *qua* sets.

[28]Note that possible worlds as construed within the theory as suggested in note 25 form a set; in fact there is a set of all propositions, since any proposition is a strict subattribute of $[x: x = \phi]$. On the possible-worlds conception of proposition,

questionable whether there is some more basic notion of attribute, not based on that of set or function, that would give rise to such an iterative construction or even satisfy the impredicative comprehension principle of higher-order modal logic.[29] In particular, taking attributes to be objects that serve as Fregean "senses" of one-place predicates seems unlikely to justify even second-order impredicative comprehension (see Essay 8, section II). If this doubt is well-founded, then the whole project of founding modal set theory on a theory of attributes is ill-conceived; indeed, what attributes there are might well depend on what *sets* there are.[30]

The theories of Kit Fine in "First Order Modal Theories I: Sets" are developed independently of any theory of attributes or propositions and do not make such strong use of quantification over possible objects as does the above attribute theory. In his two modal theories MA and MP, the rigidity principles (R $\in$), (R $\notin$), and (BF $\in$) all hold, as does the existence principle (E $\in$). Where the two theories differ is whether to admit quantification over *possibilia*. The "actualist" theory MA, in which such quantification is not given, is formulated in a logic equivalent to the free S5 of Appendix 1 with the additional logical axiom $\Diamond Et$ (according to which "possible objects" are objects that possibly *exist*). The "possibilist" theory MP adds quantifiers over the domain of possible objects; it could alternatively be viewed as a theory in fixed domain (unfree) logic

---

the existence of a set of all propositions is equivalent (given some set theory) to the existence of a set of all possible worlds.

I understand it is a matter of discussion among modal realists whether "the" possible worlds constitute a set.

[29]If one is given in advance a domain of *propositions*, another possibility is to take the higher-order entities to be functions whose values are propositions, in other words propositional functions. Cf. Church, "Russellian Simple Type Theory." It is not clear what conceptions of proposition other than the possible-worlds conception are available for this purpose. The paradox presented by Russell in Appendix B of the *Principles* (p. 527) is a problem for refined criteria of propositional identity in the context of the simple theory.

[30]If modal set theory is not to be founded on an impredicative theory of attributes, then Russell's ramified theory of types *with* the Axiom of Reducibility takes on new interest. This theory contains the *extensional* simple theory of types, but on the intensional plane the ramification is preserved. (For the significance of this for the semantical paradoxes, see Church, "Comparison.") A modal theory of this kind can be obtained by beginning with Gallin's higher-order modal logic with comprehension and extensional comprehension (notes 22 and 23 above). The types can be ramified and the (intensional) comprehension schema replaced by the predicative version. But the axiom E that yields extensional comprehension should be replaced by an axiom stating that for each relation (of a given type) there is a coextensive relation of lowest order compatible with the types of its arguments, that is rigid.

with E as a nonlogical predicate. For the details of the axiomatizations, the reader is referred to Fine's paper ("Sets," pp. 186-187); the axioms of ZFI are assumed in an essentially *de dicto* form, with quantifiers restricted to actual objects (with the exception that the schema of replacement is assumed in a form equivalent to

$$\forall a\{[\forall x \forall y \forall z \,(Axy \land Axz) \to y = z]$$
$$\to \exists b \forall z[z \in b \leftrightarrow \exists x(x \in a \land Axy)]\},$$

which in effect allows the *parameters* in the formula A to be possible objects, but the given set a has to be actual.[31]

This implies that the sets of a given world are a model of ZF but says nothing about when a set of one world will exist in another. Fine's idea for an answer to the latter question is that a set should exist if and only if all its elements do. In MP, this can be stated straight out as an axiom. But for the "actualist" MA, we can state the "only if" parts as ($E \in$). For the "if" part, we rely on the axiom of foundation and the fact that all sets are well-founded, that is, obtained by transfinite iteration of the process of forming sets of available objects, starting with individuals, i.e., every set belongs to V_α for some α, where

$$V_0(a) = a$$
$$V_{\alpha + 1}(a) = V_\alpha(a) \cup \mathfrak{P}[V_\alpha(a)]$$
$$V_\lambda(a) = \bigcup_{\alpha < \lambda} V_\alpha(a) \qquad \text{for limit } \lambda,$$

and $V_\alpha = V_\alpha(\{x\colon Ix\})$. Then one form of the desired axiom might be that if a is a set of individuals, then any set constructed from the individuals in a (i.e., in $V_\alpha(a)$ for some α) necessarily exists if a does (which of course implies that all elements of a exist), i.e.:

(1) $\qquad \forall a\{\forall x \in a\; Ix \to \forall x[\exists \alpha(x \in V_\alpha(a)\,) \to \Box(Ea \to Ex)]\}.$

However, this statement turns out to be provable with the help of a simpler "existence axiom"

[31] If this latter proviso is dropped (i.e., if '$\forall a$' is dropped or replaced by '$\forall a\Box$'), then the necessary existence of $\{x\colon x \in a\}$ becomes an immediate consequence of Separation. In the theory as it stands it is still provable, but the proof apparently requires full Replacement and Choice.

Here and for the remainder of this essay, the variables $a, b, \ldots$ will be restricted to sets (i.e., nonindividuals).

(CEA) $\forall a[\forall y \in a \ \Box(Ex \to Ey) \to \Box(Ex \to Ea)]$,

whose significance is not immediately clear but which becomes so when one reflects that it gets one across the induction step in proving (1) by transfinite induction on α, when α is a successor ordinal.

(1) implies that in a well-founded model of MA, a set exists if all its elements do, in keeping with Fine's idea. In fact, Fine is able to strengthen this observation and prove that quantifiers over possible objects are actually definable in MA and that the "possibilist" theory MP is a conservative extension of MA.[32] The idea is that the set of individuals existing at a world can be used as a code for the world itself, and then the representation of worlds within the theory can be carried out. But 'A holds at i for all possible objects' is equivalent to '$\Box \ \forall y(A$ holds at $i)$'.[33]

MA and MP avoid the untoward result of the set theory based on fixed-domain attribute theory, of giving rise to "possible sets" that do not possibly *exist*,[34] by limiting set formations to the formation of existing sets. There is an obvious translation of MP into the attribute theory, in which E is translated as E^*. If A is translated as A^*, and Πx is the quantifier over possible objects, then ΠxA has to be translated as $\forall x[\Diamond E^*x \to A^*]$, not $\forall xA^*$.

Fine's theories are clearly more stisfactory than the attribute theory and seem to be quite satisfactory theories of sets in the context of absolute metaphysical modality. The theory MA refrains from "speaking of what is not" to the extent possible in a *de re* modal theory. But in them and in the other theories we have considered so far, pure sets exist necessarily, and any set necessarily exists if all the individuals from which it is constructed do. Thus these theories do not attempt to analyze any "potential" aspects of mathematical objects themselves. The interpretation of the modal operators will not encompass the sense, discussed in the previous essay, in which a set is "potential" relative to its elements and therefore does *not* necessarily exist if its elements do. We shall now take up again the project of constructing a formal theory based on that idea.

---

[32]"Sets," p. 192.

[33]"Sets," pp. 192-193. The matter is slightly more complex if some non-set-theoretic vocabulary is added to MA and MP. Suppose, however, that this vocabulary only puts a structure on the individuals. Then we can code the worlds by structures on sets of individuals and proceed as before.

[34]Since for Fine $\Diamond Et$ is a logical axiom, if his theories did not avoid this result they would be inconsistent.

III. Theories of Potential Sets

In the previous essay I defended the thesis, based on Cantor's explanations of the notion of set, that any "multiplicity" of given objects can constitute a set. The "can" was taken seriously as indicating that relative to that of its elements, the existence of a set is potential. I proposed to develop this idea by means of a modal quantificational language. However, in that essay I took only the first steps. The purpose of the present section is to describe some formal set theories based on these ideas.

I shall describe two types of theories, first- and second-order. The idea of the second-order theories is sketched in section V of the previous essay. Various interpretations of the second-order variables are possible: Cantor's "multiplicities," Frege's concepts and relations (but without Frege's extensionality), and, for versions in which the second-order comprehension schema is restricted to insure predicativity, interpretations in terms of truth or in relative substitutional terms are also possible, along the lines of Essay 8 above.

We can also construct first-order theories; indeed the axiom expressing the idea that any multiplicity can constitute a set can be formulated naturally and convincingly as a first-order schema. However, it becomes immediately apparent that an extension of the usual first-order modal logic is needed. We might at the outset try to express the idea that any multiplicity F can constitute a set by the axiom

$$(1) \qquad \Diamond \; \exists a \; \forall x \; (x \in y \leftrightarrow Fx),$$

where, in the first-order version, 'Fx' would be replaced by any first-order formula. But it is immediately apparent that this will not do; Russell's paradox is not avoided merely by writing '$\Diamond$' in front of the naive comprehension schema. From our point of view the point is this: the objects that *are* F can constitute a set, not the objects that *would be* F if the set were formed. Suppose 'Fx' is '$x \notin x$'. We want to assume that there is potentially a set of all objects that are not elements of themselves, but that means all objects that are *actually* not elements of themselves. Once the set is formed, it will presumably not be an element of itself, but we do not expect it to belong to the set. Thus we can see (1) as suffering from a fallacy of *scope*: what is wanted is that 'Fx' should be understood as having its scope outside all modal operators within which it might occur. Then instead of (1) what we intended is

$$(2) \qquad \Diamond \; \exists a \; \forall x \; (x \in y \leftrightarrow \lfloor Fx \rfloor_0).$$

So far, the bracketing operation behaves much like the "actuality operator" that is often added to systems of modal logic.[35] In (2), we can read '$\lfloor Fx \rfloor_0$' as 'x is actually F'. But we would like to be able to say that (2) is necessarily true. But interpreting the necessitation of (2) in possible-worlds terms, what we want is that in any world i there is possibly a set of just those objects that are F *in the world i*. The actuality-operator reading would substitute 'in the actual world'. In the necessitation, we want 'Fx' to be outside the scope of the '$\Diamond$' but still within the outer '$\Box$'. Hence we need a more powerful notation for expressing scope. We will enlarge the notation of first-order modal logic by allowing formulae to be enclosed in numbered brackets; $\lfloor A \rfloor_n$, occurring within a formula, is to be interpreted (roughly) as falling within the scope of the first n nested modal operators within which it occurs but outside the scope of further modal operators. Then the natural necessitation of (2) will be

$$\Box \; \Diamond \; \exists a \; \forall x \; (x \in y \leftrightarrow \lfloor Fx \rfloor_1).$$

The first-order theories will be formulated within *scoped first-order modal logic*, which is described in more detail in Appendix 1, section 2.

Before I describe these theories further, I should say something about the intended meaning of the modal operators. In saying that a multiplicity of objects can constitute a set, I mean that they can do so without changing anything at "lower" levels, that is, without changing the structure of the individuals or of the sets that might have entered into the constitution of the objects making up the multiplicity in question. It is this strong possibility that the modal operator in (2) is meant to express. The alternate possibilities that our modal operators envisage are simply those in which possibilities of set formation have been realized that as yet have not been realized. The logic will then be somewhat like the sort of tense logic in which everything past is necessary (unalterable by later events).

· [35]See for example Hazen, "Expressive Completeness in Modal Languages," or Fine, "Postscript," pp. 142-143.

The structure of individuals and the $\in$ – structure of the sets that "already" exist will be necessary. Moreover, we do not need to allow the possibility that anything might not exist, since no object will cease to exist as a result of the formation of further sets. It follows that we do not need to distinguish the domains of free and bound variables. The possible-worlds models will be Schütte models in the sense of Appendix 1. Hence we can use ordinary rather than free quantifier rules.

The propositional logic will be S4 rather than S5: clearly the "Brouwerian" axiom B will not be valid: if it is possible, by forming more sets, to make A necessary (i.e., true "from there on"), it does not follow that A is "already" true.[36] In possible-worlds terms, the alternate possible worlds to a world i will as $\in$ – structures be *end extensions* of the structure W_i with respect to the $\in$ – relation: that is, for such a j, $W_i \subseteq W_j$, $\in \uparrow W_i$ is a subrelation of $\in \uparrow W_j$, and if $u \in W_i$ and $v \in u \cap W_j$, then $v \in W_i$.

Thus the logic of our first-order theories will be scoped unfree S4, obviously without the Barcan formula. Our rigidity axioms will be (R $\in$) and

$$(R_1 \in) \qquad \Box \forall x \, (x \in y \rightarrow \lfloor x \in y \rfloor_0).$$

Interpreted in Schütte models, these simply state the end-extension condition. (R $\in$) and (BF $\in$) then follow. Individuals are accommodated by adding a one-place predicate I and assuming axioms (R I) and ($R_1 I$) stating that I is also fully rigid. For most applications one can weaken ($R_1 I$) to (R $[x: \neg Ix]$), i.e., get by with the assumption that I is absolute.[37] '$\mathfrak{M}x$' is defined as '$\neg Ix$', as before; we assume Extensionality and

---

[36] I use the designations for schemata in modal propositional logic of Chellas, *Modal Logic*.

[37] It would be more economical to introduce a name ϕ of the empty set and define Ix as in nonmodal set theory, as $x \neq \phi \wedge \forall y(y \notin x)$. Then from the axiom $\forall y(y \notin \phi)$, the absoluteness of Ix follows just from the fact that the logic treats ϕ as a rigid designator. But this formulation conflicts with the policy that set existence is given directly only by the set-formation axiom.

($R_1 I$) (or its consequence (BF I)) is not needed to obtain the (potential) existence of the set of all *actual* individuals, but it yields the stronger statement

$$\Diamond \, \exists y \, \Box \, \forall x (x \in y \leftrightarrow Ix).$$

For the translation of ZFI into the corresponding theory of potential sets (see Appendix 2) this is needed to prove the translation of the axiom that there is a set of all individuals.

In ($R_1 \in$) and ($R_1 I$), note that $\lfloor A \rfloor_0$ is true only if its free variables have values in the domain of the world with respect to which the whole formula within which it occurs is being evaluated.

Elem $x \in y \to \mathfrak{M} \, x$.

We will consider three second-order logics, called LPS, LPE, and LIE, described in Appendix 1, section 3. The comprehension principles of LPE and LPS are predicative; LPE lacks but LPS has the scope operators. LIE has impredicative comprehension. Theories based on LPE or LPS are related to first-order theories roughly as von Neumann-Bernays-Gödel set theory NB is related to ZF; theories based on LIE are the analogue of set theories with impredicative classes. I will use the Fregean terms for second-order entities: 'concept' for monadic, 'relation' for polyadic. We can define the full rigidity of a concept F by the relativized Barcan condition of section I above:

$$Rn(F) \; =_{df} \; Abs \; (F) \wedge \square \; \forall G \; [\forall x \; (Fx \to \square \; Gx) \to \square \; \forall x \; (Fx \to Gx)].$$

In theories based on LPE or LIE, we replace $(R_1 \in)$ by $(R \notin)$ and $(BF \in)$, and similarly for $(R_1 I)$.

Let SF_0 be the first-order schema whose instances are all well-formed results of substituting for 'Fx' in (2) formulae lacking free y; let SF_1 be the closure for (2) itself by '$\forall F$'. $SF_1 E$ is the formula

$$\forall F[Rn(F) \to \Diamond \; \exists a \; \forall x(x \in a \leftrightarrow Fx)].$$

Clearly if F is fully rigid, then the idea of our set-formation axioms would motivate $SF_1 E$ (cf. Essay 10 above, section V). Therefore we can develop second-order theories in which the scope apparatus is replaced by extensional comprehension.

We now describe basic axiom systems for each logic:

MS_0:   Scoped first-order S4, $(R \in)$, $(R_1 \in)$, $(R \, I)$, $(R_1 I)$, Extensionality, Elem, Foundation, SF_0.

$MS_1 S$:   LPS, same set-theoretic axioms as MS_0 except that SF_0 is replaced by SF_1.

$MS_1 E$:   LPE, $(R \in)$, $(R \notin)$, $(BF \in)$, $(R \, I)$, $(R \; [x: \neg Ix])$, $(BF \; I)$, Extensionality, Elem, Foundation, $SF_1 E$.

$MS_1 I$:   Like $MS_1 E$ except that LPE is replaced by LIE.

Note that these theories make no assumptions of set existence except the set-formation axioms. We shall see that they are quite weak theories, but the stronger theories that we develop from them have the property that the additional assumptions needed have a certain logical character. In effect, SF_0 or SF_1 insures that any "available" objects can be formed into a set; then we can adjust the strength of the theory by different assumptions about the "availability" of objects.

If we think of sets as formed in stages, the use of S4 embodies the assumption that these stages are partially ordered. Using well-known facts about modal propositional logic, we could strengthen this assumption by strengthening the propositional logic. At the end of the previous essay, we speculated about the possibility that, beginning with a certain domain of sets (the sets formed before a certain stage), we might extend it in different ways that were incompatible, i.e., there might be "later" stages α_1 and α_2 such that we could not unite the sets at α_1 and those of α_2 into a single domain. This is ruled out if we strengthen the logic to S4.2, that is, if we add the axiom schema

G $\quad \Diamond \Box A \to \Box \Diamond A,$

since it is well known that S4.2 is sound and complete for the class of models in which the accessibility relation is *directed*, i.e., R is a partial ordering such that for any $i, j \in I$, there is a $k \in I$ such that iRk and jRk.[38] In our setting, S4.2 proves the schema

NPE$_0$ $\quad [\Box \, \forall x(\Box \, A \lor \Box \neg \, A) \land \Diamond \, \exists x A] \to \Box \, \Diamond \, \exists x A,$

which implies that if a set with a given absolute property *can* exist, then its existence cannot be made impossible by realizing *other* possibilities of set formation.

The usual picture of the iterative conception of set goes further and takes the stages to be well-ordered and therefore linearly ordered. That would justify strengthening the propositional logic still further to S4.3, by adding the axiom

L$^+$ $\quad \Box(\Box A \to B) \lor \Box(\Box B \to A).$[39]

However, it seems that in practice only S4.2 is of use in deriving set-theoretic consequences. It would be interesting if there were an axiom system in this language that had something to recommend it intuitively but was incompatible with the extension of the logic

[38]Cf. Chellas, *Modal Logic*, p. 88.
[39]Ibid., pp. 144, 185. At p. 144, Chellas helps the reader's intuition concerning this schema by giving a long list of equivalents.

to S4.2. This would be of philosophical interest to me since, as I said in the last essay, I do not find the picture of the linear ordering of the stages as evident as it should be. However, the theories considered in this essay are all compatible with such an extension. In fact, the theories we relate to ZF and NB are such that the result of adding the schema L^+ is a conservative extension for an important class of formulae.[40] However, the relations of impredicative theories (based on LIE) and standard set theories with impredicative classes have been established only by adding G to the modal theories.

To see that MS_0 and the other minimal theories are rather weak, consider how the set existence claims embodied in the axioms of ZF fare in MS_0. The null set certainly exists potentially; this is evident from SF_0 since $x \neq x$ is fully rigid. The same holds for the pair set of given objects y, z, since $x = y \lor x = z$ is also fully rigid. In fact, from the rigidity axioms for $\in$ and I it folows that any Δ_0 formula (i.e., formula each of whose quantifiers is restricted to a set) of the language of ZFI is absolute. It then follows that for a given set a, by $(R_1\in)$, that $\exists y (x \in y \land y \in a)$ is fully rigid; hence the sum set of a exists potentially. Similarly, if A is Δ_0, then $x \in a \land A$ is fully rigid, and therefore $\{x: x \in a \land A\}$ exists potentially. But for an arbitrary formula A, the most we can say is that $\{x: x \in a \land \{A\}_0\}$ exists potentially.

So far, MS_0 resembles what is sometimes called weak Zermelo set theory, that is, Zermelo set theory without the axiom of infinity and with separation restricted to Δ_0 formulae. However, we cannot show that for a given a its power set exists potentially. Even in a second-order standard model, the most that SF_1 tells us is that each subset of a exists potentially, but if a is infinite, their realization might be spread out over an unending sequence of stages so that there is no stage at which all are "available" and we can form the set of them.

One argument to motivate the power set axiom proceeds as follows: Given a set a, it could only have been formed if all its elements were already available. But since any multiplicity of available objects can constitute a set, any subset of a could equally have been formed at the same point. Hence we can suppose all subsets of a to have been formed at that stage and to be available when a is. Therefore, at the next stage we can form the set of all subsets of a. This argument can be criticized (cf. p. 277 above), but it does suggest

---

[40]See Appendix 2, Theorems 7, 9, and 10.

the following principle. Since our rigidity axioms imply that $[x: x \subseteq a]$ is absolute, that all subsets of a that could possibly be formed are available when a is can be expressed in our language as

$$PS_0 \qquad \square \ \forall x(x \subseteq a \to \lfloor x = x \rfloor_0).^{41}$$

This implies that $[x: x \subseteq a]$ is fully rigid and hence that the power set of a exists potentially. However, the latter is already implied by the assumption that PS_0 is *possible*:

$$\Diamond PS_0 \qquad \Diamond \square \ \forall x(x \subseteq a \to \lfloor x = x \rfloor_1).$$

This in fact expresses no more than that all possible subsets of a "can exist together." It is therefore a minimal way of expressing the power set axiom in our language.

$MS_0 + \Diamond PS_0$ is in a definite sense as strong as weak Zermelo set theory; in fact, if G is added, it is as strong as Zermelo set theory without Infinity. In a possible-worlds model, the rigidity axioms require a sort of transitivity of the domain of each world: if a is a set in W_i and $x \in a$, then x appears in W_j with iRj, $x \in W_j$. Since W_i is itself potentially a set, it follows that a is potentially contained in a transitive set and therefore potentially has a transitive closure. However, that every set has a transitive closure cannot be proved in Zermelo set theory.[42] A consequence is that the axiom of foundation implies in MS_0 the schema of $\in$-induction; the usual set-theoretic proof of this uses the existence of the transitive closure of a set.

To obtain a precise result relating MS_0 to a fragment of ZF, we need an appropriate translation of the language of set theory into the modal language. We do this by thinking of a formula of set theory as a negative formula of an intuitionistic language and then using the standard translation of intuitionistic logic into S4. The main points are that we treat $A \vee B$ and $\exists x A$ as abbreviations for $\neg(\neg A \wedge \neg B)$ and $\neg \forall x \neg A$ respectively; if A', B' translate A, B respectively, then $\neg A$ is translated as $\square \neg A'$, $A \to B$ as $\square (A' \to B')$,

---

[41]Note that in our logic '$\lfloor x = x \rfloor_0$' can be read as 'x actually exists'. In a free-logical version of the theory, it would be replaced by '$\lfloor Ex \rfloor_0$'.

From the fact noted in the text, that for any a there is possibly a set b of just those $x \in a$ that actually satisfy A, we can prove by PS_0 that, since b actually exists, there is actually such a b; that is, we can prove any instance of Separation. Of course $\Diamond PS_0$ does not suffice for this.

[42]Drake, *Set Theory*, p. 111.

and $\forall x A$ as $\square\ \forall x A'$. Let A^W be the translation thus obtained for a formula A of the language of ZFI. We can show that if A is a theorem of weak Zermelo set theory, the A^W is a theorem of $MS_0 + \diamond PS_0$. (See Appendix 2.)

There is an obvious translation back: Given a domain of individuals, let V_α be the set of sets of rank $< \alpha$ (as above, p. 313). For any limit ordinal λ, we can construct a model of $MS_0 + \diamond\ PS_0$ by taking the possible worlds to be the ordinals $< \lambda$ and the domain of α to be V_α. This leads to a translation of the modal language into that of ZFI such that every theorem of $MS_0 + \diamond\ PS_0$ is translated into a theorem of ZF. However, this is not the sharpest result, since the axiom of replacement is needed for the construction of the V_α. For sharper results see Appendix 2, theorems 2 and 3.

Next we consider how the axioms of Infinity, Replacement, and Choice fare in our modal theories. Let AC be some usual formulation of the axiom of choice as a closed sentence. In the first-order language, we can do little better than to postulate AC^W directly; however, we can simplify it to

AC $\diamond$ $\qquad$ $\square\ \forall a\ \diamond\ \exists x\ \forall y \in a\ \exists! z\ \{\langle y, z\rangle \in x$
$\qquad\qquad\qquad \wedge\ [\exists w(w \in y) \rightarrow z \in y]\}.$

If in the second-order theories we think of the second-order variables as ranging over quite arbitrary "multiplicities," then we might replace the potential existence of the set-relation x by the actual existence of a second-order relation H. The resulting axiom implies AC $\diamond$ and AC^W.

Infinity and Replacement are best handled in a manner suggested by axiomatizations of set theory by means of reflection principles. The situation is perhaps clearest in the second-order setting. Consider first the axiom of infinity. The question is not so much the existence of an infinite set, which we would obtain immediately if there should be infinitely many individuals, but rather that of the existence of a set of infinite rank. Of course if there are no individuals, or finitely many, it is only at infinite ranks that infinite sets appear.

A common form of the axiom infinity is the statement that there is a set containing all finite von Neumann ordinals, i.e., a set w such that $\emptyset \in w$ and for each $x \in w$, $x \cup \{x\} \in w$. One can naturally see such a set as obtained by a sort of transition from "potential" to "actual" infinity, in a way that can be formulated neatly in our modal language. Once we have a set x, we can always form the set $\{x\}$ and add it to x; in fact we can easily prove in MS_0

$$(3) \qquad \Diamond[\exists(z = \phi) \wedge \Box \forall x \Diamond \exists y(y = x \cup \{x\})].$$

However, to obtain by our set-formation axioms the set demanded by the axiom of infinity, we need to "complete" the iterations of forming $x \cup \{x\}$ from x, i.e., to have

$$(4) \qquad \Diamond[\exists z(z = \phi) \wedge \forall x \exists y(y = x \cup \{x\})].$$

The principle behind the inference from (3) to (4) could be expressed at first approximation by saying that for any F, $\Box \forall x \Diamond \exists y \, Fxy$ implies $\Diamond \forall x \exists y \, Fxy$. However, this will not hold for arbitrary F, since if for a given x we reach a "possible world" containing a y such that Fxy, it may be that for some z such that we had already reached a w such that Fzw, Fzw will have ceased to hold. We clearly want F to be *persistent*. The principle of our inference can be expressed by the statement that for any dyadic F

$$[\Box \forall x \, \forall y(Fxy \rightarrow \Box \, Fxy) \wedge \Box \, \forall x \, \Diamond \, \exists y \, Fxy] \rightarrow \Diamond \, \forall x \, \exists y Fxy.$$

However, this is equivalent to the at first sight stronger statement that for any *monadic* F

Refl$_1$ $\quad [\Box \, \forall x(Fx \rightarrow \Box \, Fx) \wedge \Box \, \forall x \, \Diamond Fx] \rightarrow \Diamond \, \forall x Fx.$[43]

This is the appropriate form of the reflection axiom for the second-order language. In the scoped first-order language, we replace it by a schema, but we have to be careful about subscripts. For any formula A, let A^1 be the result of increasing all non zero subscripts in A by one. Then the schema we want is

Refl$_0$ $\quad [\Box \, \forall x(A \rightarrow \Box A^1) \wedge \Box \, \forall x \, \Diamond A^1] \rightarrow \Diamond \, \forall x A.$[44]

---

[43]The obvious generalization to polyadic F follows, but this is an instance where in the theory with individuals we have to use (BF I).

[44]One can also formulate weaker principles of the same kind by strengthening the hypothesis on F or A, for example by assuming it to be absolute. Some of the consequences needed for the relation to standard set theory follow from these weaker versions. Although they have no clear motivation from our general point of view, they may have technical interest.

The theory resulting from MS_0 by adding $\Diamond PS_0$ and $Refl_0$ is of the same strength as ZF. We will call it PZF ('P' for potential). For its intertranslatability with ZFI, see Appendix 2, theorems 5 and 6. We can also set up second-order theories that are similarly intertranslatable with NB. Let PNBE be $MS_1E + \Diamond BF \subseteq + Refl_1$, and let PNBS be $MS_1S + \Diamond PS_0 + Refl_1$. The W–translation extends readily to second-order formulae, since the negative translation of Fx is $\neg\neg Fx$, Fx^W is $\Box \Diamond (Fx)$, and similarly for polyadic second-order variables. Theorems 8 and 9 of Appendix 2 give precise intertranslatability results for NB and PNBS and PNBE. PNBS is a direct extension of PZF; in fact the well-known theorem that NB is a conservative extension of ZF extends to this case. Since PNBE is a subsystem of PNBS, it is a conservative extension of PZF for formulae without subscripts.

IV. Comparison and Discussion

In the last section we constructed theories in a modal language embodying the idea that a set is potential relative to its elements that proved to be intertranslatable with standard set theories. In this section I will discuss some reservations one might have about these theories and compare them with Fine's theories discussed in section II. We will concentrate on an issue concerning the interpretation of the modal operators that is connected with other issues about modality and mathematics that have arisen in other essays in this book. But I shall begin with some questions that can be dealt with more briefly.

A first question arises from the fact that we aspired to a kind of reduction of the existence of sets, not eliminating it altogether but treating it as potential. But, it might be asked, does not the introduction of modality amount to trading in the existence of certain objects for the existence of possible worlds? In its most general form, this objection is certainly misguided: possible-worlds semantics is a means of interpreting modal languages in a nonintensional set-theoretic metalanguage; but if one is serious in regarding the modal notions as fundamental and developing set theory within that framework, a more fundamental metatheory would itself be done in a modal language. I have not undertaken such a task here, but there is no reason to think it impossible or even to face fundamental difficulties.

The question is more serious, however, when applied to more

powerful modal theories, since there are assumptions that can be expressed in the modal language itself that allow one to construct possible worls "internally," as we saw with the attribute theory of section II (see note 25). Then the question will arise whether one has achieved a *reduction* of the possible-worlds apparatus to something acceptable to the "modalist" or whether such a strong modal language is committed to possible worlds. In an unpublished manuscript, Allen Hazen raises in a general form the question whether a language in which the existence of sets is "modalized" can accommodate higher set theory without having the expressive power of a theory that directly quantifies over possible worlds. Hazen's argument as it stands presupposes a logic of absolute modality, which would satisfy S5. But the same suspicion is natural about our own modal logics, in view of the connection in higher-order modal logic between extensional comprehension and the coding of worlds by propositions.[45] In particular, the scoped first-order logic on which MS_0 and PZF are based appears to make it possible to keep track of the worlds indicated by outer modal operators within further modal operators, just the sort of cross-reference that is achieved by variables of quantification.[46]

On this issue, a definite result has been obtained by Harold Hodes.[47] If the propositional logic of the scoped logic is extended to S5, then it does become equivalent to a two-sorted theory with quantifiers for possible worlds. However, this is not true for the S4 logic we have actually used or for its extension to S4.2 or S4.3. It would become equivalent if we added some apparatus that would wholly or partially cancel the effect of a subscript moving the scope of a formula outside that of some of the modal operators within which it occurs. All the same, the fact that our language comes so close to simulating quantification over worlds makes one uncomfortable, and it raises a definite difficulty about a possible extension of our theories with a less restrictive interpretation of the modal operators.

This problem, whatever one's final view on it, will make more attractive another strategy for doing justice to the intuitions that our theory responded to, namely developing set theories with intuitionistic logic. In recent years, such theories have been developed and extensively analyzed, with rather satisfactory results. On the one hand, such theories have been formulated that are as strong

---

[45]See note 25 above and Gallin, *Intensional and Higher-Order Modal Logic*, §11.
[46]That this might be true was suggested to me by David Kaplan.
[47]"Modal Logic for the World Traveller."

as classical set theory, for example in the sense that an extension of Gödel's negative translation yields a proof of the consistency of ZF relative to its "intuitionistic" version.[48] On the other hand, in spite of this great logical power the theories preserve the formal properties of intuitionistic theories that express their constructive character, such as for example the "disjunction property": if $A \vee B$ is provable for closed A, B, then either A or B is provable.[49] It is even possible to extend "intuitionistic ZF" by analogues of classical large cardinal axioms.[50]

Although I do not wish to comment extensively on these theories, I shall very briefly compare the Kripke models of such a theory with the models of the theories of section III. In the intuitionistic theories, membership is not decidable, i.e., we cannot prove $\forall x \forall y \ (x \in y \vee x \notin y)$.[51] It follows that in a model, if $y \in W_i$, y may yet take on new elements in later W_j, even elements from W_i. In our modal language, although $\in$ must be persistent, it need not be fully rigid or even absolute. On the other hand, the existential axioms have the effect that the domains W_i of the model have much stronger closure properties than our axioms would require, since the rule for the existential quantifier in intuitionistic Kripke models is that $\exists x A$ is satisfied at i in a model if and only if there is an object already in W_i whose assignment to x satisfied A. Thus, for example, the axiom of pairing will force every W_i to be closed under the formation of pair sets. The axiom schema Refl_0 implies that in models of PZF, W_i with such closure properties will exist. It would be consistent with our theory to impose stronger closure conditions on all possible worlds.

Let us now turn to the problem of the rather restricted interpretation we have given to the modal operator in our theories, leading to the choice of S4 as our propositional logic. The rigidity axioms are motivated by reflections that are meant to cover other types of modality, in particular the "metaphysical" or "broadly logical" modality at stake in most of the philosophical literature on

---

[48]Powell, "Extending Gödel's Negative Interpretation to ZF."

[49]Myhill, "Some Properties of Intuitionistic Zermelo-Fraenkel Set Theory."

[50]Friedman and Ščedrov, "Large Sets."

[51]There are exceptions: the theories studied by Tharp, "A Quasi-Intuitionistic Set Theory," and Wolf, "Formally Intuitionistic Set Theories." In an intuitionistic theory, the decidability of membership is incompatible with full Separation. What holds in a model of PZF, viewed as an intuionistic Kripke model, is $\forall a \neg \neg \exists y \forall x \ [x \in y \leftrightarrow (x \in a \wedge A)]$, for each formula A.

Tharp and Wolf showed the theories they studied to be considerably weaker than ZF.

modality. One might then legitimately ask that we formulate our theory on the basis of a less restricted interpretation of the modal operators. It is likely that the propositional logic going with such an interpretaion would be S5.

However, a more general interpretation does not just lie at hand. It cannot be metaphysical modality itself, as this is generally understood. Metaphysical necessity and the necessity of our theories are such that neither implies the other. To see this, note that in Fine's theories, the existence of individuals is contingent, but given certain individuals, the existence of any sets built up from them is conditionally necessary, and hence that of pure sets is absolutely necessary. For metaphysical necessity, that is the right result. In our theories, on the other hand, the existence and structure of individuals is necessary, because we take that to be given and unaltered by the formation of sets, but the existence of a set is contingent even given that of its elements, since when the elements are given the set is initially given only *in potentia*.

Thus to formulate a more general modal set theory that will encompass both our own theories and Fine's, either we have to leave the two modalities side by side, or we must seek a third interpretation of modality, more general than both metaphysical modality and our own restricted modality. Such an interpretation is certainly not inconceivable. One's first thought might be that it should be a strictly logical modality, according to which a proposition is necessary only if it is in some sense a truth of pure logic. Our brief exploration of this conception in section I of Essay 7 turned up some anomalies that raised doubts as to whether it is very natural. The doubts all arose from the intertwined character of logical and mathematical concepts. In particular, what we wanted to call the logical possibility of a proposition amounted to the mathematical possibility of some proposition of an appropriately similar logical form.

I conclude that what we should be looking for is a form of mathematical modality. But there is an ambiguity in this notion that is not adequately attended to above in the places where mathematical necessity and possibility are discussed.[52] Along with others, I have wanted to say that the truths of pure mathematics are necessary. Although one could read this as metaphysical necessity, I view it rather as bound up with the formal character of mathematics, in the Husserlian sense mentioned in the Introduction (p. 18 above).

---

[52]See especially Essay 1, section III, and Essay 7.

It implies but is not equivalent to necessity in any "real" sense bound up with the nature of the concrete world. Whether or not this picture is correct, it does seem clear that acceptance of the necessity of mathematics need not commit us even to the meaningfulness of the notion of necessity that appears in modal realist discussions of the essence of actual objects.

On this reading of the necessity operator, of course the objects of pure mathematics necessarily exist, although probably "impure" mathematical objects such as sets built up from contingently existing individuals do not.[53] There is no obstacle to reading the necessity operator in Fine's theories in this way, although he himself seems to have been thinking of metaphysical necessity.[54] But this reading does not allow the potentiality of a set relative to its elements on which our theories of potential sets are based.

However, mathematical necessity and possibility behave differently when we are concerned with the mathematical possibility of structures, where the underlying objects may not be mathematical objects. That is what is involved when I and other writers appeal to modality in explicating the notion of existence in mathematics. There seems to be no obstacle to attributing to such modalities the formal properties of absolute modalities, i.e., validating S5. But if a structure is realized, there is not thereby any necessity for the realization of any further structure that is not internal to the given structure or others that may have been presupposed. Therefore, in this case we do not have the conditional necessity of the existence of a set given that of its elements that Fine postulates. The point is clearest if we consider the restricted situation expected in reductionist accounts of mathematical objects, where we are considering the possibility of structures whose objects are physical or at least concrete. The realization of such a structure could not possibly involve *necessarily* the existence of any large extension of it, even if such an extension is possible and even necessarily possible.

It is this latter notion of formal, mathematical modality that has the generality we are looking for. Both metaphysical modality and the modality of the theories of potential sets can then be viewed as specifications of this notion. In the latter, the specification is one in which, once a structure is given, the possibilities that are relevant

---

[53]Recall that in characterizing mathematical truth in the context of a discussion of the necessity of mathematics, we wanted statements of set theory whose truth depends on what individuals exist *not* to be mathematical truths. See Essay 7, note 34.

[54]"Sets," p. 177.

are extensions of it, but the nature of sets is in a certain way respected: we have the same set in a possible extension only if we have just the same elements. It is this that constrains the extensions to be end extensions with respect to $\in$.

Let us assume that we can formalize this general modality by a free S5 version of scoped modal logic (see Appendix 1, section 2). If the structure on the *individuals* is given by a finite number of basic relations, we can then define restricted modalities that fit the conception of our potential set theories. Suppose that, for the absolute modalities, $\in$ and I are absolute and independent of existence, and $(E \in)$ holds. Then $(\Box)\forall x(x \in y \rightarrow \{Ex \wedge x \in y\}_0)$ follows, and any possible extension will be an end extension. If we have no structure on the individuals at all, we can define $\Diamond A_R$ as

$$\Diamond \, [\Box \; \forall x\{\{Ex\}_0 \rightarrow Ex\}_1 \wedge \forall x \, (Ix \rightarrow \{Ex\}_0) \wedge A],^{55}$$

and $\Box_R$ dually, and these will be the appropriate modalities for the theories of the last section. For each basic n-place relation P on individuals (assumed to hold only of them), we would have to add the clause

$$\forall x_1 \, . \, . \; \forall x_n(Px_1 \, . \, . \, x_n \leftrightarrow \{Px_1 \, . \, . \; x_n\}_0).$$

The first conjunct within $\Diamond$ in the definiion of $\Diamond_R$ uses a device that this logic affords for simulating quantification over *possibilia*; in general, we can define $\Pi x A$ as $\Box \; \forall x\{A\}_0$, increasing subscripts within other modalities.[56]

To formulate potential set theories in this framework, we replace $\Box$ and $\Diamond$ by $\Box_R$ and $\Diamond_R$ in the modal set-theoretic axioms other than rigidity axioms. This makes the axioms rather complex, but we should note that so long as one is not concerned with individuals and their structure, the only restriction on alternate possibilities

[55]Of course subscripts must be increased if this definition is applied within further modal operators.

[56]This observation is due to Hodes, "Modal Logic for the World Traveller." Of course Πx as thus defined includes in its range only objects that possibly *exist*, but given the absolute interpretation of $\Box$ this excludes only objects that exist in no possible world.

The definability of Πx might be avoided if we required for the truth of $\{A\}_0$ that the objects assigned to its free variables *exist* in the world with respect to which zero-subscript formulæ are evaluated. Such a reading would force on us an existence-dependent treatment of membership and perhaps identity. Moreover, I do not see how we could define $\Diamond_R$.

that is actually used is that actual existences are preserved; hence for pure, individual-free set theory, the definition of $\diamond_R$ can be simplified accordingly.

Thus it appears that the problem of formulating our theories in the context of a more general interpretation of modality is solvable. The problem of formulating in this framework theories involving metaphysical modalities would give rise to problems about the non-mathematical notions the theories might contain. But we also encounter mathematical difficulties in formulating Fine's set theories in a logic that accommodates my own in the way just indicated. The problem is that Fine's existence axioms imply that all possible worlds are closed under set formation. At first sight, it seems we should try to formulate a "Fine necessity" $\Box_F$, which restricts to possible worlds closed in this way. But MS_0 already implies that there are *no* such possible worlds, since there is for any world i possibly a set of all objects existing in i that are not elements of themselves, and that set cannot exist in i.

Two more indirect ways of incorporating Fine's theories are still possible. One is to understand his own quantifiers (even over *possibilia*) in a less than absolute way: instead of considering possible worlds closed under all possible set formations, we consider worlds closed under enough set formation to be models of ZFI and perhaps of strong axioms of infinity. This corresponds to taking the quantifiers of ordinary set theory to range over a domain that looked at "from outside" is really a set (see Essays 8, 9, and 10 above). Alternatively, we can interpret Fine's quantifiers and existence in modal terms, by means of the W translation or something similar.

Rather than going into detail about this, let me mention some reservations about this "absolute" version of the theories of potential sets. First, since I have used the scoped logic strengthened to S_5, the logic is as strong as the two-sorted theory of possible objects and possible worlds; in particular, as noted above, we can express quantification over possible objects. One might reply that these results just show that possible worlds and possible objects are acceptable notions from the viewpoint with which we started, that we have in effect achieved a reduction of them.

I shall not try to adjudicate this issue. For one can press the point further and question whether the usual modal quantificational logic is appropriate for the interpretation I have in mind, where the modal operators express the most general formal mathematical modality. It is not evident that we can assume individuation of objects through all the alternate possibilities that can be relevant.

In the constructions of section III and the applications of mathematical modalities in previous essays in this volume, we avoid the problem by restrictions on the range of possibilities to be considered. In a perfectly general modal set theory, we would not be entitled to such restrictions. It follows that we must regard the problem of formulating such a theory as unsolved, and we must even leave some doubts concerning what its logical framework should be.

Appendix 1. Free Modal Logic and Modal Logic with Scope Operators

1. *Free Modal Logic.* In the usual semantics for first-order free modal logic, the "outer domain" of values of free variables and objects to which predicates can apply is fixed for all possible worlds.[57] It is technically useful and subsumes some different semantics considered in the literature if we drop this condition. Consider a first-order language, with identity and perhaps with individual constants but (to simplify exposition) without functors. Of the quantifiers, we treat only $\forall$ as primitive. We add the necessity operator $\square$, treating $\Diamond$ as an abbreviation for $\neg\square\neg$. Since we are interested in modal quantificational logics based on normal propositional logics, our models will have a set I of "possible worlds" and a binary accessibility relation R. By a *free model* we mean a sextuple $\mathfrak{M} = \langle I, R, D, F_1, F_2, V \rangle$, where I is a nonempty set, R is a binary relation on I, D is a nonempty set, F_1 and F_2 are functions from I into D such that for every $i \in I$, $F_2(i)$ is nonempty, $F_1(i) \subseteq F_2(i)$, and if iRj, $F_2(i) \subseteq F_2(j)$. V is a function that assigns to each n-place predicate P of the language and each $i \in I$ an extension $V(P, i) \subseteq [F_2(i)]^n$ and to each constant c a value $V(c, i)$ such that if iRj, then $V(c, i) = V(c, j)$.[58] We assume that the language contains a logical predicate E of existence; we can extend V by setting

$V(E, i) = F_1(i)$ and $V(=, i) = \{\langle u, v \rangle : u, v \in F_2(i) \wedge u = v\}$.

Let $\mathfrak{M}$ be such a free model. By an *i-assignment* we mean a function

---

[57]For example, the Q3 semantics of Thomason, "Modal Logic and Metaphysics" and "Some Completeness Results," and also the semantics of Fine, "Model Theory for Modal Logic, Part I," p. 129, and elsewhere.

[58]Thus we assume that individual constants are rigid designators relative to the necessity operator of the language. Of course there is no difficulty in principle in relaxing this restriction. However, to do so would complicate the axiomatization. In fact, it is useful in the presence of nonrigid constants to consider a semantics in which for each i there is an "outer outer" domain $F_3(i)$ of values for nonrigid constants and of arguments of predicates; the objects of $F_3(i) - F_2(i)$ have for practical purposes no trans-world identity.

from a set of variables of the language into $F_2(i)$. Note that if f is an i-assignment and iRj, then f is a j-assignment. That is the motivation for the inclusion condition on F_2. The motivation is not merely technical: if a formula A containing modal operators is thought of as predicating something of values assigned to its free variables, then these objects have to be "available" at the worlds at which subformulae of A are evaluated.

Given a free model $\mathfrak{M}$, $i \in I$, and an i-assignment f, $f \vDash_i A$ says that f satisfies A at i. The clauses of its inductive definition are obvious, but note that $f \vDash_k \forall x A$ iff for all $u \in F_1(i)$, $f_u^x \vDash_i A$, where $f_u^x(x) = u$ and $f_u^x(y) \simeq f(y)$ for any variable $y \neq x$; $f \vDash_i \Box A$ iff for every j such that iRj, $f \vDash_j A$.

A free model is a $Q3$ *model* if for every $i, j \in I$, $F_2(i) = F_2(j)$, i.e., the outer domain is fixed. Note that if R is symmetric, the submodel generated by any world i is $Q3$; hence free models that are not $Q3$ are of interest only in situations where we do not expect R to be symmetric, e.g., for the propositional logic $S4$. A free model is a *Schütte model* if for every $i \in I$, $F_1(i) = F_2(i)$; such models evidently suppress the free-logical aspect.[59] A *fixed domain* model is a $Q3$ model that is also a Schütte model; in such a model the inner and outer domain coincide and are fixed for all worlds.

Given a normal modal propositional logic that is sound for a class of models defined by some condition C on the accessibility relation R, the axiomatization of free modal quantificational logic given by Fine in "Model Theory for Modal Logic I," pp. 131-132, is sound for free models if the propositional logic is modified appropriately, and Fine's axiom (9), $\Diamond Et$ for any term t, is dropped. Suppose we now allow a set Θ of formulae, perhaps with free variables, as nonlogical axioms, allow generalization and necessitation in proofs headed by formulae in Θ, and add a rule of substitution of terms for free variables. Then soundness still obtains if we restrict to Θ-*models*, that is, models such that for every $A \in \Theta$, every $i \in I$, and every i-assignment f, $f \vDash_i A$.

The completeness problem can be stated in the same generalized form. We will say that the axiomatization is *complete for free variable axioms* if for every set Θ of formulae, every set Γ of formulae such that no contradiction is derivable from formulae in Γ as *premises* with formulae in Θ as *axioms* has a Θ-model. Completeness in this sense can be seen to obtain for a wide class of propositional logics,

---

[59]Schütte, *Vollständige Systeme*, p. 4; cf. Hughes and Cresswell, *Introduction*, p. 171. The concept goes back to Kripke's "Semantical Analysis of Intuitionistic Logic."

including all those with which we have been concerned in the text (i.e., S4, S4.2, S4.3, S5).

Assuming an open formula as an axiom is in general stronger than assuming its closure (see p. 300 above). Two exceptions are relevant to modal set theory: if we consider only Schütte models, replacing free by unfree quantificational logic, or if $\Box$ is absolute necessity and we assume $D = \bigcup_{i \in I} F_1(i)$, so that we have Fine's logic cited above with axiom (9). In both cases $(\Box)A \rightarrow A$ is provable for any A. It follows that in these cases completeness for free variable axioms follows from strong completeness in the usual sense.

For Q_3 models, however, the two forms of completeness diverge. For the propositional logic S4, the above axioms are not complete for free variable axioms for Q_3 models with reflexive and transitive R, but they are strongly complete in the usual sense.[60] The latter follows from completeness for free models and the fact that if every formula in Θ is closed (in particular if $\Theta = \phi$), then if a set Γ has a free Θ-model, it has a Q_3 Θ-model with the same I and R. A case of incompleteness for free variable axioms that is of interest for the present essay arises if we take Θ to consist of $(R \in)$, $(R \notin)$, $(E \in)$, and Extensionality; then there are instances of $(BF \in)$ that hold in all Q_3 Θ-models but not in all free Θ-models and hence are not provable from the axioms. We do obtain such completeness if we add Thomason's modal generalization rules R4 and R5,[61] which are in general sound for Q_3 models but not for arbitrary free models. However, they are derivable if the propositional logic contains B.[62]

2. *Scoped First-Order Modal Logic.* We now enlarge the language of first-order modal logic by allowing subformulae of a formula to be bracketed with numerical subscripts; the intended interpretation is that (in the absence of intervening brackets) $[A]_n$ is to be read as within the scope of only the first n nested modal operators within which it occurs. However, this complicates the formation rules. We define an *i-formula* as follows (assuming $\rightarrow$ to be the only primitive binary connective):

[60]Parsons, "On Modal Quantificational Logic with Contingent Domains."
[61]"Some Completeness Results," p. 63.
[62]Parsons, "On Modal Quantificational Logic with Contingent Domains." This lemma is due to Daniel Hausman. The proof proceeds much like the proof of the Barcan formula in the corresponding situation with unfree quantifier rules.

(i)   If A is an atomic formula, A is an i-formula.

(ii)   If A, B are i-formulae, so are $\neg A$ and $A \to B$.

(iii) If A is an i-formula, so is $\forall x A$.

(iv)   If A is an $i + 1$-formula, then $\Box A$ is an i-formula.

(v)   If A is a j-formula and $j \leq i$, then $\{A\}_j$ is an i-formula.

Clearly if A is an i-formula and $i \leq j$, *then A is a j-formula. A formula* is a 0-formula; if A is a formula and $\{B\}_j$ occurs in it within the scope of i nested modal operators, then $j \leq i$.

Let $\mathfrak{M} = \langle I, R, D, F_1, F_2, V \rangle$ be a free model in the sense of section 1. For $i \geq 0$, we let $[h_0 \ldots h_i]$ indicate a sequence of worlds in I such that $h_j R h_{j+1}$ for each j, $0 \leq j < i$. For an i-formula A and an h_i-assignment f, we define satisfaction of A as follows:

(a) If A is atomic, $[h_0 \ldots h_i], f \vDash A$ iff $f \vDash h_i A$.

(b) As usual if A is a negation or a conditional.

(c) If A is $\forall x\, B$, $[h_0 \ldots h_i], f \vDash A$ iff for all $u \in F_1(h_i)$, $[h_0 \ldots h_i]$, $f_u^x \vDash B$.

(d) If A is $\Box B$, $[h_0 \ldots h_i]$, $f \vDash A$ iff for every h such that $h_i R h$, $[h_0 \ldots h_i, h], f \vDash B$.

(e) If A is $\{B\}_j$, $[h_0 \ldots h_i], f \vDash A$ iff for each variable y free in B, $f(y) \in F_2(h_j)$ and $[h_0 \ldots h_j], f \vDash B$. Note here that $j \leq i$, B is a j-formula, and the condition assures that f agrees with a j-assignment for variables free in B. If A is a formula, i.e., a 0-formula, we can write $f \vDash_h A$ for $[h], f \vDash A$.

The problem of axiomatizing this logic is not simple and straightforward. However, for the theories of section III of the text, we can confine ourselves to S4 and S4.2 and Schütte models. Then we can axiomatize as follows: We add to the object language schematic predicate letters P^n, Q^n, R^n, ...for each number $n \geq 0$ of arguments. They are given widest scope. $P^n x_1 \ldots x_n$ counts as a 0-formula. To ordinary quantificational logic, we add the modal propositional logic formulated with attention to scope: instead of the axiom schemata T and 4 and the rule N of necessitation we have

T' $\Box A* \to A$

4' $\Box A* \to \Box\Box A**$

N'   From A infer $\Box A*$,[63]

---

[63]The schema K and, in the case of S4.2, G can be retained unchanged. 4' is equivalent to $\Box A \to \Box\Box A*$, with A restricted not to contain zero subscripts.

For any formula A, let A^1 result by increasing nonzero subscripts by one, except within subformulae of the form $\{B\}_0$. Then $\Box A^1 \to A$ and $\Box A \to \Box\Box A^1$ are provable by T', 4', and Subst$_0$.

"Naive" forms of these schemata can be seen to be unsound; for example if A is a 0-formula, $\{A\}_0 \leftrightarrow A$ is valid, but in general $\Box(\{A\}_0 \leftrightarrow A)$ is not.

where A^* results from A by increasing all subscripts by one.

For the schematic letters, we assume rigidity axioms going with the wide-scope interpretation

RS $\qquad Px_1\ldots x_n \to \Box\, Px_1\ldots x_n$

R$\$$ $\qquad \neg Px_1\ldots x_n \to \Box\, \neg Px_1\ldots x_n$

BFS $\qquad \forall x_1\ldots\forall x_n\,(Px_1\ldots x_n \to \Box A) \to \Box\,\forall x_1\ldots\forall x_n$

$\qquad (Px_1\ldots x_n \to A),$

and for the subscripts we have the axiom schema

A $\qquad [A]_0 \leftrightarrow A,$

and the rule

Subst$_0$ $\qquad$ From $A\,(P^n)$ infer $A\,(\,[x_1\ldots x_n \colon [A]_0]\,)$.

Then we have the

Theorem. For the propositional logic S4 or S4.2, the above axiomatization is sound and complete for Schütte models with the appropriate condition on R.

The axiomatization of free modal logic with scope operators seems not to be so straightforward. Hodes has given an axiomatization of free S5 for a slightly different language, but it is complex and unperspicuous. Still, I would conjecture that the above axiomatization modified to contain S5 and free logic becomes complete if Thomason's modal generalization rules R4 and R5 are added. The derivation of these rules in unscoped free S5 (in fact even KB) seems to break down in the scoped language.

3. *Second-Order Logic.* We can extend either the ordinary or the scoped first-order modal language to a second-order language by adding, for each $n \geq 0$, variables F^n, G^n, H^n, ... that count as n-ary predicates, and quantifiers for them. We consider here only unfree quantifier rules and Schütte models; hence we do not have E as a logical predicate. In the scoped language, second-order variables have to be distinguished from the schematic letters of section 2. Basic second-order logic is obtained by adding the obvious quantifier rules to first-order logic.[64] In a Schütte model, we add to the

---

[64]Cf. Gallin, *Intensional and Higher-Order Modal Logic*, pp. 70-71. The basic logic allows second-order universal instantiation only of variables.

structure a function D_2 that to each natural number n and each $i \in I$ assigns as domain of the n-ary second-order variables a subset of the Cartesian product $\prod_{j \in I} [F(j)]^n$, such that if iRk then $D_2(n, i) \subseteq D_2(n, k)$.[65] A model is *standard* if for every n, i, $D_2(n, i) = \prod_{j \in I} [F(j)]^n$; note that in a standard model the second-order variables have a fixed domain even if the individual variables do not.

We consider the additional axiom schemata

$$\text{C}^n \qquad \exists F^n \, \Box \, \forall x_1 \ldots \forall x_n \, (F^n x_1 \ldots x_n \leftrightarrow A)$$
$$\text{E}^n \qquad \forall G^n \, \exists F^n \, [Rn(F) \wedge \forall x_1 \ldots \forall x_n \, (Fx_1 \ldots x_n \leftrightarrow Gx_1 \ldots x_n)]$$
$$\text{EC}^n \qquad \exists F^n \, [Rn(F) \wedge \forall x_1 \ldots \forall x_n \, (Fx_1 \ldots x_n \leftrightarrow A)],$$

where in C^n and EC^n, F does not occur free in A. C is the schema of Comprehension, EC of Extensional Comprehension. For a given A, EC^n evidently follows from C^n and E^n.[66]

The logic LPS (p. 318 above) is obtained from scoped first-order S4 by adding, for each n, all instances of C^n in which A lacks bound second-order variables (what would be called relative first-order comprehension). In LPS, E^n for each n is provable; hence so is EC^n if A lacks bound second-order variables.

The logic LPE is formulated by adding to unfree first-order S4 the second-order apparatus but not the scope indicators, with relative first-order comprehension and E^n for each n. It is evidently a subsystem of LPS.

The logic LIE is the extension of LPE by adding all instances of C^n; it is thus the impredicative second-order logic. In it all instances of EC are provable. The scope apparatus can be introduced by contextual definition; hence it is a conservative extension of LPS.

Each of these logics can be seen to be sound and complete for general models satisfying the relevant comprehension axioms.

Appendix 2. Relation of Theories of Potential Sets to Standard Set Theory[67]

We have described above (pp. 321-322) a translation of classical set theory into the modal language of set theory. Given a formula

---

[65]Since we are speaking of a Schütte model, 'F' can denote indifferently F_1 or F_2. If $n = 0$, $D_2(n, i)$ is a subset of I, that is, a proposition.

[66]Cf. Gallin, *Intensional and Higher-Order Modal Logic*, p. 78.

[67]Some results in this Appendix were announced in "Modal Set Theories."

A of set theory, let A' be its negative translation, and for any B let B' be its standard translation (as an intuitionistic formula) into S4.[68] Then we can define A^W as $(A')^I$. We can then view the truth of A^W at a world in a model of the modal set theory as the forcing of A' at that world in the model considered as an intuitionistic Kripke structure. As an interpretation of the classical language, it is a weak version of the infinite forcing introduced by Abraham Robinson.[69] We can take advantage of our rigidity axioms to simplify the translation; for atomic A A^W is absolute and equivalent by these axioms to A, and this extends to bounded formulae (i.e., Δ_0 formulae). We have

Theorem 1. If A is a theorem of weak Zermelo set theory, then A^W is a theorem of $MS_0 + \Diamond PS_0$.

Proof. Straightforward from the facts about the development of MS_0 indicated above (pp. 320-322), together with the observation that if A is Δ_0, then A^W is absolute and equivalent to A.

If we add NPE_0 to our logic, then the absoluteness of A^W obtains for all first-order formulae. We then obtain

Theorem 2. If A is a theorem of Zermelo set theory without Infinity, then A^W is a theorem of $MS_0 + \Diamond PS_0 + NPE_0$ (or G). If Choice is included, AC $\Diamond$ needs to be added.

Proof. To prove the translation of all instances of Separation, it suffices to show that for any nonmodal A, $[x: x \in a \wedge A^W]$ is fully rigid. This follows from the following lemma, which is easy to prove by the rigidity axioms:

Lemma. If for a (modal) formula A, $[x: A]$ is absolute, then $[x: x \in a \wedge A]$ is fully rigid.

We obtain a reverse translation by considering the following notion of $\in$-model for the modal theory: let K be a class of transitive sets, partially ordered by inclusion.[70] We take them to be possible

---

[68]Schütte, *Vollständige Systeme*, p. 35.

[69]"Infinite Forcing in Model Theory," §2. This connection was pointed out by the above-mentioned anonymous referee. It raises some mathematical questions that I do not pursue here.

[70]Consider a Schütte model satisfying Extensionality and the scoped rigidity axioms. Let I be its set of possible worlds; for each $i \in I$, we have a structure $\langle W_i, E_i \rangle$ such that if iRj, $\langle W_j, E_j \rangle$ is an end extension of $\langle W_i, E_i \rangle$. Using the Mostowski collapsing lemma, let W_i' be a transitive set such that $\langle W_i', \in \rangle$ is isomorphic to $\langle W_i, E_i \rangle$. Then it is easy to see that if iRj, $W_i' \subseteq W_j'$; hence we have an isomorphic model of the sort considered above (p. 317).

What we now consider is more permissive than the above concept of $\in$-model in allowing a proper class of worlds, but it is also more restrictive; it amounts to assuming that if $W_i \subseteq W_j$, then iRj.

It should be remarked that the second-order Barcan condition insures full rigidity

worlds and domains for a model; $\in$ and I are interpreted by themselves. We use the notation $[z_0 \ldots z_n]$ for a sequence of transitive sets in K such that $z_0 \subseteq \ldots \subseteq z_n$. Then for an n-formula of the language of MS_0, we define a formula $A_{[z_0 \ldots z_n]}$ of set theory that says that A is satisfied in the model at that sequence of worlds.

(a) If A is atomic, $A_{[z_0 \ldots z_n]}$ is A.

(b) If A is $\neg B$ or $B \to C$, then $A_{[z_0 \ldots z_n]}$ is $\neg B_{[z_0 \ldots z_n]}$ or $B_{[z_0 \ldots z_n]} \to C_{[z_0 \ldots z_n]}$, respectively.

(c) If A is $\forall x B$, then $A_{[z_0 \ldots z_n]}$ is $\forall x \in z_n \; B_{[z_0 \ldots z_n]}$.

(d) If A is $\Box B$, then $A_{[z_0 \ldots z_n]}$ is
$$\forall z \in K \; [(Trans \, (z) \wedge z_n \subseteq z) \to B_{[z_0 \ldots z_n, z]}].$$

(e) If A is $\{B\}_i$, then $A_{[z_0 \ldots z_n]}$ is
$x_1 \ldots x_n \in z_i \wedge B_{[z_0 \ldots z_i]}$, where $x_1 \ldots x_n$ are all free variables of A.

Now let Z^* be Zermelo set theory without Infinity but with an axiom saying that every set is contained in a transitive set. Let Z_0^* be the same without Power Set. If we let K be the class of all transitive sets, we can see in Z_0^* that the axioms of MS_0 hold in the model. For ($R_1 \in$), consider transitive z_0 and let z_1 be transitive with $z_0 \subseteq z_1$. If $a \in z_0$, $x \in z_1$, and $x \in a$, then $x \in z_0$ by transitivity of z_0, so $\{x \in a\}_0$ holds at $[z_0, z_1]$. For SF_0, let $x_1 \ldots x_m$ be the free variables of A other than x, and consider $x_1 \ldots x_m \in z_0$. Then by Separation there is a set a such that

$$\forall x[x \in a \leftrightarrow (x \in z_0 \wedge A_{[z0]}) \,].$$

Now let z be the transitive closure of $z_0 \cup \{a\}$, so that $z_0 \subseteq z$ and $a \in z$. Since clearly $a \subseteq z_0$, if $x \in a$, then $\{A\}_0$ holds at $[z_0, z]$. Conversely, if $(\{A\}_0)_{[z_0, z]}$ holds, then $x \in z_0$ and $A_{[z_0]}$, whence $x \in a$. Hence $\forall x \, (x \in a \leftrightarrow \{A\}_0)$ holds at $[z_0, z]$, and therefore $\Diamond \; \exists a \, \forall x \, (x \in a \leftrightarrow \{A\}_0)$ holds at $[z_0]$. It is also evident that G holds in this model. We have

Theorem 3. If A is a closed theorem of $MS_0 + G$, $\forall z \, (Trans \, (z) \to A_{[z]})$ is a theorem of Z_0^*. If A is a closed theorem of $MS_0 + \Diamond \, PS_0 + G$, then $\forall z \, (Trans \, (z) \to A_{[z]})$ is a theorem of Z^*.

Theorems 2 and 3 yield

Theorem 4. $MS_0 + \Diamond \, PS_0 + G$ and Z^* are intertranslatable, also with the addition of AC $\Diamond$ and AC respectively.

---

only in standard models; hence the end-extension condition may not be satisfied in all general models of our theories based on LPE or LIE. The condition is needed to establish the above isomorphism.

We now turn to the relation of ZF and PZF (see p. 324). As we have noted, NPE_0 or G is no longer needed to prove the translation of full Separation.

Theorem 5. If A is a theorem of ZF, then A^W is a theorem of PZF. If A is a theorem of ZFC, then A^W is a theorem of PZF + AC $\Diamond$.

Proof. Given theorem 1, we need to handle the cases of Infinity, full Separation, and Replacement.

The case of Infinity is clear from (4) (p. 323): one simply formalizes the argument that according to (4) there is a possible world in which ϕ and for each x, $x \cup \{x\}$ exist; since the domain of that world is potentially a set a, it will satisfy the condition

$$\phi \in a \wedge \forall x \in a \; \exists y \in a \; (y = x \cup \{x\}),$$

which is Δ_0.

Now let C be the formula $\forall x \, [x \in b \leftrightarrow (x \in a \wedge A)]$, so that $\exists b \, C$ is the axiom of separation for the formula A and set a. To prove $(\exists b \, C)^W$, i.e., $\Box \Diamond \; \exists b \, C^W$, it would suffice to show that $[x: x \in a \wedge A^W]$ is fully rigid. (It was here that we needed NPE_0 for theorem 2.) Thus we can prove

$$Rn \, ([x: x \in a \wedge A^W]) \to \Diamond \; \exists b \, C^W$$

(weakening the conclusion). But then, to prove $\Diamond \; \exists b \, C^W$ and hence $(\exists b \, C)^W$, it suffices to prove

(5) $\Diamond \; Rn \, ([x: x \in a \wedge A^W]).$

But this follows from $Refl_0$. For if for a modal formula B, $[x: B]$ is persistent, then $\Box \forall x \Diamond \, (\Box B \vee \Box \neg B)$ holds, and by reflection $\Diamond \; \forall x \, (\Box B \vee \Box \neg B)$ follows. But $\forall x \, (\Box B \vee \Box \neg B)$ implies that $[x: x \in a \wedge B]$ is fully rigid, with the help of (BF $\in$). Since $[x: A^W]$ is provably persistent, (5) is also provable, and so is $(\exists b \, C)^W$.

Now for Replacement, we can use the fact that it follows from full Separation and the axiom schema of Collection

$$\forall x \in a \; \exists y \, A \to \exists b \; \forall x \in a \; \exists y \in b \, A.$$

If we assume $\Box\ \forall x \in a\ \Diamond\ \exists y\, A^W$, by the persistence of $\exists y\, A^W$ and Refl_0, $\Diamond\ \forall x \in a\ \exists y\, A^W$. Again, the domain of this world is potentially a set, and since by $(\text{BF} \in)\ \forall x \in a\ \exists y\, A^W$ is persistent, we can infer $\Diamond\ \exists b\ \forall x \in a\ \exists\ y \in b\, A^W$, which implies the desired conclusion. This proves theorem 5.

We can translate PZF back into ZF using either the obvious model of V_α's or the more general type of model with a class of transitive sets. The former is a special case of the latter, but it is more convenient to take the ordinals as possible worlds and the V_α as their domains. I use the notation $[\alpha_0 \dots \alpha_n]$ for a nondecreasing sequence of ordinals. The translation given above still applies, but note that the quantifier and necessity clauses now take the form

If A is $\forall x\, B$, then $A_{[\alpha_0 \dots \alpha_n]}$ is $\forall\, x \in V_{\alpha_n}\, B_{\alpha_0 \dots \alpha_n]}$.
If A is $\Box\, B$, then $A_{[\alpha_0 \dots \alpha_n]}$ is $\forall \beta \geq \alpha_n B_{[\alpha_0 \dots \alpha_n,\, \beta]}$.

Theorem 6. If A is a closed theorem of PZF, then $\forall \alpha\, A_{[\alpha]}$ is a theorem of ZF. Similarly for PZF + AC $\Diamond$, ZFC.
Proof. Extensionality, the rigidity axioms, and SF_0 follow much as they do in the proof of theorem 3. To handle Refl_0, let $x_1 \dots x_n$ be the free variables of A other than x. We need to show that for any $x_1 \dots x_n \in V_\alpha$,

$$\{\forall \beta \geq \alpha\ \forall x\ \in V_\beta(A_{[\alpha,\beta]} \rightarrow \forall \gamma \geq \beta\ A^1_{[\alpha,\beta,\gamma]})$$
$$\wedge\ \forall \beta \geq \alpha\ \forall x \in V_\beta\ \exists \gamma \geq \beta\ A^1_{[\alpha,\beta,\gamma]}) \rightarrow \exists\ \beta \geq \alpha\ \forall\ x \in V_\beta\, A_{[\alpha,\,\beta]}$$

If we set $\beta_0 = \alpha$, $\beta_{n+1} = sup_{x \in\ V\beta_n}\ \mu\alpha \geq \beta_n\, A_{[\alpha,\gamma]}$, it is not difficult to see that to satisfy the consequent, we can take $\beta = sup\ \beta_n$.[71] The first conjunct of the antecedent (i.e., the persistence of A) insures that if for a given $x \in V_{\beta_n}$ we find a γ such that $A_{[\alpha,\,\gamma]}$, then $A_{[\alpha,\beta_{n+1}]}$ will still hold.

It is quite straightforward to prove by induction on the construction of a nonmodal formula A that if all its free variables have values in V_α, then $(A^W)_{[\alpha]} \leftrightarrow A$ holds. It then follows from theorems 5 and 6 that A is provable in ZF if and only if A^W is provable in PZF. But now note that since the ordinals are linearly ordered, the model by the V_α satsifies the propositional logic S4.3. Therefore if A^W is provable in PZF with the addition of L^+ (and a fortiori with

---

[71]Cf. the proofs of first-order reflection principles in ZF, e.g., Drake, *Set Theory*, chap. 3, §6. Note that since A^1 can contain subformulae with subscript 1 only within subformulae with subscript 0, $A^1_{[\alpha,\beta,\gamma]}$ is equivalent to $A_{[\alpha,\gamma]}$.

the addition of G or NPE$_0$), then A^W is already provable in PZF. We have

Theorem 7. PZF + L$^+$ is a conservative extension of PZF for formulae of the form A^W.

Much the same arguments show that the relations between ZF and PZF established by theorems 5 and 6 also obtain between NB and either PNBS or PNBE (p. 324).

Theorem 8. If A is a theorem of NB, then A^W is a theorem of PNBE and therefore of PNBS.

Theorem 9. If A is a closed theorem of PNBS + L$^+$, then $\forall \alpha\, A_{[\alpha]}$ is a theorem of NB.

Proof. Much as before. But for the translation into NB reflecting the model with the V_α as domains, we have to add to each second-order variable F an extra argument for the ordinal α_n in $(Fx_1 \ldots x_n)_{[\alpha_0 \ldots \alpha_n]}$ and restrict the variables to relations on V_{α_n}. But we can then carry out the translation for the theory PNBS, again even with the addition of L$^+$. Moreover, we can still prove in NB $(A^W)_{[\alpha]} \leftrightarrow A$ for each closed A.

Theorem 10. PNBS + L$^+$ is a conservative extension of PNBE for formulae of the form A^W.

Now let KM be the extension of NB that admits impredicative classes, called Kelley-Morse or Bernays-Morse set theory. Then a similar relation obtains between KM and the extension of PNBE that results by replacing LPE by LIE, i.e., assuming full comprehension. KM can be extended further in interesting ways, for example by adding a natural second-order reflection schema of Bernays that implies the existence of strongly inaccessible and Mahlo cardinals.[72] A reflection schema with the same purport can be formulated very naturally in our (second-order) language. An interesting further strengthening of KM is by the stronger "extendibility schema" of Reinhardt discussed in the last essay, which implies the existence of many measurable cardinals.[73] This again can be formulated, somewhat more awkwardly, in an extension of our modal language. However, in both these cases the intertranslatability has been shown only if G is added to the modal propositional logic, which allows the replacement of the W-translation by another in which second-order variables are interpreted as ranging over absolute concepts and relations.

---

[72]"Zur Frage der Unendlichkeitsschemata," p. 25; cf. Drake, *Set Theory*, pp. 122-123.
[73]Reinhardt, "Remarks on Reflection Principles," p. 196; cf. Essay 10 above, end of section III.

Bibliography

The following bibliography includes all works cited in this book but is not intended as a comprehensive bibliography of the subject. Works are cited by author and title only, the latter often abbreviated. Certain principles govern the citation:

1. Works of an author are listed alphabetically by title (ignoring only the definite and indefinite article) rather than chronologically to facilitate locating the entry from the references in the text and footnotes.

2. Books are cited in the latest edition listed in this bibliography, unless the contrary is explicitly stated.

3. For articles, reprintings in collections of the author's own papers are listed, but reprintings in anthologies are not, with the exception of two that are indispensable for workers in this field: Paul Benacerraf and Hilary Putnam's *Philosophy of Mathematics: Selected Readings* and Jean van Heijenoort's *From Frege to Gödel*. The "latest edition" principle is interpreted to mean that page references to an article reprinted in a book by the same author are to the reprint. Thus a reference to Quine, "New Foundations," pp. 90-91, is not to the original publication of that article in the *American Mathematical Monthly* but to *From a Logical Point of View*, 2d ed. (1961).

4. Concerning works originally published in foreign languages, I have tried to observe the principle that any commentary on a text should contain a reference enabling the reader to locate it in the original language. This principle is all too often breached in contemporary American writing on the history of philosophy.

5. For a few classic authors, I have observed special conventions:

Cantor. All references are to *Gesammelte Abhandlungen*, which is abbreviated as *GA*.

Frege. His collected works now exist in a convenient (though expensive) German edition, consisting of the two collections edited by Angelelli, *Begriffsschrift und andere Aufsätze* and *Kleine Schriften*, the Olms reprints of *Grundlagen* and *Grundgesetze*, and the two volumes of the surviving *Nachlass, Nachgelassene Schriften* and *Wissenschaftlicher Briefwechsel*. For works published in Frege's lifetime I have nevertheless given the pagination of the *original* publication, which is given in the margins of the Angelelli collections and agrees with that of the two other reprints and of Austin's translation of the *Grundlagen*. It is also given in the margin of some other translations, particularly the Geach and Black collection and Furth's *Basic Laws*.

Kant. The *Critique of Pure Reason* is cited according to the original editions in the standard way (A for first edition [1781], B for second edition [1787]). This pagination is given in most modern editions and in Norman Kemp Smith's translation, which is used for all quotations (though sometimes altered). Other writings are cited according to the standard edition, *Gesammelte Schriften*, edited by the Prussian Academy of Sciences and its successors, abbreviated as *Ak*. When possible I have given section numbers and other edition-invariant information.

Ackermann, Wilhelm. "Zum Hilbertschen Aufbau der reellen Zahlen." *Mathematische Annalen* 99 (1928), 118-133. Translation in van Heijenoort, pp. 493-508.

Anscombe, G. E. M., and Peter Geach. *Three Philosophers*. Oxford: Blackwell, 1961.

Axiomatic Set Theory. Proceedings of Symposia in Pure Mathematics, vol. 13. Providence, R.I.: American Mathematical Society, Part I, 1971; Part II, 1974.

Bar-Hillel, Yehoshua; E. I. J. Poznanski; M. O. Rabin; and Abraham Robinson, eds. *Essays on the Foundations of Mathematics*. Dedicated to A. A. Fraenkel. Jerusalem: Magnes Press, 1961.

Barwise, Jon, ed. *Handbook of Mathematical Logic*. Amsterdam: North Holland, 1977.

Belnap, Nuel D., Jr. "Gupta's Rule of Revision of Theory of Truth." *Journal of Philosophical Logic* 11 (1982), 103-116.

Benacerraf, Paul. "Logicism: Some Considerations." Ph.D. dissertation. Princeton University, 1960.

——. "What Numbers Could Not Be." *Philosophical Review* 74 (1965), 47-73.

Benacerraf, Paul, and Hilary Putnam, eds. *Philosophy of Mathematics: Selected Readings*. Englewood Cliffs, N.J.: Prentice-Hall, 1964. 2d ed. Cambridge University Press, 1983. Citations from first edition.

Bernays, Paul. *Abhandlungen zur Philosophie der Mathematik*. Darmstadt: Wissenschaftliche Buchgesellschaft, 1976.
——. "Mathematische Existenz und Widerspruchsfreiheit." In *Etudes de philosophie des sciences, en hommage à F. Gonseth à l'occasion de son soixantième anniversaire*, pp. 11-25. Neuchâtel: Griffon 1950. Reprinted in *Abhandlungen*, pp. 92-106.
——. "On Platonism in Mathematics." Translated by C. D. Parsons in Benacerraf and Putnam, pp. 274-286.
——. "Sur le platonisme dans les mathématiques." *L'enseignement mathématique* 34 (1935), 52-69. English translation "On Platonism." (The German version in *Abhandlungen*, pp. 62-78, is a translation from the French.)
——. "Zur Frage der Unendlichkeitsschemata in der axiomatischen Mengenlehre." In Bar-Hillel et al., pp. 3-49.
——. See also Hilbert and Bernays.
Beth, E. W. "Über Lockes 'allgemeines Dreieck.'" *Kant-Studien* 48 (1956-1957), 361-380.
Boolos, George. "The Iterative Conception of Set." *Journal of Philosophy* 68 (1971), 215-231.
Boolos, George, and Richard C. Jeffrey. *Computability and Logic*. Cambridge University Press, 1974. 2d ed., 1981.
Brittan, Gordon G., Jr. *Kant's Theory of Science*. Princeton: Princeton University Press, 1978.
Brouwer, L. E. J. *Collected Works*. Vol. I: *Philosophy and Intuitionistic Mathematics*, edited by Arend Heyting. Amsterdam: North-Holland, 1975.
——. "Consciousness, Philosophy, and Mathematics." *Proceedings of the Tenth International Congress of Philosophy, Amsterdam 1948*, I, Fasc. 2, pp. 1235-1249. Amsterdam: North Holland, 1949. Reprinted in *Collected Works*, I, 480-494.
——. "Intuitionism and Formalism." Translated by A. Dresden. *Bulletin of the American Mathematical Society* 20 (1913), 81-96. Reprinted in Benacerraf and Putnam, pp. 66-77, and in *Collected Works*, I, 123-138.
——. *Intuitionisme en formalisme*. Groningen: Noordhoff, 1912. Reprinted in *Wiskunde, waarheid, werkelijkheid*. Groningen: Noordhoff, 1919. English translation "Intuitionism and Formalism."
——. "Mathematik, Wissenschaft, und Sprache." *Monatshefte für Mathematik und Physik* 36 (1929), 153-164. Reprinted in *Collected Works*, I, 417-428.
——. "Willen, weten, spreken." *Euclides* 9 (1933), 177-193. (Only a short excerpt is translated in *Collected Works*, I, 443-446.)
Buchholz, Wilfried, Solomon Feferman, Wolfram Pohlers, and Wilfried Sieg. *Iterated Inductive Definitions and Subsystems of Analysis*. Lecture Notes in Mathematics, 897. Berlin: Springer, 1981.
Burge, Tyler. "The Liar Paradox: Tangles and Chains." *Philosophical Studies* 41 (1982), 353-366.
——. "Semantical Paradox." *Journal of Philosophy* 76 (1979), 169-198. Reprinted in Martin, *Recent Essays*.

Bibliography

Butts, R. E., and Jaakko Hintikka, eds. *Logic Foundations of Mathematics, and Computability Theory.* Dordrecht: Reidel, 1977.

Cantor, Georg. *Gesammelte Abhandlungen mathematischen und philosophischen Inhalts.* Edited by Ernst Zermelo. Berlin: Springer, 1932. Reprinted Hildesheim: Olms, 1961. Cited as *GA*.

Carnap, Rudolf. "Formalwissenschaft und Realwissenschaft." *Erkenntnis* 5 (1935-1936), 30-36. Translation in Herbert Feigl and May Brodbeck, eds., *Readings in the Philosophy of Science*, pp. 123-128. New York: Appleton, 1953.

Cartwright, Richard L. "Propositions." In R. J. Butler, ed., *Analytical Philosophy*, pp. 81-103. Oxford: Blackwell, 1962.

Chellas, Brian F. *Modal Logic: An Introduction.* Cambridge University Press, 1980.

Chihara, Charles. *Ontology and the Vicious-Circle Principle.* Ithaca, N.Y.: Cornell University Press, 1973.

——. "The Semantic Paradoxes: A Diagnostic Investigation." *Philosophical Review* 88 (1979), 590-618.

Church, Alonzo. "Comparison of Russell's Resolution of the Semantical Antinomies with That of Tarski." *Journal of Symbolic Logic* 41 (1976), 747-760. Reprinted in Martin, *Recent Essays*.

——. *Introduction to Mathematical Logic.* Princeton: Princeton University Press, 1956.

——. "Russellian Simple Type Theory." *Proceedings and Addresses of the American Philosophical Association* 47 (1973-1974), 21-33.

Crossley, J. N., and Michael Dummett, eds. *Formal Systems and Recursive Functions.* Amsterdam: North Holland, 1965.

Davidson, Donald. "On Saying That." *Synthese* 19 (1968), 130-146; also in Davidson and Hintikka, pp. 158-174.

——. "True to the Facts." *Journal of Philosophy* 66 (1969), 748-764.

——. "Truth and Meaning." *Synthese* 17 (1967), 304-323.

Davidson, Donald, and Gilbert Harman, eds. *Semantics of Natural Language.* Dordrecht: Reidel, 1972.

Davidson, Donald, and Jaakko Hintikka, eds. *Words and Objections.* Dordrecht: Reidel, 1969. Rev. ed. 1975.

Dedekind, Richard. *Was sind und was sollen die Zahlen?* Braunschweig: Vieweg, 1888, 3d ed., 1911. English translation in *Essays on the Theory of Numbers*. Chicago: Open Court, 1901.

Drake, F. R. *Set Theory: An Introduction to Large Cardinals.* Amsterdam: North-Holland, 1974.

Dummett, Michael. *Elements of Intuitionism.* Oxford: Clarendon Press, 1977.

——. *Frege: Philosophy of Language.* London: Duckworth, 1973. 2d ed., 1981.

——. "Nominalism." *Philosophical Review* 65 (1956), 491-505. Reprinted in *Truth and Other Enigmas*, pp. 38-49.

——. "The Philosophical Basis of Intuitionistic Logic." In Rose and Shepherdson, pp. 5-40. Reprinted in *Truth and Other Enigmas*, pp. 215-247.

——. *Truth and Other Enigmas.* London: Duckworth, 1978.

Edwards, Paul, *The Encyclopedia of Philosophy.* 8 vols. New York: Free Press–Macmillan, 1967.

Enderton, Herbert. *Elements of Set Theory.* New York: Academic Press, 1977.

Feferman, Solomon. "Gödel's Incompleteness Theorems and the Reflective Closure of Theories." *Journal of Symbolic Logic,* forthcoming.

——. "A More Perspicuous Formal System for Predicativity." In Kuno Lorenz, ed., *Konstruktionen versus Positionen,* I, 68-93. Berlin: de Gruyter, 1978.

——. "Predicative Provability in Set Theory." *Bulletin of the American Mathematical Society* 72 (1966), 486-489.

——. "Systems of Predicative Analysis." *Journal of Symbolic Logic* 29 (1964), 1-30.

——. "Toward Useful Type-Free Theories, I." *Journal of Symbolic Logic,* forthcoming. Reprinted in Martin, *Recent Essays.*

Fenstad, J. E., ed. *Proceedings of the Second Scandinavian Logic Symposium.* Amsterdam: North-Holland, 1971.

Field, Hartry. *Science without Numbers: A Defense of Nominalism.* Princeton: Princeton University Press, 1980.

——. "Tarski's Theory of Truth." *Journal of Philosophy* 69 (1972), 347-375.

Fine, Kit. "First-Order Modal Theories I: Sets." *Noûs* 15 (1981), 177-205.

——. "Model Theory for Modal Logic, Part I—the *de re/de dicto* Distinction." *Journal of Philosophical Logic* 7 (1978), 125-156.

——. "Postscript: Prior on the Construction of Possible Worlds and Instants." In Prior and Fine, pp. 116-168.

——. "Properties, Propositions, and Sets." *Journal of Philosophical Logic* 6 (1977), 135-191.

Fitch, Frederic B. "Comments and a Suggestion." In Martin, *The Paradox of the Liar,* pp. 75-78.

Føllesdal, Dagfinn. "Interpretation of Quantifiers." In van Rootselaar and Staal, pp. 271-281.

Fraenkel, A. A., Yehoshua Bar-Hillel, and Azriel Lévy. *Foundations of Set Theory.* Amsterdam: North-Holland, 1973. Revised edition of a work by the first two authors published in 1958.

Frege, Gottlob. *The Basic Laws of Arithmetic: Exposition of the System.* Translated and edited with an introduction by Montgomery Furth. Covers Introduction, sections 0-52, portions of sections 54, 55, 91, and Appendix to vol. II of *Grundgesetze.* Berkeley and Los Angeles: University of California Press, 1964.

——. *Begriffsschrift, eine der arithmetischen nachgebildete Formelsprache des reinen Denkens.* Halle: Nebert, 1879. Reprinted with original pagination in *Begriffsschrift und andere Aufsätze.* English translation in van Heijencort, pp. 1-82.

——. *Begriffsschrift und andere Aufsätze.* Edited by Ignacio Angelelli. Hildesheim: Olms, 1964.

Bibliography

——. *The Foundations of Arithmetic*. Translated by J. L. Austin. Contains the German text. Oxford: Blackwell, 1950. 2d ed., 1953.

——. *Funktion und Begriff*. Jena: Pohle, 1891. Reprinted in *Kleine Schriften*. Translation in Geach and Black, pp. 21-41.

——. *Grundgesetze der Arithmetik, begriffsschriftlich abgeleitet*. Jena: Pohle, vol. I, 1893; vol. II, 1903. Partial English translation in *Basic Laws*.

——. *Die Grundlagen der Arithmetik*. Breslau: Koebner, 1884. Reprinted, Hildesheim: Olms, 1961. English translation *Foundations*.

——. *Kleine Schriften*. Edited by Ignacio Angelelli. Hildesheim: Olms, 1967.

——. *Nachgelassene Schriften*. Edited by Hans Hermes, Friedrich Kambartel, and Friedrich Kaulbach. Hamburg: Meiner, 1969.

——*Translations from the Philosophical Writings of Gottlob Frege*. See Geach and Black.

——. "Über Begriff und Gegenstand." *Vierteljahrsschrift für wissenschaftliche Philosophie* 16 (1892), 192-205. Reprinted in *Kleine Schriften*. Translation in Geach and Black, pp. 42-55.

——. *Wissenschaftlicher Briefwechsel*. Edited by Gottfried Gabriel, Hans Hermes, Friedrich Kambartel, Christian Thiel, and Albert Veraart. Hamburg: Meiner, 1976.

Friedman, Harvey. "The Consistency of Classical Set Theory Relative to a Set Theory with Intuitionistic Logic." *Journal of Symbolic Logic* 36 (1971), 315-319.

Friedman, Harvey, and Andrej Ščedrov. "Large Sets in Intuitionistic Set Theory," forthcoming.

Furth, Montgomery. "Two Types of Denotation." In *Studies in Logical Theory*. American Philosophical Quarterly Monograph Series no. 2, pp. 9-45. Oxford: Blackwell, 1968.

Gallin, Daniel. *Intensional and Higher-Order Modal Logic*. Amsterdam: North-Holland, 1975.

Geach, Peter. "Class and Concept." *Philosophical Review* 64 (1955), 561-570. Reprinted in *Logic Matters*, pp. 226-235.

——. *Logic Matters*. Oxford: Blackwell, 1972.

——. See also Anscombe and Geach.

Geach, Peter, and Max Black, eds. *Translations from the Philosophical Writings of Gottlob Frege*. Oxford: Blackwell, 1952. 3d ed., 1980.

Gerber, Harvey. "Brouwer's Bar Theorem and a System of Ordinal Notations." In Myhill, Kino, and Vesley, pp. 327-338.

Gödel, Kurt. *The Consistency of the Continuum Hypothesis*. Princeton: Princeton University Press, 1940.

——. "On an Extension of Finitary Mathematics Which Has Not Yet Been Used." Unpublished English translation, with additional notes by the author, of "Über eine bisher noch nicht benützte Erweiterung des finiten Standpunktes." This is not the translation published in *Journal of Philosophical Logic* 9 (1980), 133-142.

——. "Russell's Mathematical Logic." In P. A. Schilpp, ed., *The Philosophy of Bertrand Russell*, pp. 125-153. Evanston: Northwestern University, 1944. 3d ed., New York: Tudor, 1951. Reprinted in Benacerraf and Putnam, pp. 211-232.

———. "Über eine bisher noch nicht benützte Erweiterung des finiten Standpunktes." *Dialectica* 12 (1958), 280-287. For translations see "On an Extension."

———. "What Is Cantor's Continuum Problem?" *American Mathematical Monthly* 54 (1947), 515-525. Revised and expanded version in Benacerraf and Putnam, pp. 258-273. Citations are from the latter version.

Goodman, Nelson. *Problems and Projects*. Indianapolis: Bobbs-Merrill, 1972.

Goodman, Nelson, and W. V. Quine. "Steps toward a Constructive Nominalism." *Journal of Symbolic Logic* 12 (1947), 105-122. Reprinted in Goodman, *Problems and Projects*, pp. 173-198.

Goodman, Nicolas D. "A Genuinely Intensional Set Theory." in Gumb and Shapiro.

Gottlieb, Dale. *Ontological Economy: Substitutional Quantification and Mathematics*. Oxford: Clarendon Press, 1980.

Grattan-Guinness, Ivor. "The Correspondence between Georg Cantor and Philip Jourdain." *Jahresbericht der deutschen Mathematiker-Vereinigung* 73 (1971), 111-130.

Gréco, Pierre, Jean-Blaise Grize, Seymour Papert, and Jean Piaget. *Problèmes de la construction du nombre*. Etudes d'épistémologie génétique, XI. Paris: Presses Universitaires de France, 1960.

Gumb, Raymond, and Stewart Shapiro, eds. *Intensional Mathematics*. Flushing, N.Y.: Haven, forthcoming.

Gupta, Anil. "Truth and Paradox." *Journal of Philosophical Logic* 11 (1982), 1-60. Reprinted in Martin, *Recent Essays*.

Hambourger, Robert. "A Difficulty with the Frege-Russell Definition of Number." *Journal of Philosophy* 74 (1977), 409-414.

Harman, Gilbert. "An Introduction to Translation and Meaning: Chapter Two of *Word and Object*." *Synthese* 19 (1968), 14-26. Also in Davidson and Hintikka, pp. 14-26.

———. "Logical Form." *Foundations of Language* 9 (1972), 38-65.

———. "Quine on Meaning and Existence." *Review of Metaphysics* 21 (1967), 124-151, 343-367.

———. Review of Wilfrid Sellars, *Philosophical Perspectives*. *Journal of Philosophy* 66 (1969), 133-144.

Hazen, Allen P. "Expressive Completeness in Modal Languages." *Journal of Philosophical Logic* 5 (1976), 25-46.

Heidegger, Martin. *Kant und das Problem der Metaphysik*. Bonn: Cohen, 1929. Later printings Frankfurt: Klostermann.

Henrich, Dieter. *Identität und Objektivität: Eine Untersuchung zu Kants transcendentaler Deduktion*. Heidelberg: Carl Winter Universitätsverlag, 1976.

Herzberger, Hans G. "Dimensions of Truth." *Journal of Philosophical Logic* 2 (1973), 535-556.

Bibliography

——. "Naive Semantics and the Liar Paradox." *Journal of Philosophy* 79 (1982), 479-494.
——. "New Paradoxes for Old." *Proceedings of the Aristotelian Society* N.S., 81 (1980-1981), 109-123.
——. "Notes on Naive Semantics." *Journal of Philosophical Logic* 11 (1982), 61-102. Reprinted in Martin, *Recent Essays*.
——. "Paradoxes of Grounding in Semantics." *Journal of Philosophy* 67 (1970), 145-167.
——. "The Truth-Conditional Consistency of Natural Languages." *Journal of Philosophy* 64 (1967), 29-35.
——. "Truth and Modality in Semantically Closed Languages." In Martin, *The Paradox of the Liar*, pp. 25-46.
Higginbotham, James. Review of Martin, *The Paradox of the Liar. Journal of Philosophy* 69 (1972), 398-401.
——. "Some Problems in Semantics and Radical Translation." Ph.D. dissertation, Columbia University, 1973.
Hilbert, David, and Paul Bernays. *Grundlagen der Mathematik.* Berlin: Springer, vol. I, 1934; vol. II, 1939; 2d ed., vol. I, 1968; vol. II, 1970.
Hintikka, Jaakko. "Are Logical Truths Analytic?" *Philosophical Review* 74 (1965), 178-203. Reprinted in *Knowledge and the Known*, pp. 135-159.
——. "Kant on the Mathematical Method." *The Monist* 51 (1967), 352-375. Reprinted in *Knowledge and the Known*, pp. 160-183.
——. "Kantian Intuitions." *Inquiry* 15 (1972), 341-345.
——. "Kant's 'New Method of Thought' and His Theory of Mathematics." *Ajatus* 27 (1965), 37-47. Reprinted in *Knowledge and the Known*, pp. 126-134.
——. *Knowledge and the Known: Historical Perspectives in Epistemology.* Dordrecht: Reidel, 1974.
——. *Logic, Language-Games, and Information.* Oxford: Clarendon Press, 1973.
——. "On Kant's Notion of Intuition (*Anschauung*)." In Terence Penelhum and J. J. MacIntosh, eds., *Kant's First Critique*, pp. 38-53. Belmont, Calif.: Wadsworth, 1969.
——. See also Butts and Hintikka; Davidson and Hintikka.
Hodes, Harold. "Modal Logic for the World Traveller." Unpublished manuscript, 1982.
Howell, Robert. "Intuition, Synthesis, and Individuation in the *Critique of Pure Reason.*" *Noûs* 7 (1973), 207-232.
Hughes, G. E., and M. J. Cresswell. *An Introduction to Modal Logic.* London: Methuen, 1968.
Husserl, Edmund. *Ideen zu einer reinen Phänomenologie und phänomenologischen Philosophie.* Erstes Buch. Edited by Walter Biemel. Husserliana, vol. 3. The Hague: Nijhoff, 1950. First published Halle: Niemeyer, 1913.
——. *Philosophie der Arithmetik, mit ergänzenden Texten* (1890-1901). Edited by Lothar Eley. Husserliana, vol. 12. The Hague: Nijhoff, 1970. First published Halle: Pfeffer, 1891.

Jeffrey, Richard C. *Formal Logic: Its Scope and Limits*. New York: McGraw-Hill, 1967. 2d ed., 1981.

——. See also Boolos and Jeffrey.

Jubien, Michael, "Ontological Commitment to Particulars." *Synthese* 28 (1974), 513-532.

——. "Ontology and Mathematical Truth." *Noûs* 11 (1977), 133-150.

Kant, Immanuel. *Critique of Pure Reason*. Translated by Norman Kemp Smith. London: Macmillan, 1929. 2d ed., 1933.

——. *Gesammelte Schriften*. Edited by the Preussische Akademie der Wissenschaften, Berlin; later by the Deutsche Akademie der Wissenschaften, Berlin; later by the Akademie der Wissenschaften zu Göttingen. Berlin: Reimer, later de Gruyter, 1902-. Cited as *Ak*.

——. *Inaugural Dissertation and Early Writings on Space*. Translated by John Handyside. Chicago: Open Court, 1929.

——. *Kritik der reinen Vernunft*. Edited by Raymund Schmidt. Leipzig: Meiner, 1926. 2d ed., 1930.

——. *Prolegomena to Any Future Metaphysics*. Edited by Lewis White Beck. New York: Liberal Arts Press, 1950. Later printings, Indianapolis: Bobbs-Merrill.

Kaplan, David. "S5 with Quantifiable Propositional Variables." Abstract. *Journal of Symbolic Logic* 35 (1970), 355.

Kitcher, Philip. "The Plight of the Platonist." *Noûs* 12 (1978), 119-136.

Kleene, S. C. *Introduction to Metamathematics*. New York: Van Nostrand, 1952.

Kleene, S. C., and R. E. Vesley. *The Foundations of Intuitionistic Mathematics*. Amsterdam: North-Holland, 1965.

Kneale, William. "Propositions and Truth in Natural Languages." *Mind* N.S., 81 (1972), 225-243.

Kreisel, G. "Ordinal Logics and the Characterization of Informal Concepts of Proof." In *Proceedings of the International Congress of Mathematicians, Edinburgh 1958*, pp. 289-299. Cambridge University Press, 1960.

Kripke, Saul A. "Naming and Necessity." In Davidson and Harman, pp. 253-355, 763-769. 2d ed.: *Naming and Necessity*. Cambridge, Mass.: Harvard University Press, 1980.

——. "Outline of a Theory of Truth." *Journal of Philosophy* 72 (1975), 690-716. Reprinted in Martin, *Recent Essays*.

——. "Semantical Analysis of Intuitionistic Logic." In Crossley and Dummett, pp. 92-129.

——. "Semantical Considerations on Modal Logic." *Acta Philosophica Fennica* 16 (1963), 83-94.

——. "A Theory of Truth I, II." Abstracts. *Journal of Symbolic Logic* 41 (1976), 556-557.

Krivine, Jean-Louise. *An Introduction to Axiomatic Set Theory*. Translated by David Miller. Dordrecht: Reidel, 1971.

Leibniz, Gottfried Wilhelm. *Leibnizens mathematische Schriften*. Edited by C. J. Gerhardt. Halle, 1849-1863. Reprinted, Hildesheim: Olms, 1962.

——. *Nouveaux essais sur l'entendement humain*. Vol. 5 of *Die philosophischen Schriften von G. W. Leibniz*, edited by C. J. Gerhardt. Berlin, 1875-1890. Reprinted, Hildesheim: Olms, 1965.

Lévy, Azriel. *Basic Set Theory*. Berlin: Springer, 1979.

——. See also Fraenkel, Bar-Hillel, and Lévy.

Lewis, David. "Counterpart Theory and Quantified Modal Logic." *Journal of Philosophy* 65 (1968), 113-126.

——. "General Semantics." *Synthese* 22 (1971), 18-67. Also in Davidson and Harman, pp. 169-218.

Marcus, Ruth Barcan. "Classes and Attributes in Extended Modal Systems." *Acta Philosophica Fennica* 16 (1963), 123-136.

——. "Classes, Collections, and Individuals." *American Philosophical Quarterly* 11 (1974), 227-232.

——. "Modalities and Intensional Languages." *Synthese* 13 (1961), 303-322.

Martin, Donald A. Review of Quine, *Set Theory and Its Logic*, 2d ed. *Journal of Philosophy* 67 (1970), 111-114.

——. "Sets versus Classes." Unpublished comment on "Sets and Classes" (Essay 8 of this volume).

Martin, Gottfried. *Arithmetik und Kombinatorik bei Kant*. Berlin: de Gruyter, 1972. Expanded version of a 1934 dissertation, printed Itzehoe, 1938.

——. *Kant's Metaphysics and Theory of Science*. Translated by P. G. Lucas. Manchester: Manchester University Press, 1953.

——. *Klassische Ontologie der Zahl*. Kant-Studien Ergänzungsheft 70. Cologne: Kölner Universitätsverlag, 1956.

Martin, Robert L. "Are Natural Languages Universal?" *Synthese* 32 (1976), 271-291.

——, ed. *The Paradox of the Liar*. New Haven: Yale University Press, 1972. 2d ed., Reseda, Calif.: Ridgeview, 1978.

——, ed. *Recent Essays on Truth and the Liar Paradox*. New York: Oxford University Press, forthcoming.

Martin, Robert L., and Peter W. Woodruff. "On Representing 'True-in-L' in L." *Philosophia* 5 (1975), 213-217. Also in Asa Kasher, ed., *Language in Focus*, pp. 113-117. Dordrecht: Reidel, 1976. Reprinted in Martin, *Recent Essays*.

Martin-Löf, Per. "Hauptsatz for Iterated Inductive Definitions." In Fenstad, pp. 179-216.

——. "An Intuitionistic Theory of Types: Predicative Part." In Rose and Shepherdson, pp. 73-118.

Mates, Benson. *Elementary Logic*. New York: Oxford University Press, 1965. 2d ed., 1972.

Mendelson, Elliott. *Introduction to Mathematical Logic*. Princeton: Van Nostrand, 1964. 2d ed., 1979.

Michael, Emily. "Peirce's Paradoxical Solution of the Liar's Paradox." *Notre Dame Journal of Formal Logic* 16 (1975), 369-374.

Mirimanoff, Dimitri. "Les antinomies de Russell et de Burali-Forti et le problème fondamental de la théorie des ensembles." *L'enseignement mathématique* 19 (1917), 37-52.

Montague, Richard. *Formal Philosophy: Selected Papers.* Edited by Richmond H. Thomason. New Haven: Yale University Press, 1974.

———. "Pragmatics and Intensional Logic." *Synthese* 22 (1971), 69-94. Also in Davidson and Harman, pp. 142-168; *Dialectica* 24 (1970), 277-302; reprinted in *Formal Philosophy*, pp. 119-147.

———. "Universal Grammar." *Theoria* 36 (1970), 373-398. Reprinted in *Formal Philosophy*, pp. 222-246.

Morton, Adam. "Denying the Doctrine and Changing the Subject." *Journal of Philosophy* 70 (1973), 503-510.

Moschovakis, Y. N. "Predicative Classes." In *Axiomatic Set Theory*, Part I, pp. 247-264.

Myhill, John. "Intensional Set Theory." In Gumb and Shapiro.

———. "Some Properties of Intuitionistic Zermelo-Fraenkel Set Theory." In *Cambridge Summer School in Mathematical Logic, Proceedings 1971*, pp. 206-231. Lecture Notes in Mathematics, 337. Berlin: Springer, 1973.

Myhill, John, Akiko Kino, and R. E. Vesley, eds. *Intuitionism and Proof Theory.* Amsterdam: North-Holland, 1970.

Papert, Seymour. "Problèmes épistémologiques et génétiques de la récurrence." In Gréco et al., pp. 117-148.

———. "Sur le réductionnisme logique." in Gréco et al., pp. 97-116.

Parsons, Charles. "Brouwer, Luitzen Egbertus Jan." In Edwards, I, 399-401.

———. "Hierarchies of Primitive Recursive Functions." *Zeitschrift für mathematische Logik und Grundlagen der Mathematik* 14 (1968), 357-376.

———. "Intensional Logic in Extensional Language." *Journal of Symbolic Logic* 47 (1982), 289-328.

———. "Mathematical Intuition." *Proceedings of the Aristotelian Society* N.S., 80 (1979-1980), 142-168.

———. "Mathematics, Foundations of." In Edwards, V, 188-213.

———. "Modal Set Theories." Abstract. *Journal of Symbolic Logic* 46 (1981), 683-684.

———. "Objects and Logic." *The Monist* 65 (1982), 491-516.

———. "On Modal Quantificational Logic with Contingent Domains." Abstract. *Journal of Symbolic Logic* 40 (1975), 302.

———. "On Translating Logic." *Synthese* 27 (1974), 405-411.

———. "Sets and Possible Worlds." Abstract of a lecture at Columbia University, December 5, 1974.

———. "Some Remarks on Frege's Conception of Extension." In Schirn, I, 265-277.

———. "Substitutional Quantification and Mathematics (Review of Gottlieb, *Ontological Economy*)." *British Journal for the Philosophy of Science* 33 (1982), 409-421.

———. "Was ist eine mögliche Welt?" *Kant-Studien* 65 (1974), 378-396.

Bibliography

Peirce, Charles Sanders. *The Collected Papers of Charles Sanders Peirce*, edited by Charles Hartshorne and Paul Weiss. 5 vols. Cambridge, Mass.: Harvard University Press, 1931-34.

Plaass, Peter. *Kants Theorie der Naturwissenschaften*. Göttingen: Vandenhoeck & Rupprecht, 1965.

Plantinga, Alvin. *The Nature of Necessity*. Oxford: Clarendon Press, 1974.

Poincaré, Henri. *La science et l'hypothèse*. Paris: Flammarion, 1902.

——. *Science et méthode*. Paris: Flammarion, 1908.

Powell, William C. "A Completeness Theorem for Zermelo-Fraenkel Set Theory." *Journal of Symbolic Logic* 41 (1976), 323-327.

——. "Extending Gödel's Negative Interpretation to ZF." *Journal of Symbolic Logic* 40 (1975), 221-230.

——. "Set Theory with Predication." Ph.D. dissertation, State University of New York at Buffalo, 1972.

Prior, A. N., and Kit Fine. *Worlds, Times, and Selves*. London: Duckworth, 1976.

Putnam, Hilary. "Is Logic Empirical?" In *Boston Studies in the Philosophy of Science* V, 216-241. Dordrecht: Reidel, 1969. Reprinted as "The Logic of Quantum Mechanics," in *Mathematics, Matter, and Method*.

——. *Mathematics, Matter, and Method: Collected Philosophical Papers*. Vol. I. Cambridge University Press, 1975. 2d ed. 1980.

——. "Mathematics without Foundations." *Journal of Philosophy* 64 (1967), 5-22. Reprinted in *Mathematics, Matter, and Method*, pp. 43-59.

——. "The Thesis That Mathematics Is Logic." In Ralph Schoenman, ed., *Bertrand Russell: Philosopher of the Century*, pp. 273-303. London: Allen & Unwin, 1967. Reprinted in *Mathematics, Matter, and Method*, pp. 12-42.

——. "What Is Mathematical Truth?" In *Mathematics, Matter, and Method*, pp. 60-78.

——. See also Benacerraf and Putnam.

Quine, W. V. "Existence and Quantification." In *Ontological Relativity and Other Essays*, pp. 91-113.

——. *From a Logical Point of View*. Cambridge, Mass.: Harvard University Press, 1953. 2d ed., 1961. The fourth printing (1980), though labeled "second edition, revised," has a new Foreword and one other small revision.

——. *Mathematical Logic*. New York: Norton, 1940. Rev. (2d) ed., Cambridge, Mass.: Harvard University Press, 1951.

——. *Methods of Logic*. New York: Henry Holt, 1950. Rev. (2d) ed., 1959. 3d ed., New York: Holt, Rinehart, and Winston, 1972. 4th ed. Cambridge, Mass.: Harvard University Press, 1982.

——. "Necessary Truth." In *The Ways of Paradox*, pp. 68-76.

——. "New Foundations for Mathematical Logic." *American Mathematical Monthly* 44 (1937), 70-80. Reprinted with revisions in *From a Logical Point of View*, pp. 80-101.

——. "Ontological Relativity." *Journal of Philosophy* 65 (1968), 185-212. Reprinted in *Ontological Relativity and Other Essays*, pp. 26-68.

——. *Ontological Relativity and Other Essays*. New York: Columbia University Press, 1969.

——. *Philosophy of Logic*. Englewood Cliffs, N.J.: Prentice-Hall, 1970.

——. "Replies." In Davidson and Hintikka, pp. 292-352.

——. "Reply to D. A. Martin." *Journal of Philosophy* 67 (1970), 247-248.

——. "Reply to Professor Marcus." *Synthese* 13 (1961), 323-330. Reprinted in *The Ways of Paradox*, pp. 177-184.

——. *The Roots of Reference*. La Salle, Ill.: Open Court, 1974.

——. *Set Theory and Its Logic*. Cambridge, Mass.: Harvard University Press, 1963. 2d ed. 1969.

——. "Three Grades of Modal Involvement." In *Proceedings of the XIth International Congress of Philosophy*, XIV, 65-81. Amsterdam: North-Holland, 1953. Reprinted in *The Ways of Paradox*, pp. 158-176.

——. "Truth and Disquotation." In *Proceedings of the Tarski Symposium*, pp. 373-384. *Proceedings of Symposia in Pure Mathematics*, XXV. Providence: American Mathematical Society, 1974. Reprinted in *The Ways of Paradox*, pp. 308-321. (Not in 1st ed.)

——. *The Ways of Paradox and Other Essays*. New York: Random House, 1966. 2d ed., enlarged, Cambridge, Mass.: Harvard University Press, 1976.

——. *Word and Object*. Cambridge, Mass.: Technology Press (now MIT Press), and New York: Wiley, 1960.

——. See also Goodman and Quine.

Reinhardt, W. N. "Remarks on Reflection Principles, Large Cardinals, and Elementary Embeddings." In *Axiomatic Set Theory*, Part II, pp. 189-206.

——. "Satisfaction Definitions and Axioms of Infinity in a Theory of Properties with Necessity Operator." In A. I. Arruda, R. Chuaqui, and N.C.A. da Costa, eds., *Mathematical Logic in Latin America*, pp. 267-304. Amsterdam: North-Holland, 1980.

——. "Set Existence Principles of Shoenfield, Ackermann, and Powell." *Fundamenta Mathematicae* 84 (1974), 5-34.

Resnik, Michael D. *Frege and the Philosophy of Mathematics*. Ithaca, N.Y.: Cornell University Press, 1980.

——. "Frege's Methodology." Ph.D. dissertation, Harvard University, 1963.

Robinson, Abraham. "Infinite Forcing in Model Theory." In Fenstad, pp. 317-340. Reprinted in *Selected Papers, Vol. 1: Model Theory and Algebra*, edited by H. J. Keisler, pp. 243-266. New Haven: Yale University Press, 1979.

Rose, H. E., and J. C. Shepherdson, eds. *Logic Colloquium '73*. Amsterdam: North-Holland, 1975.

Russell, Bertrand. *Essays in Analysis*. Edited by Douglas Lackey. New York: Braziller, 1973.

——. *Logic and Knowledge*. Edited by Robert C. Marsh. London: Allen & Unwin, 1956.

——. "Mathematical Logic as Based on the Theory of Types." *American Journal of Mathematics* 30 (1908), 222-262. Reprinted in *Logic and Knowledge*, pp. 57-102, and in van Heijenoort, pp. 150-182.

——. *The Principles of Mathematics.* Cambridge University Press, 1903. 2d ed., London: Allen & Unwin, 1937.

——. See also Whitehead and Russell.

Schirn, Matthias, ed. *Studien zu Frege.* 3 vols. Stuttgart: Frommann, 1976.

Schultz, Johann. *Prüfung der Kantischen Kritik der reinen Vernunft.* Königsberg: Hartung, vol. I, 1789; vol. II, 1792.

Schütte, Kurt. *Beweistheorie.* Berlin: Springer-Verlag, 1960.

——. "Eine Grenze für die Beweisbarkeit der transfiniten Induktion in der verzweigten Typenlogik." *Archiv für mathematische Logik und Grundlagenforschung* 7 (1965), 45-60.

——. "Predicative Well-Orderings." In Crossley and Dummett, pp. 280-303.

——. *Proof Theory.* Revised and expanded translation of *Beweistheorie.* Berlin: Springer-Verlag, 1977.

——. *Vollständige Systeme modaler und intuitionistischer Logik.* Berlin: Springer, 1968.

Schwichtenberg, Helmut. "Proof Theory: Applications of Cut-Elimination." In Barwise, pp. 867-986.

Sellars, Wilfrid. "Abstract Entities." *Review of Metaphysics* 16 (1962-1963), 627-671. Reprinted in *Philosophical Perspectives*, pp. 229-269.

——. "Classes as Abstract Entities and the Russell Paradox." *Review of Metaphysics* 17 (1963-1964), 67-90. Reprinted in *Philosophical Perspectives* pp. 270-290.

——. *Philosophical Perspectives.* Springfield, Ill.: Charles C Thomas, 1967.

Shamoon, Alan. "Kant's Logic." Ph.D. dissertation, Columbia University, 1979.

Sharvy, Richard. "Why a Class Can't Change Its Members." *Noûs* 2 (1968), 303-314.

Shoenfield, J. R. "Axioms of Set Theory." In Barwise, pp. 321-344.

——. *Mathematical Logic.* Reading, Mass.: Addison-Wesley, 1967.

Skyrms, Brian. "Notes on Quantification and Self-Reference." In Martin, *The Paradox of the Liar*, pp. 67-74.

——. "Return of the Liar: Three-Valued Logic and the Concept of Truth." *American Philosophical Quarterly* 7 (1970), 153-161.

Smith, Norman Kemp. *Commentary to Kant's 'Critique of Pure Reason.'* London: Macmillan, 1918. 2d ed., 1923.

Steiner, Mark. *Mathematical Knowledge.* Ithaca, N.Y.: Cornell University Press, 1975.

Strawson, P. F. *Logico-Linguistic Papers.* London: Methuen, 1971.

——. "Singular Terms and Predication." *Journal of Philosophy* 58 (1961), 393-412. Reprinted in Davidson and Hintikka, pp. 97-117, and *Logico-Linguistic Papers*, pp. 53-74. Citations from the reprint in Davidson and Hintikka.

Suppes, Patrick. *Axiomatic Set Theory*. Princeton: Van Nostrand, 1960.

Tait, W. W. "Constructive Reasoning." In van Rootselaar and Staal, pp. 185-198.

——. "Finitism." *Journal of Philosophy* 78 (1981), 524-546.

Takeuti, Gaisi. *Proof Theory*. Amsterdam: North-Holland, 1975.

Takeuti, Gaisi, and Wilson M. Zaring. *An Introduction to Axiomatic Set Theory*. Berlin: Springer, 1971.

Tarski, Alfred, Andrzej Mostowski, and Raphael M. Robinson. *Undecidable Theories*. Amsterdam: North-Holland, 1953.

Tharp, Leslie H. "Necessity, Apriority, and Provability." Abstract. *Notices of the American Mathematical Society* 21 (1974), A-320.

——. "A Quasi-Intuitionistic Set Theory." *Journal of Symbolic Logic* 36 (1971), 456-460.

——. "Three Theorems of Metaphysics." Unpublished manuscript, 1974.

Thomason, Richmond H. "Modal Logic and Metaphysics." In Karel Lambert, ed., *The Logical Way of Doing Things*, pp. 119-146. New Haven: Yale University Press, 1969.

——. "Some Completeness Results for Modal Predicate Calculi." In Karel Lambert, ed., *Philosophical Problems in Logic: Some Recent Developments*, pp. 56-76. Dordrecht: Reidel, 1970.

Thompson, Manley. "Singular Terms and Intuitions in Kant's Epistemology." *Review of Metaphysics* 26 (1972-1973), 314-343.

Troelstra, A. S. *Principles of Intuitionism*. Lecture Notes in Mathematics, 95. Berlin: Springer, 1969.

Tymoczko, Thomas. "The Four-Color Problem and Its Philosophical Significance." *Journal of Philosophy* 76 (1979), 57-83.

van Fraassen, Bas C. "Inference and Self-Reference." *Synthese* 21 (1970), 425-438. Also in Davidson and Harman, pp. 695-708.

——. "Presupposition, Implication, and Self-Reference." *Journal of Philosophy* 65 (1968), 136-152.

——. "Rejoinder: On a Kantian Conception of Language." In Martin, *The Paradox of the Liar*, pp. 59-66.

——. "Truth and Paradoxical Consequences." In Martin, *The Paradox of the Liar*, pp. 13-23.

van Heijenoort, Jean, ed. *From Frege to Gödel: A Source Book in Mathematical Logic, 1879-1931*. Cambridge, Mass.: Harvard University Press, 1967.

van Rootselaar, B., and J. F. Staal, eds. *Logic, Methodology, and Philosophy of Science III*. Amsterdam: North-Holland, 1968.

Vleeschauwer, H. J. de. *La déduction transcendentale dans l'oeuvre de Kant.* 3 vols. Antwerp: "De Sikkel," 1934-1937.

Waismann, Friedrich. *Introduction to Mathematical Thinking*. Translated by Theodore J. Benac. New York: Ungar, 1951.

Wang, Hao. *From Mathematics to Philosophy*. London: Routledge & Kegan Paul, 1974.

——. "Large Sets." In Butts and Hintikka, pp. 309-334.

——. "Process and Existence in Mathematics." in Bar-Hillel et al., pp. 328-351.

Weyl, Hermann. "Der *circulus vitiosus* in der heutigen Begründung der Analysis." *Jahresbericht der deutschen Mathematiker-Vereinigung* 28 (1919), 85-92. Reprinted in *Gesammelte Abhandlungen*, II, 43-50.

——. *Gesammelte Abhandlungen*. Edited by K. Chandrasekharan. Berlin: Springer, 1968.

——. *Das Kontinuum. Kritische Untersuchungen über die Grundlagen der Analysis*. Leipzig: Veit, 1918.

——. "Über die neue Grundlagenkrise der Mathematik." *Mathematische Zeitschrift* 10 (1921), 39-79. Reprinted in *Gesammelte Abhandlungen*, II, 143-180.

White, Nicholas P. "What Numbers Are." *Synthese* 27 (1974), 111-124.

Whitehead, Alfred North, and Bertrand Russell. *Principia Mathematica*. 3 vols. Cambridge University Press, 1910-1913. 2d ed., 1925-1927.

Wittgenstein, Ludwig. *Bemerkungen über die Grundlagen der Mathematik*. Frankfurt: Suhrkamp, 1974.

——. *Lectures on the Foundations of Mathematics*, Cambridge 1939. Edited by Cora Diamond. Ithaca, N.Y.: Cornell University Press, 1976.

——. *Remarks on the Foundations of Mathematics*. Translated by G.E.M. Anscombe. Oxford: Blackwell, 1956. 3d ed., 1978. The first edition contains the German: the third edition does not and contains an additional text. For the original of the third edition, see the Suhrkamp *Bemerkungen*.

Wolf, Robert S. "Formally Intuitionistic Set Theories with Bounded Formulas Decidable." Ph.D. dissertation, Stanford University, 1974.

Wolff, Robert Paul. *Kant's Theory of Mental Activity*. Cambridge, Mass.: Harvard University Press, 1963.

Zermelo, Ernst. "Über Grenzzahlen und Mengenbereiche." *Fundamenta Mathematicae* 16 (1930), 29-47.

Index

359

Library of Congress Cataloging in Publication Data

Parsons, Charles, 1933–
    Mathematics in philosophy.

    Bibliography: p.
    Includes index.
        1. Mathematics—Philosophy—Addresses, essays,
lectures.   I. Title.
QA8.6.P37   1983   510'.1   83-45153
ISBN 0-8014-1471-7   (Alk. paper)